Anna Maria Gil-Lafuente

Fuzzy Logic in Financial Analysis

Studies in Fuzziness and Soft Computing, Volume 175

Editor-in-chief
Prof. Janusz Kacprzyk
Systems Research Institute
Polish Academy of Sciences
ul. Newelska 6
01-447 Warsaw
Poland
E-mail: kacprzyk@ibspan.waw.pl

Further volume of this series
can be found on our homepage:
springeronline.com

Vol. 160. K.K. Dompere
Cost-Benefit Analysis and the Theory of Fuzzy Decisions – Fuzzy Value Theory, 2004
ISBN 3-540-22161-1

Vol. 161. N. Nedjah, L. de Macedo Mourelle (Eds.)
Evolvable Machines, 2005
ISBN 3-540-22905-1

Vol. 162. R. Khosla, N. Ichalkaranje, L.C. Jain
Design of Intelligent Multi-Agent Systems, 2005
ISBN 3-540-22913-2

Vol. 163. A. Ghosh, L.C. Jain (Eds.)
Evolutionary Computation in Data Mining, 2005
ISBN 3-540-22370-3

Vol. 164. M. Nikravesh, L.A. Zadeh, J. Kacprzyk (Eds.)
Soft Computing for Information Prodessing and Analysis, 2005
ISBN 3-540-22930-2

Vol. 165. A.F. Rocha, E. Massad, A. Pereira Jr.
The Brain: From Fuzzy Arithmetic to Quantum Computing, 2005
ISBN 3-540-21858-0

Vol. 166. W.E. Hart, N. Krasnogor, J.E. Smith (Eds.)
Recent Advances in Memetic Algorithms, 2005
ISBN 3-540-22904-3

Vol. 167. Y. Jin (Ed.)
Knowledge Incorporation in Evolutionary Computation, 2005
ISBN 3-540-22902-7

Vol. 168. Yap P. Tan, Kim H. Yap, Lipo Wang (Eds.)
Intelligent Multimedia Processing with Soft Computing, 2005
ISBN 3-540-22902-7

Vol. 169. C.R. Bector, Suresh Chandra
Fuzzy Mathematical Programming and Fuzzy Matrix Games, 2005
ISBN 3-540-23729-1

Vol. 170. Martin Pelikan
Hierarchical Bayesian Optimization Algorithm, 2005
ISBN 3-540-23774-7

Vol. 171. James J. Buckley
Simulating Fuzzy Systems, 2005
ISBN 3-540-24116-7

Vol. 172. Patricia Melin, Oscar Castillo
Hybrid Intelligent Systems for Pattern Recognition Using Soft Computing, 2005
ISBN 3-540-24121-3

Vol. 173. Bogdan Gabrys, Kauko Leiviskä, Jens Strackeljan (Eds.)
Do Smart Adaptive Systems Exist?, 2005
ISBN 3-540-24077-2

Vol. 174. Mircea Negoita, Daniel Neagu, Vasile Palade
Computational Intelligence: Engineering of Hybrid Systems, 2005
ISBN 3-540-23219-2

Vol. 175. Anna Maria Gil-Lafuente
Fuzzy Logic in Financial Analysis, 2005
ISBN 3-540-23213-3

Anna Maria Gil-Lafuente

Fuzzy Logic in Financial Analysis

Springer

Professor Dr. Anna Maria Gil-Lafuente
Universidad de Barcelona
Facultad de Ciencias
Económicas y Empresariales
Dept. de Economía y Organización de Empresas
Av. Diagonal, 690
08034 Barcelona
Spain
E-mail: amgil@ub.edu

ISSN print edition:1434-9922
ISSN electronic edition:1860-0808
ISBN-10 3-540-23213-3 Springer Berlin Heidelberg New York
ISBN-13 978-3-540-23213-1 Springer Berlin Heidelberg New York

Library of Congress Control Number: 2004117077

Springer is a part of Springer Science+Business Media
springeronline.com

Printed in The Netherlands

Typesetting: by the author and TechBooks using a Springer LaTeX macro package
Cover design: E. Kirchner, Springer Heidelberg
Printed on acid-free paper 89/3141/jl- 5 4 3 2 1 0

Preface

With ever increasing frequency, financial studies in business are acquiring greater importance in the field of management as a consequence of the complexity of the problems arising in present day society. The old treaties that included certain aspects on the raising and placing of payment means have been relegated to history, leaving the way open to monographic studies that include and treat in depth specific aspects of financial activities in business.

The work we are presenting intends to set out an ordered set of basic elements on which to base financial analysis in business in the XXI century. In it are included a series of previous works which serve as the basis and foundation to the new contributions that are being made in the field of financial economy and provide business with certain instruments and models that are suitable for the treatment of the new economic context.

But a work such as this would not have any sense if it were not to be oriented towards the solution of the problems arising in current reality. It is well known that we are immersed in times that are characterised by certain social and economic changes that are extremely fast and profound and which have no precedent whatsoever throughout the whole history of mankind. We are immersed in a world in which every event occurs and develops with such speed that it becomes practically impossible to know with any degree of exactitude what the future holds in store for us. All the events and circumstances we can expect are charged with a high degree of uncertainty.

In order to be able to tackle the problems of a general economic or business nature arising from this uncertainty, which is implicit in all future events, knowledge based on formal logic is no longer sufficient. The so-called modern mathematics based on mechanistic schemes, as well as Boolean algebra, are no use for explaining and foreseeing those actions which we should carry out in the future and they become impotent when faced with the new manner of acting of our society.

In order to be able to adequately attend to the requirements, which very often are brought up relative to a certain type of forecast or opinion on the nature of future events, it has become essential to use a new set of techniques based on uncertainty. This manner of operation has as its objective the expression in the results arrived at with all the subtleness and subjectivity that characterises human thought. Unlike these new techniques, up to just a few years ago what were used were certain models based some times on certain data, at others on probabilistic elements the data of which was measurable or verifiable. The first of these is not applicable today, because the uncertainty that characterises the future events of our society does not permit us to work with certain data. Neither are the second usable for the same reasons, since to be able to apply a probability it is necessary to have, in the first place, a succession of phenomena that are repeated under determined circumstances and, in the second, be able to apply the results arrived at on another phenomenon submitted to the same conditions as the former.

If we place ourselves in the sphere of business of our day and age, there is no doubt whatsoever of the importance of financial decisions. An incorrect decision can easily cause the disappearance of the business from the market. In this context of uncertainty, the decisions to be taken are increasingly more complex due to the changing circumstances relative to legal provisions, the influence of other sectors and countries, prospects of changes in monetary systems, etc. This causes the existing problems to be multiple and varied: expansion, modernisation, financial balance, quality of the work place, etcetera, is only a very small sample. If we only concentrate on one aspect, technology, it can be seen that on a daily basis more models appear aimed at savings in energy, effort and space, causing ever increasing requirements for investment in order to attain greater expansion and profits, precisely due to the uncertainty relative to what will happen the next day.

But, in spite of all the technological advances, in spite of the fact that increasingly mankind will have a greater need to be supported by the so-called "artificial intelligence" in order to be able to handle a larger amount of information at one and the same time, in spite of all this, there is something available to us all and which makes us superior to any mechanical apparatus: that is imagination, a quality which allows us to be creative, to think and decide.

The financier, in the exercise of the activity, is obliged to take decisions for an uncertain tomorrow, all of which are subject to very rapid and substantial changes. In order for the business to reach the desired objectives it is necessary, increasingly more frequently, to resort to the aid of experts. These, based on their knowledge and experience, should be able to orient the business activity and attempt to decrease the range of possibilities for erroneous decisions. To base the studies of financial analysis on the opinions of experts instead of doing this on past data or on probabilistic forecasts, constitutes a radical change, which we hope will bring fruitful results.

But the individualised opinion of an expert is not exempt from a certain degree, more or less high, of subjectivity. It is for this reason that a good part of this work is dedicated to the problem of aggregation of the opinion of experts in the financial field, with the object of limiting, wherever possible, the subjective component of the opinions and making sure that the decisions have the best guarantee of reaching the desired objectives.

The future faced by the businessman, financier, is uncertain and full of risks, although also very full of great prospects. Fortunately current science allows for the drawing up of techniques, previously unthinkable, which are very suitable for the treatment of the problems of our society, of our businesses, of our finances. All that is required is that advantage be taken of all the tools available to the executive to be able to advance through the competitive jungle arising from the new economic conglomerates.

February 2005 *Anna Maria Gil-Lafuente*

Contents

1 The Financial Activity of Businesses

1.1 The Changeability of Financial Phenomena

The mission of science does not end with the grasping by man of real or ideal entities and their relationships, but it attempts to act on the same by creating new forms. Within the framework in which business management moves a certain activity takes place, the most important characteristic of which is the raising and placement of payment means that are necessary for carrying out the objectives of the business. This field, which is called financial, is the object of a process of formalisation, which varies as the phenomena that it must apprehend change. This task is not easy, as also it is not easy in any other field, because as stated by Ortega y Gasset, "he who discovers a new scientific truth beforehand had to pulverize nearly all that was learnt before and arrives at the new truth with bloodied hands because of having to strangle innumerable common places"[1].

Financial analysis in business has been considered as a part of financial management in business, and includes the study of the connections existing between monetary facts and phenomena that take place internally in the same, as well as of the results arising from the same.

The understanding of all the distorted mosaic of facts, phenomena and connections, which the raising, treatment and placing of monetary means by the business introduce, gives rise to certain problems which this part of Business Economy aspires to garner. The systemisation of this knowledge at times requires a long process of development without which it is impossible to attain the desired results.

It is unquestionable that social surroundings in which businesses carry out their activity have at all times an obvious influence on their financial structures as they do also on their future evolution.

Financial problems vary over time and with them can be seen an evolution of the techniques that are used for their resolution. As a consequence of this each phase of the interesting business history has been the origin of the birth of different instruments that have been considered as suitable for giving a reply to the problems that arose. Now then, in the same way that, according

[1] Ortega y Gasset J (1958) La rebellion de las masas (in Spanish), 14th edition. (Ed) Espasa-Calpe, Madrid, p. 131.

Anna Maria Gil-Lafuente: *Fuzzy Logic in Financial Analysis*, StudFuzz **175**, 1–11 (2005)
www.springerlink.com

to Nicol, "below their natural specialisations all the different sciences constitute a unit"[2] it is also possible to study under a certain unitary situation the diverse elements that are necessary for financial analysis in business.

Now, over latter years we have seen that these variations have been occurring with greater frequency, in such a way that we are now faced with a situation, the most marked characteristic of which is mutability. In fact, the rapid evolution taking place in the environment of business activity ensures that financial problems undergo continuous modifications, which go hand in hand with the pressing need for adapting the instruments that are used for their treatment in this permanently changing reality. Therefore, neither in the formal field nor in the material field does the resistance to change which occurs in other aspects of economic science, exist.

In short, it can be stated that the irruption of new techniques into the sphere of financial management in business has been caused both by the changes that have taken place, take place and will take place in business financial phenomenology and by the modification taking place in the social sphere surrounding it and in which it is immersed. It is well known that one and the same monetary mass can receive different treatment and that one and the same financial requirements can be resolved by resorting to different products and sources. But in current reality, and we believe much more so in the future, the range of possibilities placed before the businessman will vary with greater speed. Faced with this perspective of rapid and permanent change it will be necessary to have available certain instruments, the dominant characteristics of which must be flexibility and adaptability.

1.2 Business within the Framework of Monetary Circulation

From a very general aspect, it can be considered that the businessman, in order to be able to continue functioning, requires a certain amount of monetary means by which payments can be made over time. As a result of sales a certain amount of money flows into the business which, in an economy in expansion, is not always sufficient for making certain payments in the short, medium and long-term. For this reason recourse has to be made to other means from different sources.

Therefore, the businessman receives from consumers, as payment for goods delivered or services rendered a certain monetary mass that, in a stationary economic system and therefore, among other things, without any increase in activity and no inflation, should be greater than the amounts handed over to production factors. However, in reality these conditions are not complied with, and from this situation stems the important role played

[2] Nicol Eduardo (1965) Los principios de la ciencia (in Spanish). (Ed) Fondo de Cultura Económica, México, p. 9.

by the credit institutions, which provide businessmen with certain monetary masses that allow for maintaining a continuous process for creating income.

Over the years credit institutions have assumed different roles. Thus, at the time when industry was developing, or even prior to this, in the Middle Ages, those rudimentary banks were limited to collecting legal residual payment means in order to place them at the disposal of merchants and negotiators. Currently this idea only reflects the function they carry out in a very incomplete way, for two reasons:

1. Businessmen, as a group, do not have available sufficient payment means for financing production.
2. Credit institutions do not exactly and exclusively carry out this function of intermediaries, which consists in receiving payment means from the public and placing them at the disposal of businessmen.

The characteristic trait of modern production lies in the fact that the businessman counts on in house financial means, which are employed "in principal" for acquiring fixed assets, totally or in part. Obtaining long-term credits also reinforces the permanent financial structure of the business. And finally working capital is financed to a greater degree by means of external short-term financing.

In a modern economy it is therefore, unthinkable that businesses can carry out their activity and attain their objectives exclusively with their own financial means and this is the reason for the existence of credit institutions. Within the sphere of the activity of a country, the latter occupy a central position from which they carry out the function of maintaining and lubricating the extensive networks which, like the blood circulation system, feeds the cells of the economic fabric, providing financial means to those economic units that are in need of it.

If we commence with a general scheme, obviously simplified, the role played by credit institutions can be placed within the monetary circulation of a system (see Fig. 1.1).

For this we can start out from the fact that businesses overall make payments arising from the cost of labour, and for the acquisition of raw materials, semi-manufactured products and services. This is why from business arises a current which goes as far as consumers and another current that takes place internally that promotes a circulation of money from one business to another. If we carry out a process of general abstraction this inter-business monetary flow need not be considered and we could assume a simplified system in which all businesses form a unitary group. From this first current stems another, which leads to the Government and constitutes a reduction for covering taxes on the work yield of the people.

When businesses obtain a profit what is expected from this in the first place is a current, retribution of capital in the form of dividends and interest which ends up with the consumers, and from which it is hoped there will be a sub-current which also is destined for the Government: the tax on capital

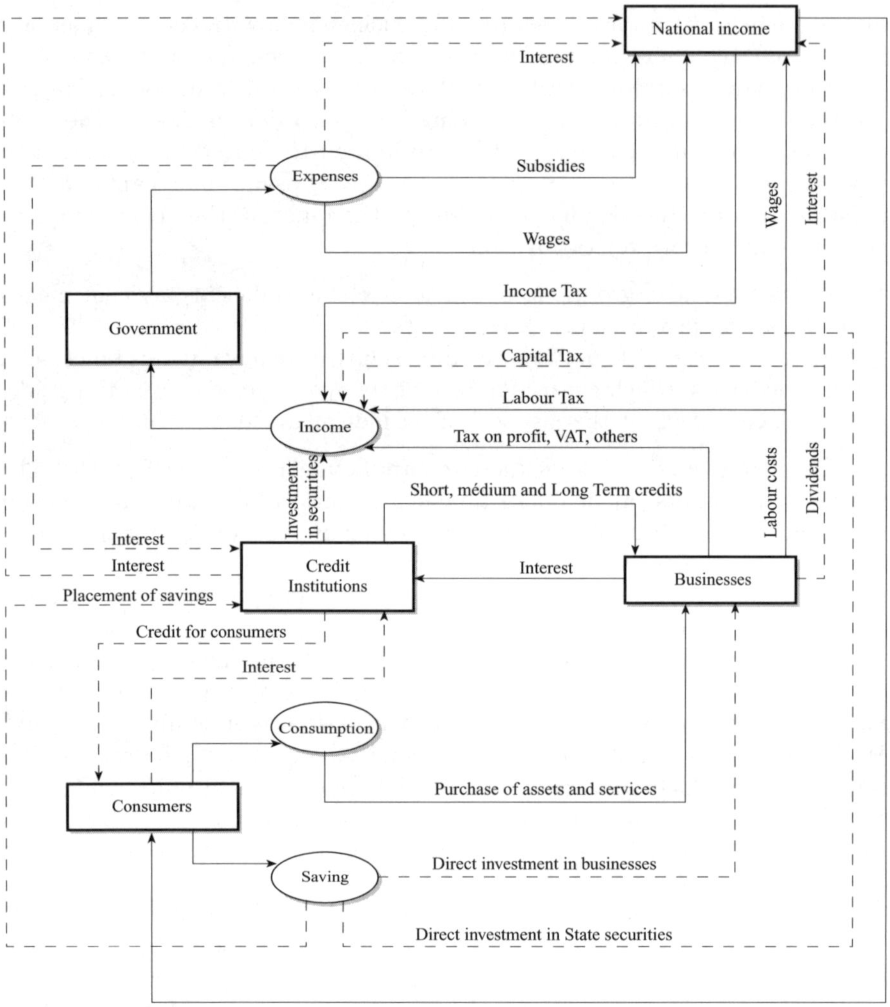

Fig. 1.1. The monetary circulation of a system

gains. Independently of this current there is another formed as a consequence of the tax on the overall profits of business, which also ends up in the Government. Also to be taken into account, included in the Government income, is the current generated by the tax on overall income of the consumers.

As a counterpart to this income a new current appears corresponding to Government expenses. These are constituted, in a first approximation, by two groups: wages and subsidies. In a much wider diagram consideration could also be given to the current relative acquisitions by the Administration from businesses. To these two items, subsidies and wages, it is customary to add a third: interest for loans received.

With this description it is possible to see the formation of a concept belonging to the macro-economic sphere: that of national income. This basically includes, the interest of businesses and credit institutions, wages for labour, both that stemming from business and Government, subsidies and interest also from the Government.

The aggregation of these flows will give rise to the monetary mass reaching the consumers, with which they will have to solve their own financial problem. For this the monetary mass is separated into two parts: that which is consumed and therefore forms the current flowing towards business, and the other part, that which constituters savings, which forms one of the fundamental cores of business financing.

In fact, the savings of consumers can obviously be directed at business as direct investment and also as an investment in State securities, which would increase Government income, which would accrue as compensation certain interest, mentioned before, which form a part of national income. But there is a third current which consists in the placement of savings in credit institutions, that is, banks, savings banks, etc., be they public, semi-public or private financing entities. This placement of savings will give rise to a current of interest from credit institutions to consumers and payment on these savings, that is, as a counterpart to the use of these monetary masses.

From the point of view of credit institutions their problem resides in the manner of placing the funds, which have been deposited with them. The first possibility consists in an investment in State securities, which will accrue a certain interest, which will give rise to a current from Government towards credit institutions. The second takes place when credits are granted to private citizens, either for consumption or for increasing patrimony. The compensation occurs with the return of the capital plus interest and other charges.

Finally a monetary current can arise from credit institutions towards businesses, as a consequence of the granting of short, medium or long-term credits, which will mean another current in the opposite direction due to the return of these capitals and interest and other payments that businesses must make to credit institutions.

Financial analysis is occupied principally with the study of these two currents, of their incidence on the economic-financial structures of business and of the influence that these exercise on the objectives of the same. It is for this reason that three fundamental aspects appear: the raising of monetary resources, their internal treatment and their placing.

In fact, it is obvious that in order to aid in attaining its objectives it is necessary for a business to acquire certain fixed assets, stocks, and also requires a complement of working capital with which to cover ordinary payments. In order for it to carry out these acquisitions it is necessary that financial means be available. A part of these are provided by the owners of the business, another stems from the difference obtained by the sale of goods and/or services, and a third appears as a support from financial institutions to business activity Obviously these financial resources generate a cost: this is what is called capital cost. The business, in order to be able to continue with its activity and reach its objectives must make sure that the profitability it obtains on placing these financial means is higher than the cost of obtaining the same.

The notion of capital cost constitutes the core around which the financial activity of businesses revolves, because included in the same are not only the problems of "raising" but also the problems of "investment". It is difficult then to totally separate the problems of financing from those of investment, therefore it is customary to include both in a single body called "business financing". Anglo-Saxon literature is accustomed to include business financial problems, considered in the very widest sense, within the term capital budgeting. Also the term capital allocation is often used.

If we should want to specify the fundamental problems of capital budgeting in a single term than we could say that this can be specified as the obtaining the greatest profitability from the capital invested. For this reason the business, when it acts in a rational manner, should give preference to those investments that imply the highest profitability, and from this the existence of a function of marginal profitability with a decreasing nature can be considered. On the other hand, capital costs acquire increasing forms, since the business will resort in the first place to the cheapest financing sources in order later to resort to those sources that require a higher cost. In this way it can be said that the marginal cost of capital is increasing. Under these hypotheses the instruments of economic analysis point out that the balance takes place when the marginal cost of capital is equal to marginal profitability on investments.

Obviously this approach, which is useful for understanding a problem, lacks all operational possibilities. It is for this reason that it is necessary to separate from this aggregate the necessary elements for correct financial management of business, placing at the disposal of the businessman a set of techniques with a wide range of possibilities for use in very differentiated situations from the rich financial reality of today.

1.3 Classical and Modern Aspects of Financial Studies

Studies carried out relative to financial activities in business have undergone considerable changes throughout the last century as a consequence of modifications both with reference to the sphere of study of business financial problems, and in the use of techniques which have varied in accordance with the needs at each instant, in an attempt to resolve problems that financial phenomena have caused.

Initially, studies which we could call classical referred basically to one objective: the raising of the financial resources required by the business at each point in time and the means necessary for the same to flow towards the business under the best possible conditions. Obviously if this objective was sufficient to justify studies on financial administration at certain points in time, current reality does not allow for limiting oneself strictly to the problem of raising funds, but that requirements are greater.

In fact, at the beginning of the century, when the birth of large businesses occurs, when concentrations take place, when internal pressures such as the development of union activity cause certain businesses to have serious problems due to the fact of their deficient financial structure (due mainly to the need for resorting to outside financial sources, which gave rise to very high financial costs), the closing of many industries and the failure of important business projects occurs. Likewise, when a policy of excessive dividends, provoked under-capitalisation of business with the consequent lack of liquidity, an environment was generalised when the need became normal to seek new paths for obtaining the monetary masses required for covering these requirements.

Later on, as a consequence of the crisis of 1929, moving into first place are the problems of the lack of liquidity due to the phenomena of the accumulation of stocks and the lack of resources to cover the most pressing payments taking into account the fact that one of the normal sources of financing, the sale of products, had become reduced in a drastic and spectacular manner. Mixed in with these problems was the need that arose for studying the causes and consequences brought on by the economic depression, arising from the liquidation of business through bankruptcy, suspension of payments, etc.

After the end of the Second World War, new financial problems appear, but this time with different characteristics, arising from the need to rebuild an economic world destroyed by the disaster of the second great war and by the need to reconvert a wartime industry into an production industry for peace.

Currently, the evolution of events leads us to a new context in which financial problems have varied substantially. To make an attempt to determine the moment at which these problems arose would be, in the very best of cases, risky. Nevertheless, it can be pointed out that when economists become aware of the fact that a new financial framework exists, is perhaps during the late fifties or early sixties. During this period an important change occurs in the financial structure of businesses, which leads to the fact that the classical problems of raising means are practically relegated to a second level and that they then constituted a consequence of the study of the internal financial structure of businesses. What then occurs is the raising of funds is done by means of resorting to the different sources of financing of the business and, obviously, from the economic-financial system in which the business carries out its activities. The study of the financial structure allows for looking at the two fundamental aspects of the financial problem that is the raising of resources and their placing under the best possible conditions, from a different angle.

In this way studies on the cost of capital and as a consequence of this the attempts, by theoreticians, for determining the optimum financial structure, enter the scene. Obviously both problems, which go to make up the basic core of financial activity, are not dissociated, but are very closely related.

Accompanying these two basic aspects other elements appear in financial studies. On the one hand the concern for attaining a certain degree of liquidity and on the other, to control the degree of liabilities. Forming a framework to both these aspects appears the problem of profitability.

Finally, the dividend policy will be the element to be taken into account in the cash analysis of the business. It has an influence on liquidity and in the degree of self-financing of the same.

There can be no doubt whatsoever that all these aspects are all interrelated. If classical studies were basically centred on the financial structure of the balance sheet, currently it is considered that this aspect is one of the elements that it is necessary to consider, if it is required to know the relationship existing between the cost of financing on the one hand and profitability of assets on the other.

In short, the passage of classical to modern studies relative to this particular subject, has meant that financial studies have become more extended with the incorporation of the study of financial phenomena that occur within business and their relation outside the same, both those that imply raising funds and those that are as a result of placing financial means, by establishing certain criteria for the assignment of the financial resources obtained.

With regard to the techniques used in the study of financial problems three very different phases should be highlighted. The first of these coincides with the classical studies of financial activity, in which the techniques used are fundamentally descriptive. No use is made, in this case, of mathematical instruments, since the objective does not require any direct quantification. This is the era in which "the ritual of the demand for credit" becomes a myth. The sources of financing are established, and along with these the paths to be followed for obtaining the necessary means of payments, and only as an afterthought do certain comparisons appear relative to the overall cost of the financial resources for each of the possible areas of raising funds.

In the second phase, which coincides with the commencement of internal financial analysis in business, traditional mathematics is used, based on purely mechanistic logic. This is the time when any probability is quantified by means of measurements. Models are drawn up based on certainty or probability. What this achieves is a convenient formulation, and in a certain way models that show reality with a certain approximation, a reality that at that time did not give rise to neither important nor rapid changes.

But the economic-financial situation of business has gone on changing substantially. During latter times profound changes have taken place that, and this is the most important fact, in the immediate future they will be far more intense and accelerated. Therefore, current financial reality is characterised by change, and to talk of change means, when faced with the future, to place problems within the sphere of uncertainty. A great deal of effort has been required in order to attempt to draw up certain schemes that were suitable for showing and treating this permanently changing reality. At this time certain new instruments are being drawn up that will be used for the treatment of financial problems in an attempt to draw models that will show this new reality.

We will return to this important aspect further on.

1.4 The Economic-Financial and Dividend Policy

Business, through its activities, seeks determined objectives the study of which has given rise to extensive literature, which has even gone much further than the limits of mere management study. On tackling this section we have no intention of increasing the number of works that exist on this subject, but faced with this fact and starting out from the assumption that these objectives are multiple, we have attempted to draw up a scheme.

We feel then that the actions of the businessman are aimed at obtaining not only the maximum profit, or the enrichment of the owners of the business, but that, also they seek other goals, which can be found on a level of equality or be secondary objectives or even intermediary objectives. In any event, they are all necessary for the "satisfaction" of the shareholders.

Thus, a whole series of concepts can be listed that represent, in some way, the objectives sought by business. Merely as an indication, and without intending that these be exclusive, the following can be considered:

1. Primary objectives:
   - Increase the value of the business.
   - Increase the value of the shares.
   - Obtain permanence in economic life.
2. Secondary objectives:
   - Obtain political-social power.
   - Consolidate prestige in the market.
   - Strengthen the position for agreements with other businesses.
3. Intermediary objectives:
   - Facilitate obtaining credits
   - Favour commercial expansion.

On the other hand, a whole set of elements exist that can have a certain consequence on these objectives. To list them would be too long and to be certain incomplete, however much we attempted to add headings to those already existing. For this reason and taking into account the need for synthesis that permits agility, some of the "causes" having a consequence on the objectives and which belong to the different areas of business activity, can be mentioned. For example:

1. Economic-financial aspect:
   - Annual profits.
   - Invoicing.
   - Degree of liquidity.
2. Labour aspect:
   - Number of employees.
   - Union activity.
3. Productive aspect:
   - Stock levels.
   - Degree of technology.
4. Commercial aspect:
   - Range of products in the market.
   - Quality and presentation of the products.
5. Governing co-ordinating aspect:
   - Management team.

As can be seen, both in the objectives for the business as in the causes that have been described, there appear elements that form a part directly of the financial area of the same. It is obvious that from the interconnection existing between the different areas in which the business moves, financial aspects will have an incidence or will be affected also by the other areas of activity of the business, such as labour, production, commercial, etc.

We should point out here that currently business financial studies do not now start out from the hypothesis according to which the sole objective of the business resides on the optimisation of profit, but that normally attention is placed more on the "optimisation of the value of the business for the shareholders", which is equivalent to saying that what is being attempted is to increase the quotation of the shares.

It cannot escape the notice of anybody that the increase in the value of the business is intimately tied in with the value of the shares and it becomes obvious that high annual profits imply in some way an increase in the quotation of the shares.

Now then, when talking about the relationship existing between profits of the business and the quotation of the shares of the same another important aspect should be taken into account: is the dividend policy. It is particularly well known that the decision to distribute a percentage of the profits acquired by the business to the shareholders is one of the important elements made by the businessman from a financial point of view, since the value of the company will depend on this.

In fact, profits can be separated into two parts: one of these leaves the sphere of the business in order to reach the shareholders and the other may stay in the business in the concept of reserves and these constitute an increase in the auto-financing of the same. The quotation of the shares depends both on the profits that are distributed and on the fundamental value of the business. Therefore, it can be stated that the decision will be the better from an economic point of view when the market value of the shares is increases as much as possible.

Thus, it will be necessary to establish a dividend policy such that it allows for equilibrium between the internal requirements of the business, that is to say, the amount of profit retained as reserves, and the amount for paying the shareholders for their participation in the capital of the same.

Dividends, this compensation to the shareholders for their deposit of a monetary mass in the company, constitute an essential element for the value of the business itself in the share market to be maintained or to improve.

Nevertheless, in reality, what has true importance is the net annual profit of the business, which constitutes through its evolution, year after year, the indication of the health of the business itself.

Obviously, the concept of profit is not a monetary mass with an absolute, overall, definite value but it depends, to a greater degree, on the different internal policies such as the investment policy, the renewal policy, the depreciation policy, etc. The relativity of profit from one year can be eliminated if a sufficiently large period of time is considered and the political-financial prospects of the business for the future are analysed.

To summarise, it can be stated that there exists in the business an economic-financial policy that attempts to attain certain objectives that are tied to that which another type of policy seeks, which the theoreticians in financial administration like to separate: that is the policy of dividends. Obviously, although it may be convenient to consider these types of policies for clarification of the financial study of the business, it must be taken into account that both are intimately tied and inter-linked in such a way that a reciprocal influence exists between them.

2 Basic Elements for the Treatment of Uncertainty

2.1 From Binary Logic to Multivalent Logic

Human thought, although charged with a high degree of subjectivity, seeks to find the objective. In the seeking process, scientific knowledge has been developed over history. The combination of logical reasoning and sensations makes up the behaviour in the varied spheres in which mankind carries out its activity. It is, although not exclusively, the former which allow for calculation, thought and selection. These acts constitute the central hub around which economic life revolves and on which financial activity is carried out.

The formalisation of financial phenomena, as occurred with the construction of economic science, has been traditionally carried out with the aid of certain instruments, which were useful in other sciences. Another's success led to their being used in the specific field. This was, in a first phase, the incorporation of mechanist mathematics that required the consideration of a being, Homo oeconomicus, lacking all feeling, which prevented giving to his acts the slightest hint of subjectivity. For this aid to the logical treatment of thought was later added binary logic that constitutes the necessary support for the operation of computers. This logic, in which the principal came into being of the excluded middle principle, has remote origins which are to be found in Aristotle and Crisipides, although its efficiency is due to the form that it acquired thanks to the work of George Boole with the publication in 1853 of his work, Laws of Thought. The results, whilst being used in the relation man-machine, have been spectacular, and science owes much of its advances to the possibilities of treatment that Boolean algebra allows.

Nevertheless, in thought not all is binary, as included in it, from perception to decision, are nuances, and frequently human logic is imprecise, vague, fuzzy. It then becomes necessary to pass over from binary logic to multivalent logic.

We have entered a new era in which, the association between man and machine is going to be possible and fruitful. The one intelligent and imperfect, the other theoretically perfect but requiring programming, will allow for the treatment of the most complex problems that will arise in this highly intricate world of today and the future.

It is generally accepted that knowledge takes place starting out with models, that is to say, of representations that are nearly always schematic that

Anna Maria Gil-Lafuente: *Fuzzy Logic in Financial Analysis*, StudFuzz **175**, 13–36 (2005)
www.springerlink.com

the brain stores in the memory as images, sensations, numerical data, formulae, relations of causality... Scientific knowledge is constructed principally from these models, which are acquired by man by means of education and teaching. The vast quantity of models existing in a certain manner makes it necessary to seek a balance between their efficiency and their simplicity. The same happens with logic. It can be neither universal nor perfectly objective, but it must be easily understood and simple to use in order to avoid an excess of sophisms. Aristotelian logic is not a universal logic, but one among many. For a very wide variety of languages, very varied logic exists.

The excluded middle principle, and that of no contradiction forms a part of our mental patrimony. We say that a proposition cannot be at the same time true and false, but that it is "always" true or false. Nevertheless, existing in thought, between truth and falseness, is an infinity of nuances. The need to seek a fluid and simple communication has led to the custom of using binary values for truth. The introduction of computers, constructed by binary elements, has done no less than reinforce this tendency. In the field of electronics, the binary system is the cheapest and easiest form of treatment. With 0 and 1 it is possible to express any number, concept, operation, although the number of digits (position 0 or 1) can be enormous. Mathematicians call this algebra, Boolean algebra, in honour of the person who formulated the binary rules of thought.

Two symbols are used for truth: V that means true and F that means false, that can be substituted for any other pair the meaning of which must correspond explicitly with V and F. In a practically universal manner the pair 0,1 has been chosen, where 1 corresponds to truth and 0 to false, in the same way as in the theory of sets membership is expressed by 1 and non membership by 0. As happens with normal language, "operators" or "connectors" are also used: for "and" Δ is used Δ, for "or/and" ∇ is used, for denial or negation $\neg$ or else $^{-}$, for equality $=$. Now as in binary logic 0 and 1 can be put into an order $1 > 0$, instead of using Δ and ∇ symbols $\wedge$, "minimum", and $\vee$ "maximum". As far as negation goes, this becomes "complementation", in such a way that $\bar{1} = 1 - 1 = 0$, $\bar{0} = 1 - 0 = 1$. In this way the following tables can be constructed:

$\wedge$	0	1
0	0	0
1	0	1

$\vee$	0	1
0	0	1
1	1	1

$^{-}$	0	1
	1	0

It is also frequent, above all in data processing, to use $(\cdot)$ product, for representing "and", as well as $(\dotplus)$ to represent "or/and", which should not be confused with the addition symbol $+$ of common algebra, since $1 + 1 = 2$, but $1 \dotplus 1 = 1$.

One of the most important mechanisms of logic is the "relation of inference", which is one of the functions of the two binary variables that are most used presented under the form of "if X ... then Y", for example "if the interest rate decreases, then ask for the credit". When an inference is such that P infers Q, this is written as $P \to Q$ and this can be expressed by saying that "if P is true then Q is true", which in binary logic would give rise to $\bar{P} \nabla Q$ $(\neg P \nabla Q)$ and by the binary variables $\bar{a} \dot{+} b$. Therefore the following tables can be established:

$\bar{P} \nabla Q$ (P \ Q)	F	V
F	V	V
V	F	V

$\bar{a} \dot{+} b$ (a \ b)	0	1
0	1	1
1	0	1

The value truth of a proposition P is normally expressed by $v(P)$, therefore when $P = V$, we have $v(P) = 1$, while if $P = F$, then it will be $v(P) = 0$, which is the same as writing $v(P) = a$, taking into account that a can only be equal to 1 or 0.

In the sphere of Boolean logic an inference is the same $v(P \to Q) = 1$ and, therefore, $v(\bar{P} \nabla Q) = 1$. Let us take a look at what happens when starting out from a and it is established that $\bar{a} \dot{+} b = 1$:

1. If: $a = 0, \bar{a} \dot{+} b = 1 \dot{+} b = 1$, therefore b can be either 0 or 1, without a possible decision, "undecidable".
2. If: $a = 1, \bar{a} \dot{+} b = 0 \dot{+} b = 1$, therefore b can only be 1.

When starting out from b and it is established that $\bar{a} \dot{+} b = 1$:

1. If: $b = 0, \bar{a} \dot{+} b = \bar{a} \dot{+} 0 = 1$, therefore a can only be 0.
2. If: $b = 1, \bar{a} \dot{+} b = \bar{a} \dot{+} 1 = 1$, here the result is that a can be 0 or 1, possible decision, "undecidable".

It is for this reason that only two "modes" arise in order for an inference to be usable: the modus ponens, in which the premises are $v(P) = 1, v(P \to Q) = 1$, and the conclusion $v(Q) = 1$, and the modus tollens the premises of which are $v(Q) = 0, v(P \to Q) = 1$, and the conclusion $v(P) = 0$.

The problem that arises in binary logic consists in determining in a sequence of simple or compound propositions, which are the true, the false, the undecidable or impossible. Let us assume that in T intervene operators $\Delta, \nabla, \neg$ or even none (simple proposition). The following can be found:

- $v(T) = 1, T$ is true
- $v(T) = 0, T$ is false
- $v(T) = \{0,1\}$, undecidable
- $v(T) = 1$ and $v(\bar{T}) = 1$, impossible or not able to satisfy

It is obvious that when a single impossible proposition appears, all the sequence is also impossible.

Up to now an attempt has been made to bring to light how to express human thought by means of binary logic. Nevertheless, the human brain carries out its activity in a different way from binary mechanisms. Between truth and falseness there is a very wide range of nuances in the same way as there are infinite shades of greys between black and white.

In real financial problems, the logical treatment, be this numerical or not, resorts to information of a very diverse nature. Some of this information comes to light by means of formal or probable measurements, others are simply the result of the opinion of experts and, as such, of a subjective nature. Estimating a value of the truth of a proposition cannot be limited to values 1 (truth) or 0 (Falseness). Language is full of adjectives and adverbs that provide manifestations that are between the truth and the false. In certain cases we express an idea with precision, as we say in this way that "the inter-bank interest rate is 13.5%", in other instances, in imprecise terms, such as when it is said that "working capital must be positive" the use of precise and imprecise data is a constant in our communications according to the level of knowledge and as a consequence of the inevitable subjectivity of our reasoning. This is the reason why "human treatment of information is neither bivalent nor binary, but multivalent. Between the truth and falseness intermediary positions can be accepted in order to arrive at a transcription that is closer to reality. Between what is false (0) and what is true (1) intermediary positions will be admitted in which the values will be frequently located between 0 and 1. In this way, logical values will be accepted positioned between 0 and 1, for which it is convenient to define the corresponding semantics. This correspondence as with estimating will be subjective and will vary between one individual and another"[1].

Therefore, in multivalent logical reasoning a proposal is accepted with a "level of truth" by taking any number between 0 and 1, obviously including the extremes 0 and 1. With the object of providing a greater degree of freedom of opinion, it can also be expressed by means of "confidence intervals". If P is a simple or compound proposal it can be assigned a value of truth $v(P)$ which can be either a number between 0 and 1, or else a confidence interval $[v_1, v_2]$ taken in [0,1], which indicates that the opinion can be found between v_1 and v_2 degrees of truth located in the segment [0,1] and where $0 \leq v_1 \leq v_2 \leq 1$.

Faced with a proposition such as:

P: "in the next economic accounting period there will be gross profits of 300 million"

the following estimates of the degree of truth can be assigned:

- Fairly true: $v(P) = 0{,}7$
- [more false than true, fairly true] $= [v_1, v_2] = [0{,}4,\ 0{,}7]$

[1] Kaufmann A (1988) Les Logiques humaines et artificielles. (In french) (ed) Hermes, Paris, p. 35.

Obviously the opinion of a single expert cannot be considered as exact, given the degree of subjectivity of any individual estimate. Procedures exist, but this is another problem, of the aggregation of opinions of several experts. At the right time we will insist on this important aspect.

In binary reasoning a great variety of functions exist that are linked by operators such as $\Delta, \nabla, \neg$. These same operators are also used in multivalent logic. When using numerical values the operator or connector "and" is symbolised by $\wedge$, "minimum", "and/or" by $\vee$, "maximum" and negation by the line over the symbol. Nevertheless, the use of $(\cdot)$ logical product and of $(\dot{+})$ logical sum must be avoided. Let us take a look at an example:

If:

$$v(P) = 0{,}4 \quad v(Q) = 0{,}7$$

we have:

$$\begin{aligned} v(P) \wedge v(Q) &= 0{,}4 \quad v(P) \wedge v(Q) = 0{,}7 \\ v(\bar{P}) &= \overline{0{,}4} = 1 - 0{,}4 = 0{,}6 \\ v(\bar{Q}) &= \overline{0{,}7} = 1 - 0{,}7 = 0{,}3 \end{aligned}$$

There also exists a very simple algebra for confidence intervals:

If:

$$\begin{aligned} [v_1, v_2] &= [0{,}3, 0{,}6] \\ [w_1, w_2] &= [0{,}2, 0{,}7] \end{aligned}$$

We have:

$$\begin{aligned} [0{,}3, 0{,}6] \wedge [0{,}2, 0{,}7] &= [0{,}2, 0{,}6] \\ [0{,}3, 0{,}6] \vee [0{,}2, 0{,}7] &= [0{,}3, 0{,}7] \end{aligned}$$

The logical multivalent inferences are an extension of binary inferences in which qualification of the reasoning is permitted. As with binary inference, in multivalent inferences $v(P \to Q)$ is a piece of information, and when $v(P)$ is known, $v(Q)$ must be arrived at, while when $v(Q)$ is known, $v(P)$ must be found. Now, if only one binary inference exists, on the other hand an infinity of multivalent inferences can be conceived. In fact, if the binary structure, in numerical values, can be described by the following table, which was seen before:

$\bar{a} \vee b$		b: 0	b: 1
a	0		
	1		

in the event of multivalent logic inferences, Table 2.1[2], in which the grading of 0 to 1 has been done in tenths, that is to say, 0, 0,1, 0,2, . . . , 0,9, 1. This then is what we have in Table 2.1:

Table 2.1.

v (P) \ v (P→Q)	0	0,1	0,2	0,3	0,4	0,5	0,6	0,7	0,8	0,9	1
0	1	1	1	1	1	1	1	1	1	1	1
0,1											1
0,2											1
0,3											1
0,4											1
0,5											1
0,6											1
0,7											1
0,8											1
0,9											1
1	0										1

In this direction the value of truth never diminishes (upward, right side of table)

In this direction the value of truth never diminishes (left to right, below table)

Two fundamental rules must be respected in the sense of left to right and from below to above: the values must never diminish, that is to say they can be the same or higher. And on the other hand for value 0 and 1 the multivalent inferences must provide the same result as the binary inference. If these two conditions are complied with, as many multivalent inferences can be drawn up as are desired. It is sufficient to fill in the boxes of the previous table with numbers from the segment [0,1]. It is obvious then that the essential problem that arises is how to assign a number to each empty box. Certain authors have proposed determined inferences by using mathematical relations adapted to the requirements or conveniences that have arisen.

From among the many inferences that appear in specialised works on the subject, we will mention as the most used the Lee inference and the Lukaciewicz inference, which normally are presented as follows:

[2] This table has been taken from the already mentioned work: Kaufmann A (1988) Les Logiques humaines et artificielles. (In french) (ed) Hermes, Paris, p. 40.

$$\text{Inference of Lee:} \quad v(P \to Q) = v(\bar{P}) \vee v(Q) = \bar{a} \vee b$$
$$\text{Inference of Lukaciewicz:} \quad v(P \to Q) = 1 \wedge (v(\bar{P}) + v(Q)) = 1 \wedge (\bar{a} + b)$$

But others exist which a very adequately adaptable to a determined number of problems. The is the case of the Gödel inference and the Goguen inference:

$$\begin{array}{lll} \text{Gödel inference:} & a \to b = 1\,, & \text{when } a \le b \\ & a \to b = b\,, & \text{when } a > b \\ \text{Goguen inference:} & a \to a = 1\,, & \text{when } a \le b \\ & a \to a = b/a\,, & \text{when } a > b \end{array}$$

As will be seen in later chapters, the result arrived at by using one or the other is different. This lack of uniqueness in the solutions in our understanding does not constitute an inconvenience but all to the contrary, it means the possibility of a better treatment of real problems on using, in each case, the inference that is better adapted to the specific case. With this what is arrived at is that human "logic" is susceptible to reproduction for providing a better treatment to the complex reality of our world.

2.2 Fuzzy Logic and Fuzzy Sub-Sets

The theory of fuzzy sub-sets constitutes a very wide context in which to situate multivalent logic. Their origin can be found in the works, which in 1965 were developed by L. Zadeh, professor at the University of California, and today constitutes a mathematical theory constructed in all rigor that allows for the treatment of subjectivity and/or uncertainty. Its development has brought up an epistemological problem in the sense that is it better to use a certain model, which is unlikely to represent reality, or a fuzzy model that constitutes a valid reflection. In our understanding, it is necessary to observe economic and financial phenomena and determine their nature. It will be when they are presented in a fuzzy, vague manner, with limits that it will be necessary to use fuzzy mathematics. But we should not fall into the temptation of converting into fuzzy that which is not, but neither should we qualify as certain that which appears as fuzzy.

Knowledge of the fact, persons and things is situated at different levels the specification of which is difficult. Between perfect knowledge of a phenomenon and total ignorance, knowledge that is more or less imprecise can be found.

Let us now take a look at some of the basic elements of the theory of fuzzy sub-sets, starting out from the most elementary mathematical knowledge. Thus, from a purely intuitive point of view, a group of objects, different from each other and perfectly specified is called a set. If the six elements that form set E are designated by a, b, c, d, e, f, the following could be written:

$$E = \{a, b, c, d, e, f\}$$

If instead of considering all the elements only some of them are considered, then a so-called "sub-set" is formed, for example:

$$A = \{a, c, e, f\}$$

When using 1 and 0 for representing membership or non-membership of the elements of the referential to the sub-set, this could also be expressed:

	a	b	c	d	e	f
A =	1	0	1	0	1	1

where the 0 and 1 define the characteristic function of membership. That is to say that in this presentation a 1 is assigned if an element that is a member of A and a 0 if not being a member of A. The theory of sets is principally based in this way on the idea that an element of the referential set is a member or not of a sub-set of this set.

This approach, the use of which in many aspects has been sufficiently proven, does not reflect in depth the functioning of our brain, because it does not include an activity as important and continuous as nuances. For a population a property could be considered, such as "to be young". But as in the case of many other properties, this is vague, imprecise, "fuzzy", because it is difficult to catalogue the major part of the population as young or not young. Although no doubt exists that a newborn is young (a 1 is assigned to the function characteristic of membership), and as old a person of 80 years (assign a 0 to the function characteristic of membership), what can be said of a person of 30, 45 or 55 years old? In reality, we get older from the very day we are born. Therefore, the sub-set of young people of this set cannot be represented only by 0 (is not young) or by 1 (is young), but will require a grading from 0 to 1. Then it will be said that the sub-set of young people of the population formed by the referential is a "fuzzy sub-set". The degree or "level" of youth assigned to each person of the referential will obviously be subjective.

If we consider the same referential, a fuzzy sub-set of this referential E would be for example:

	a	b	c	d	e	f
$\underset{\sim}{B}$ =	0,6	0	0,3	0,7	0,1	1

which indicates that a is a member of $\underset{\sim}{\mathbf{B}}$ with a degree or "level" estimated at 0,6, b with a level 0 (not a member of $\underset{\sim}{\mathbf{B}}$, etc. In order to show the nature of fuzziness of the sub-set we place a swung dash under the capital letter (in this case B).

Fuzzy sub-sets posses the same properties as regular sub-sets, excepting for the "excluded middle principle" and "non contradiction", that is:

$$\underset{\sim}{\mathbf{B}} \cap \bar{\underset{\sim}{\mathbf{B}}} \neq \emptyset$$

$$\underset{\sim}{\mathbf{B}} \cup \bar{\underset{\sim}{\mathbf{B}}} \neq E$$

where $\cap$ is the intersection, $\cup$ the union, $\bar{\underset{\sim}{\mathbf{B}}}$ the complementation of $\underset{\sim}{\mathbf{B}}$, $\emptyset$ the vacant sub-set and E the referential.

As with the classical theory of sets, also in the field of fuzziness a resort can be made to operators such as the intersection, union, complementation, among others. If we start out from two fuzzy sub-sets of referential E such as:

	a	b	c	d	e	f
$\underset{\sim}{B}_1 =$	0,2	0,9	1	0,5	0	0,7

	a	b	c	d	e	f
$\underset{\sim}{B}_2 =$	0	0,4	0,8	0,1	0	0,6

The intersection $\cap$ corresponding to the "and" is done by choosing for each element the lowest value of the function characteristic of membership. Thus, a is a member of $\underset{\sim}{\mathbf{B}}_1$ up to a level of 0,2, whilst in $\underset{\sim}{\mathbf{B}}_2$ at a level of 0, therefore the smallest value is taken, which is 0; b is a member of $\underset{\sim}{\mathbf{B}}_1$ to a level of 0,9, and of $\underset{\sim}{\mathbf{B}}_2$ to a level of 0,4, therefore 0,4 is chosen; and so on successively. We then arrive at:

	a	b	c	d	e	f
$\underset{\sim}{B}_1 \cap \underset{\sim}{B}_2 =$	0	0,4	0,8	0,1	0	0,6

The union or reunion $\cup$ that corresponds to "and/or" (the one or the other or both) is done by choosing, for each element, the highest value of the function characteristic of membership: We would have:

	a	b	c	d	e	f
$\underset{\sim}{B}_1 \cup \underset{\sim}{B}_2 =$	0,2	0,9	1	0,5	0	0,7

The complementation, $\bar{\ }$, finally, is arrived at, by assigning to each element of the referential the complement to the unit. Therefore:

	a	b	c	d	e	f
$\bar{\underset{\sim}{B}}_1 =$	0,8	0,1	0	0,5	1	0,3

	a	b	c	d	e	f
$\bar{\underset{\sim}{B}}_2 =$	1	0,6	0,2	0,9	1	0,4

In the theory of fuzzy sub-sets these operators are used, that are different from the analysers described, such as semantic operators, among which we could mention the relative, superlatives, etc. There are as many semantic operators as there are in normal language, some are easy to transfer to the theory of fuzziness, other present certain difficulties that are not impossible to resolve. In this manner, the theory of fuzzy sub-sets is associated to fuzzy logic in the same way as the classical theory is associated to binary logic.

2.3 Imprecise Numbers

There exists a type of fuzzy sub-set that warrants special attention. We are referring to fuzzy numbers.

The fuzzy number is a fuzzy sub-set of the referential of the real R, or the referential of the naturals N, or the referential of the integers Z, which posses the properties of convexity or normality.

It is said that a fuzzy sub-set is convex when the values of the function characteristic of membership "does not increase" when we move to the left or to the right of the highest value of this function (this highest value is called "maximum presumption". Let us see some examples in the referential of the integer numbers:

In the field of the real numbers and in the hypothesis of continuity, convexity or non-convexity can be represented by means of the graphs shown in Figs. 2.1 and 2.2.

	4	5	6	7	8	9	10	
$\underset{\sim}{A} =$	0	0,4	1	1	0,7	0,2	0	, convex

	4	5	6	7	8	9	10	
$\underset{\sim}{B} =$	0	0,2	1	0,8	0,9	0,3	0	, not convex

	4	5	6	7	8	9	10	
$\underset{\sim}{C} =$	0	0,3	0,6	0,8	0,2	0,1	0	, convex

A fuzzy sub-set is normal if it posses at least one element the characteristic function of which has as its value the unit. Thus, in the field of the integer numbers:

	4	5	6	7	8	9	10	
$\underset{\sim}{D} =$	0	0,4	0,2	1	0,7	0,2	0	, normal

	9	10	11	12	13	14	15	
$\underset{\sim}{E} =$	0	0,6	0,8	0,9	0,3	0,1	0	, no normal

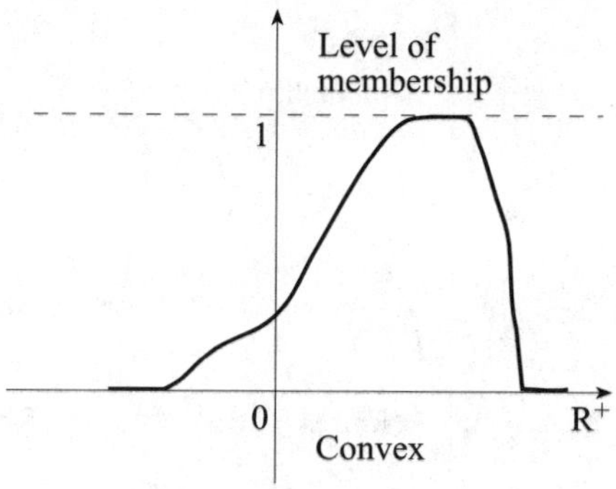

Fig. 2.1.

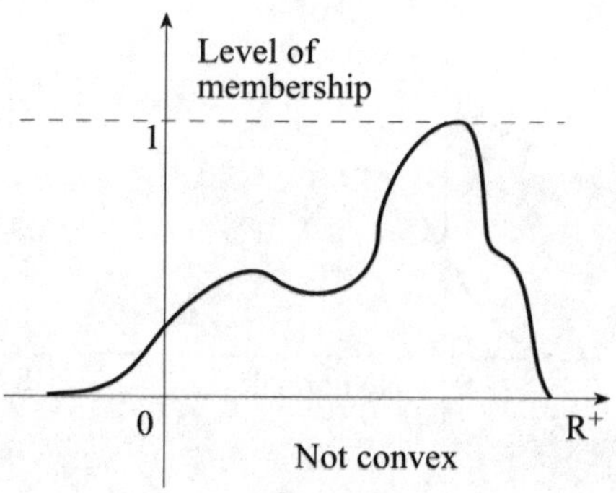

Fig. 2.2.

In the hypothesis of continuity this can be shown as seen in figures below. These therefore will be fuzzy numbers (see Figs. 2.5 and 2.6):

	−3	−2	−1	0	1	2	3	4
$\underset{\sim}{G}$ =	0	0,4	0,7	1	0,8	0,5	0,3	0

	2	3	4	5	6	7	8	9
$\underset{\sim}{H}$ =	0	0,6	0,8	1	1	1	0,4	0

A particular case of fuzzy numbers is that of fuzzy triangular numbers. In this case the maximum value of presumption is unique and the development of the function characteristic of membership is linear both in the direction of the lower extreme and the upper extreme. In other words, the level of membership or level of presumption diminishes at the same rhythm for decreasing values and also diminishes at the same rhythm for increasing values of the referential. The rhythm of decrease of the left can be different from that of the right. In

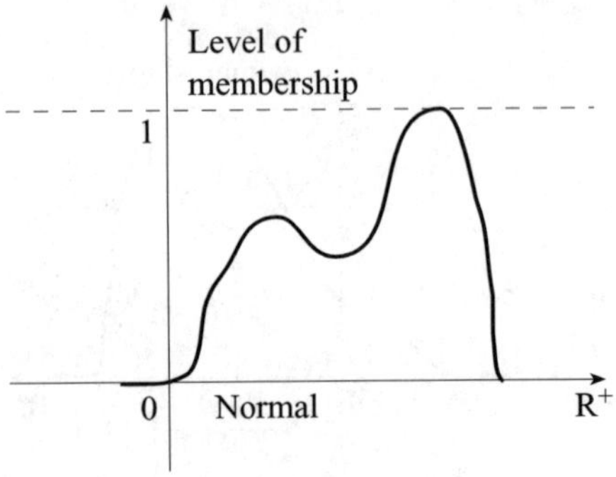

Fig. 2.3.

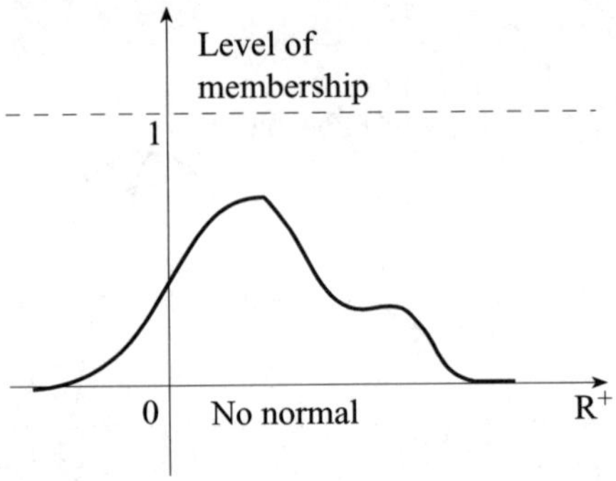

Fig. 2.4.

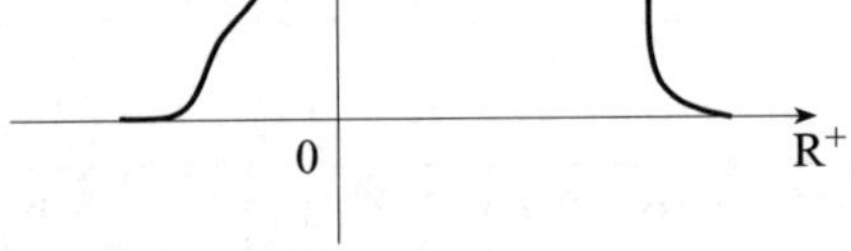

Fig. 2.5.

this case, the fuzzy number can be drawn in a graph by means of a triangle and can be represented in the ternary (by means of three numbers) form. Therefore the fuzzy triangular number $\underset{\sim}{\mathbf{X}} = (x_1, x_2, x_3)$ can be represented in a graph as is shown in Fig. 2.7.

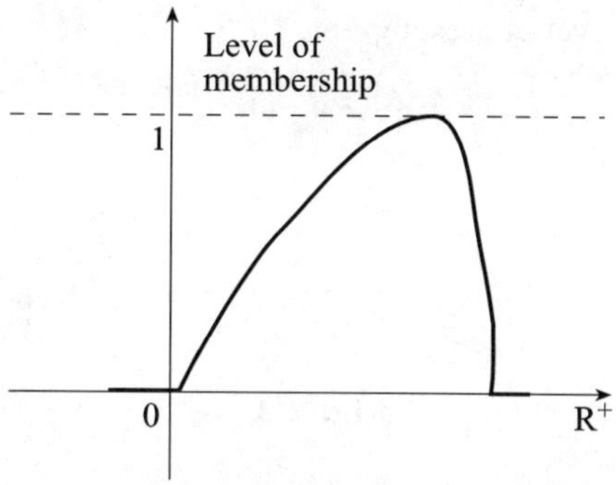

Fig. 2.6.

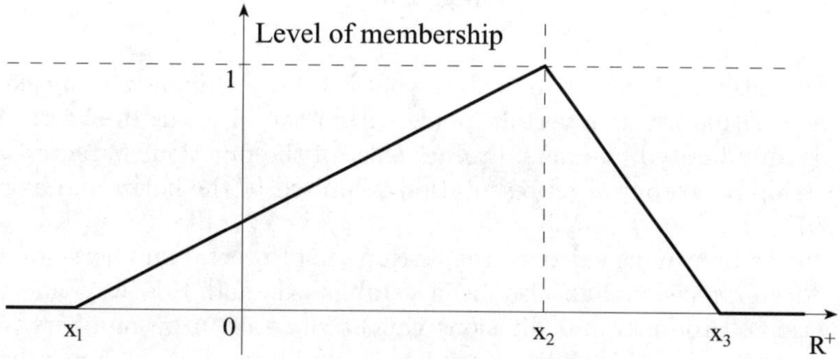

Fig. 2.7.

In the case of fuzzy triangular numbers it is assumed that the value of the function characteristic of membership has been estimated for all the values of the referential. Therefore, information is possessed – normally subjective – of any value between x_1 and x_2 and between x_2 and x_3. When only possessing information on the extremes x_1, x_3 and the maximum presumption x_2, we are faced with a "confidence triplet". In confidence triplets the values of the function characteristic of membership are totally unknown for the intermediary values of the referential. Thus, the triplet $Y = (y_1, y_2, y_3)$ can be represented graphically as is shown in Fig. 2.8.

This is equivalent to saying, that the estimate of a determined value, for example, is reduced to stating that this will not have any values lower than y_1, nor higher than y_3, and that the maximum presumption (it is assumed with the maximum intensity), will be y_2.

In the last instance the case can be considered in which the information is reduced to stating that a phenomenon or value will not be less than a determined number z_1 not higher than z_2. In this case the confidence interval $T = [t_1, t_2]$ appears. The upper and lower extremes are known between

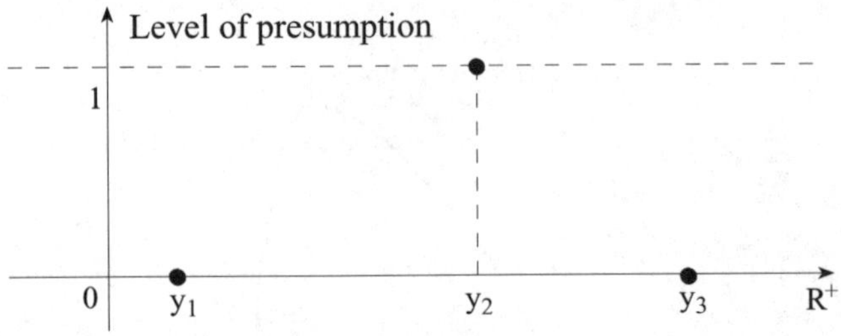

Fig. 2.8.

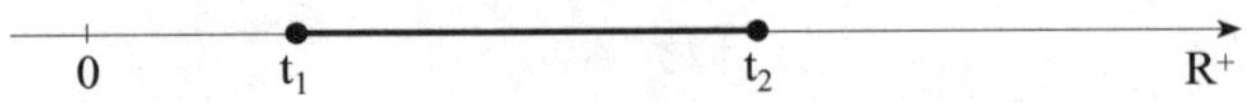

Fig. 2.9.

which the estimated value will be, but what is totally unknown is any element that could situate it at a certain point within the interval. In short, uncertainty is only limited, and as no values exist of the function characteristic of membership its graphical representation is limited to the horizontal axis (see Fig. 2.9).

In the same way as exists in the arithmetic of precise numbers, an arithmetic for imprecision has also been established, and this way additions, subtractions, products and divisions can be done of fuzzy numbers, triangular fuzzy numbers, confidence triplets, confidence intervals and other instruments. In successive chapters we will have the opportunity of using a varied range of operations for resolving the problems created by uncertainty in financial phenomena. We will limit ourselves at this point to the most elementary operations with the easiest instrument for tackling uncertainty: confidence intervals. We have already pointed out how to arrive at the maximum and minimum so we will now look at the possibility of doing other operations, with the condition of remaining within $[0; 1]$.

For addition, it is sufficient to add the lower extreme of one interval with the lower of another (or others) with the upper extreme of one with the upper extreme of the other (or others). Therefore:

$$[a_1, a_2]\,(+)\,[b_1, b_2] = [a_1 + b_1, a_2 + b_2]$$

Using numbers instead of letters:

$$[2,4]\,(+)\,[-3,6] = [2 + (-3), 4 + 6] = [-1,10]$$

In subtraction[3], from the lower extreme of the minuend the upper extreme of the subtrahend must be subtracted and from the upper extreme of the

[3] Here we have described the normal subtraction of intervals. There are other types of subtraction, such as that of Minkowski, which consists in finding the difference

minuend the lower extreme of the subtrahend must be subtracted, that is to say that a cross-over takes place of the extremes. We will have then:

$$[a_1, a_2]\,(-)\,[b_1, b_2] = [a_1 - b_2, a_2 - b_1]$$

and with a numerical example:

$$[2{,}4]\,(-)\,[-3{,}6] = [2 - 6{,}4 - (-3)] = [-4{,}7]$$

When dealing with the product, and the intervals are formed by non negative numbers, the lower extreme of an interval is multiplied by the lower extreme (or extremes) of the other and the upper extreme with the upper extreme or (extremes) of the other, When there are negative extremes it may be necessary to invert (or cross over) the extremes in order for the general condition, that in the resulting interval all possible solutions are present, to be complied with:

$$\begin{aligned} R^+ : [a_1, a_2](\cdot)[b_1, b_2] &= [a_1 \cdot b_1, a_2 \cdot b_2] \\ R : [a_1, a_2](\cdot)[b_1, b_2] &= [\mathrm{Min}(a_1 \cdot b_1, a_1 \cdot b_2, a_2 \cdot b_1, a_2 \cdot b_2), \\ &\qquad \mathrm{Max}(a_1 \cdot b_1, a_1 \cdot b_2, a_2 \cdot b_1, a_2 \cdot b_2)] \end{aligned}$$

and with numerical examples:

$$\begin{aligned} [3{,}4](\cdot)[5{,}8] = [3.5, 4.8] &= [15{,}32] \\ [-5, -1](\cdot)[-2{,}4] &= [\mathrm{Min}(-5 \cdot (-2), -5.4, -1 \cdot (-2), -1.4), \\ &\qquad \mathrm{Max}(-5 \cdot (-2), -5.4, -1 \cdot (-2), -1.4)] \\ &= [-20, 10] \end{aligned}$$

With regard to the quotient, a distinction must also be made between non-negative and negative numbers. In the former case, when we are in R^+ the lower extreme will be the result of doing the division between the lower extreme of the dividend and the upper extreme of the divisor, and the upper extreme the result of dividing the upper extreme of the dividend by the lower extreme of the divisor. If one or more negative numbers appear, then it will be necessary to calculate all the possible quotients with the extremes and select the lowest result as the lower and the highest as the upper:

$$R^+\text{: } [a_1, a_2](:)[b_1, b_2] = \left[\frac{a_1}{b_2}, \frac{a_2}{b_1}\right]$$

$$R\text{: } [a_1, a_2](:)[b_1, b_2] = \left[\mathrm{Min}\left(\frac{a_1}{b_1}, \frac{a_1}{b_2}, \frac{a_2}{b_1}, \frac{a_2}{b_2}\right), \mathrm{Max}\left(\frac{a_1}{b_1}, \frac{a_1}{b_2}, \frac{a_2}{b_1}, \frac{a_2}{b_2}\right)\right]$$

---

between the lower extremes on the one hand and between the upper extremes on the other, that is without crossing over the extremes. It is used for solving equations and for showing determined financial operations, as will be seen in later chapters.

As an example:

$$[2,5](:)[3,6] = \left[\frac{2}{6}, \frac{5}{3}\right]$$

$$[4,8](:)[-3,5] = \left[\mathrm{Min}\left(\frac{4}{-3}, \frac{4}{5}, \frac{8}{-3}, \frac{8}{5}\right), \mathrm{Max}\left(\frac{4}{-3}, \frac{4}{5}, \frac{8}{-3}, \frac{8}{5}\right)\right] = \left[-\frac{8}{3}, \frac{8}{5}\right]$$

We are going to leave arithmetical operations for now[4] and move on to a matter that acquires special importance in the treatment of problems of uncertainty. We are referring to the "inclusion" of two confidence intervals. Two types of inclusion are normally considered: 1) inclusion called of sets, 2) inclusion by decreasing values.

It is said that an interval $[a_1, a_2]$ is included in the sense of the sets in another interval $[b_1, b_2]$ if, and only if, $a_1 \geq b_1$ and $a_2 \leq b_2$. Numerically speaking we will have then, [3,7] is included in the sense of sets in [2,9] since $3 \geq 2$ and $7 \leq 9$.

On the other hand, it is said that $[a_1, a_2]$ is smaller than or equal to $[b_1, b_2]$ if, and only if, $a_1 \leq b_1$ and $a_2 \leq b_2$. The result of this is that [1,5] is smaller than [4,8] since $1 < 4$ and $5 < 8$.

These two classes of inclusion also allow us to define two essential manners of "fitting" confidence intervals: fitting called of sets, and fitting called by decrease. Let us take a look at the two types of "fitting" by means of some examples:

Confidence intervals [0,10], [2,8], [3,7], [4,7], [5,6] have their fitting of sets represented in the Fig. 2.10.

While confidence intervals [0,4], [2,5], [3,7], [3,9], [5,9] have their fitting by decrease represented in Fig. 2.11.

An essential consideration can now be formulated: a fuzzy sub-set of a referential of real numbers constructed by fitting of sets gives rise to a fuzzy number in the real numbers.

In the graph in Fig. 2.12 a fuzzy number is shown of the referential of the real numbers constructed by continuous correspondence of the confidence intervals. It can be seen that the confidence interval corresponding to level 0,3, which in this case is [4,10] contains to that obtained at level 0,7, that corresponds in the figure to [5, 8,5].

We will see in later chapters that the logical operations that are done with confidence intervals are extendable, level by level, to fuzzy numbers.

[4] For further information on the arithmetical operations we have described see: Kaufmann A and Gil Aluja J: (1987) Técnicas operativas de gestión para el tratamiento de la incertidumbre (in Spanish). Hispano Europea, Barcelona, Chaps. 3–5.

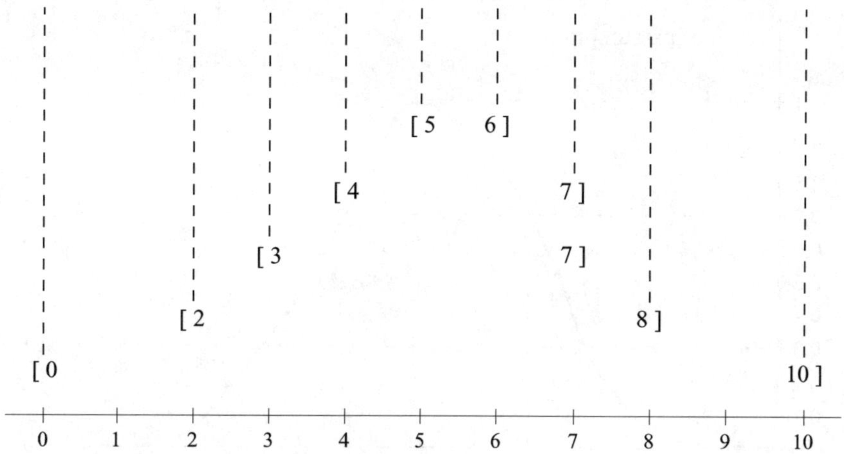

Fig. 2.10. Fitting of sets

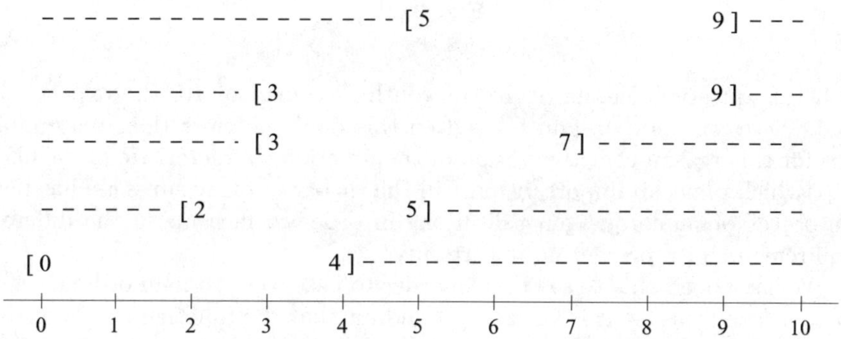

Fig. 2.11. Fitting of decrease

2.4 With Regard to the Problem of Subjectivity

The assignment of values to the function characteristic of membership constitutes one of the aspects that has raised the most controversy in the use of the techniques for the treatment of uncertainty in management problems. When an expert takes on the responsibility of assigning a number to the function characteristic of membership, what he is showing in a number is his or her feeling, which is charged with a component filled with a higher or lower element of subjectivity. Thus, when putting a number of 0,7 to the degree of youth of a person of 40 years old, what it is not removed from this opinion of the very expert are: the age, the physical condition and the opinion regarding youth. Also on many occasions of providing a number in [0,1] and only one, restricts the liberty of opinion of the expert, who finds it very difficult to select between two or three numbers in order to assign a value to the function characteristic of membership. It is possible that faced

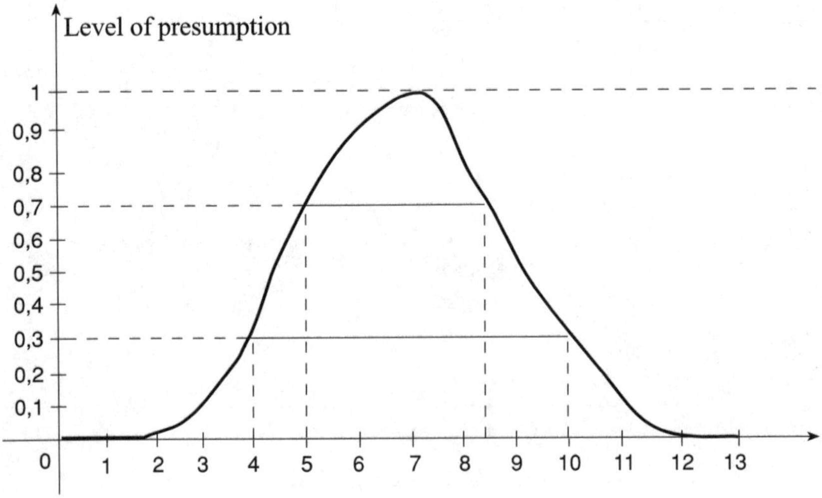

Fig. 2.12.

with this question on the degree of youth of a man of 40, the expert may doubt between 0,5, 0,6, and 0,7. Often this doubt is more than reasonable. It is for this reason that the notion of the function characteristic of membership, which plays an important part in the theory of fuzzy sub-sets, has been the object of numerous generalisations in order to face up to the different requirements of representing uncertainty.

We have been able to see that in order to pass over from an ordinary sub-set to a fuzzy sub-set it is necessary to admit that the function characteristic of membership instead of having its values in the set {0,1}, takes these from the segment [0; 1]. Let us assume now that this function of membership takes its values in the form of confidence intervals $[a_1, a_2]$ always contained within [0; 1].

A sub-set of elements of a referential the values of the function characteristic of membership of which are intervals in [0,1] is called "ϕ-fuzzy sub-set"[5].

Let us assume a finite referential of:

$$E = \{a, b, c, d, e\}$$

A ϕ-fuzzy sub-set of E could be

	a	b	c	d	e
$\underset{\sim}{A}$ =	[0,2, 0,5]	[0,6, 0,7]	0,4	[0,9, 1]	[0,3, 0,8]

[5] The concept of ϕ-fuzzy sub-set is owed to Sambuc A (December 1975), Functions ϕ-flous. Applications au diagnostic en pathologie thyroïdienne (in French). Doctoral thesis in Medicine, Medical Faculty of Marseille, Marseille, France.

It will be seen that in the case of the ϕ-fuzzy sub-set a broken $\sim$ has been placed under the capital letter (in this case A), instead of a diacritic $\sim$.

We should point out that the crisp number is a special case of an interval in which the extremes are so close that they become confused in a single number. Thus $[0{,}4;\ 0{,}4] = 0{,}4$. For this reason it is said that a fuzzy sub-set is a special case of a ϕ-fuzzy sub-set.

The same operations can be carried out with ϕ-fuzzy sub-sets as can be done with fuzzy sub-sets, although taking the due precautions into account relative to the properties of the confidence intervals, such as the intersection (by means of the minimum), union (through the maximum) and complementation ($1 - a_2$ for the lower extreme and $1 - a_1$ for the upper extreme).

ϕ-fuzzy sub-sets constitute a very interesting extension of the theory of fuzzy sub-sets since it allows us to show, in a far more authentic manner, the perception of a phenomenon when faced with the lack of an objective measurement we are obliged to allow intervention of the subjective opinion of an expert intervene.

The liberty of expressing an opinion that is allowed by the ϕ-fuzzy sub-set, does not prevent the fact that on determined occasions the opinion of a single expert is considered as insufficient. In these cases resort can be made to the opinion of several experts and the problem then arising is how to aggregate these opinions. There are several procedures that are more or less favourable, for aggregating opinions. Just for its interest we are going to develop one of these which gives rise to "random fuzzy sub-sets", the function characteristic of membership of which is a random variable: that is to say, that for a same value of the referential a probability is established for all the values considered of the function characteristic of membership. Below is an example of the process for arriving at this:

Starting out from a referential:

$$E = \{a, b, c, d\}$$

And we are going to request a number of experts, let us say 12, to give an opinion relative to each one of the elements of the referential. After gathering the opinions we can draw up the table as shown in Table 2.2.

We now do a statistical chart of the replies given by the experts and arrive at the corresponding probabilities[6] (see Table 2.3).

For operative effects, it is convenient to arrive at the law of complementary probability, by doing the accumulation of probabilities starting out from level 1 to level 0. In this way we arrive at a random fuzzy sub-set of referential E (see Table 2.4). In order to designate a random fuzzy sub-set it is customary to place below the capital letter (in this case A) a diacritic with a point, as can be seen in the example.

[6] We are conscious of the fact of the abuse of language when talking about probabilities with a population of 12 units, but didactic requirements oblige us to consider a reduced number of experts.

Table 2.2.

Expert	a	b	c	d
1	0,6	0,5	0,8	0,4
2	0,4	0,6	0,3	0,7
3	0,3	0	0,4	0,1
4	0	0,3	0,6	0,5
5	1	0,9	0,5	0,8
6	0	0,2	0,3	0,1
7	0,1	0,7	0,6	0,7
8	0,9	1	0,5	0,8
9	0,6	0,8	0,9	0,5
10	0,5	1	1	0,7
11	0	0,4	0,2	0,1
12	0,4	0,5	0,1	0,4

Several operations can be done with random fuzzy sub-sets, which make them very useful for the treatment of a large number of problems, as will be seen in later developments relative to the financial area.

An important and fruitful extension of random fuzzy sub-sets has given rise to "expertons", the idea and development of which is due to A. Kaufmann[7]. The advance represented by expertons relative to other instruments for treatment of uncertainty is shown by the fact that they allow at one and the same time for a good aggregation of the opinion of several experts and that these express their opinions with the freedom given by confidence intervals. Let us look very briefly at how to construct an experton.

Let us assume a new referential.

$$E = \{a, b, c, d\}$$

And the opinion of 14 experts, on determined characteristics of a, b, c, d of a phenomenon expressed by means of confidence intervals (see Table 2.5).

The valuations expressed by the experts warrant certain considerations. Thus it can be seen that expert 8 on assigning $[0; 1]$ relative to element a indicates that he cannot give any opinion relative to this characteristic of the phenomenon under consideration. On the other hand, totally opposing opinions can be seen between a same element, as occurs with experts 4 and

[7] Kaufmann A (1987) Les expertons. Hermès, Paris.

Table 2.3.

	a	b	c	d
0	3	1		
0,1	1		1	3
0,2		1	1	
0,3	1	1	2	
0,4	2	1	1	2
0,5	1	2	2	2
0,6	2	1	2	
0,7		1		3
0,8		1	1	2
0,9	1	1	1	
1	1	2	1	

Times that the expert have asigned the same valuation

	a	b	c	d
0	0,250	0,083		
0,1	0,083		0,083	0,250
0,2		0,083	0,083	
0,3	0,083	0,083	0,166	
0,4	0,166	0,083	0,083	0,166
0,5	0,083	0,166	0,166	0,166
0,6	0,166	0,083	0,166	
0,7		0,083		0,250
0,8		0,083	0,083	0,166
0,9	0,083	0,083	0,083	
1	0,083	0,166	0,083	

Divided by 12: probability of each valuation

Table 2.4.

$\underset{\sim}{A} =$

	a	b	c	d
0	1	1	1	1
0,1	0,750	0,916	1	1
0,2	0,666	0,916	0,916	0,750
0,3	0,666	0,833	0,833	0,750
0,4	0,583	0,750	0,666	0,750
0,5	0,416	0,666	0,583	0,583
0,6	0,333	0,500	0,416	0,416
0,7	0,166	0,416	0,250	0,416
0,8	0,166	0,333	0,250	0,166
0,9	0,166	0,250	0,166	0
1	0,083	0,166	0,083	0

7 relative to characteristic a. It can also be seen that there are experts who give their opinion by means of a number, such as the case of expert 2 for element a, or 13 for element b, c and d for example. As has been stated this does not break the harmony of the process, since a crisp number is a special

Table 2.5.

Expert	a	b	c	d
1	[0,3, 0,6]	[0,6, 0,8]	[0,2, 0,5]	0,8
2	0,7	[0,4, 0,9]	[0,1, 0,6]	[0,7, 1]
3	[0,2, 0,4]	0,5	[0,2, 4]	[0,4, 0,9]
4	1	[0,6, 0,8]	0	[0,5, 0,6]
5	[0,1, 0,8]	[0,5, 0,7]	[0,3, 0,7]	0,8
6	[0,4, 0,5]	[0,1, 0,8]	[0,4, 0,5]	[0,7, 1]
7	0	[0,6, 0,7]	[0,1, 0,8]	1
8	[0, 1]	[0,4, 0,9]	[0,2, 0,3]	[0,4, 0,5]
9	[0,5, 0,9]	[0,2, 0,5]	[0,1, 0,7]	[0,5, 0,9]
10	[0,4, 0,6]	[0,6, 0,7]	[0, 0,3]	0,7
11	0,8	[0,8, 1]	[0,2, 0,8]	[0,4, 0,8]
12	[0,2, 0,5]	[0,7, 0,9]	[0,1, 0,3]	[0,3, 0,6]
13	[0,7, 0,9]	0,8	0,4	0,7
14	[0,3, 0,7]	[0,4, 0,6]	0,2	0,9

case of a confidence interval. For the corresponding calculations it will suffice to take this number as the lower extreme and as the upper extreme.

In the same way as was done for random fuzzy sub-sets, we now arrive at a statistic for the lower extremes and the upper extremes of each element. In this way we will arrive at the number of times that the experts have expressed the same opinion as lower extreme and as upper extreme for one and the same element (see Table 2.6).

With the consideration we made before, on referring to random fuzzy sub-sets, relative to the validity of the term probability we now arrive at the corresponding "probabilities". We arrive at the result, Table 2.7, by dividing each row by 14 (the number of experts consulted).

Finally, the experton is arrived at by doing the accumulation of probabilities (see Table 2.8). In order to designate an experton we place a hooked line with a point below the capital letter (in this case A), as can be seen in the example.

With the object of providing a simplified representation of an experton, resort can be made to obtaining the mathematically expected value for each extreme of each element, which would give us a confidence interval for each one, which, in short will give rise to a Φ-fuzzy sub-set. In order to arrive at the mathematically expected value of an experton, for each extreme, the

Table 2.6.

	a		b		c		d	
0	2	1			2	1		
0,1	1		1		4			
0,2	2		1		5	1		
0,3	2				1	3	1	
0,4	2	1	3		2	2	3	
0,5	1	2	3	2		2	2	1
0,6		2	3	1		1		2
0,7	2	2	1	3		2	4	2
0,8	1	2	2	4		2	2	3
0,9		2		3			1	3
1	1	2		1			1	3

Table 2.7.

	a		b		c		d	
0	0,142	0,071			0,142	0,071		
0,1	0,071		0,071		0,285			
0,2	0,142		0,071		0,357	0,071		
0,3	0,142				0,071	0,214	0,071	
0,4	0,142	0,071	0,214		0,142	0,142	0,214	
0,5	0,071	0,142	0,214	0,142		0,142	0,142	0,071
0,6		0,142	0,214	0,071		0,071		0,142
0,7	0,142	0,142	0,071	0,214		0,142	0,285	0,142
0,8	0,071	0,142	0,142	0,285		0,142	0,142	0,214
0,9		0,142		0,214			0,071	0,214
1	0,071	0,142		0,071			0,071	0,214

accumulated probabilities are added (except for level 0) and the result is divided by the number of levels (excepting 0). In our example we would have:

	a	*b*	*c*	*d*
$E(\underset{\sim}{A}) =$	[0,399, 0,670]	[0,506, 0,756]	[0,178, 0,463]	[0,628, 0,799]

Table 2.8.

		a		b		c		d	
	0	1	1	1	1	1	1	1	1
	0,1	0,857	0,928	1	1	0,857	0,928	1	1
	0,2	0,785	0,928	0,928	1	0,571	0,928	1	1
	0,3	0,642	0,928	0,857	1	0,214	0,857	1	1
	0,4	0,500	0,928	0,857	1	0,142	0,642	0,928	1
$\underset{\sim}{A}$ =	0,5	0,357	0,857	0,642	1	0	0,500	0,714	1
	0,6	0,285	0,714	0,428	0,857	0	0,357	0,571	0,928
	0,7	0,285	0,571	0,214	0,785	0	0,285	0,571	0,785
	0,8	0,142	0,428	0,142	0,571	0	0,142	0,285	0,642
	0,9	0,071	0,285	0	0,285	0	0	0,142	0,428
	1	0,071	0,142	0	0,071	0	0	0,071	0,214

Now we could ask ourselves why the mathematically expected value is not arrived at initially in confidence intervals, since the extremes are probabilities. The answer is twofold: on the one hand, as a consequence of the fact that the logical operations are not always linear operators, and, on the other, in order to maintain "all" the information right up to the end of the process. It is at this point when it is convenient to obtain indicators that are susceptible to being interpreted by the human operator.

Expertons constitute an important generalisation of the instruments for the treatment of uncertainty and subjectivity. In fact, a random fuzzy sub-set is a special case of the experton, in the same way as the Φ-fuzzy sub-set, a fuzzy sub-set and a common sub-set. The possibility of applying multivalent logic to expertons converts this into an element of great value for modern techniques for the treatment of financial problems.

We are conscious of the fact that the contents of this chapter does not include in a exhaustive manner all the elements necessary for carrying out the study of the problems of management in uncertainty. Throughout each chapter new concepts and new techniques will be incorporated which will make the study of the problems and solutions included for resolution easier. All we have done here is to set the bases on which to found the later development of our work.

3 Accountancy and Decision Techniques

3.1 Financial Statements, Representative Hub of the Financial Situation

The importance acquired by the quantification of economic phenomena for business management brings to light the interest the data has that accountancy provides for the control and measurement of the economic-business day to day affairs. But the use of this data is not limited to the analysis of facts that have already occurred, but knowledge of the same allows for the taking of decisions relative to the future.

Among those instruments common to accounting, the use of which is more fruitful in business management, the balance sheet occupies a prominent position. As occurs with other instruments used in business management, (as, for example, the operating accounts, accounting statements, profit and loss accounts, etc.), the concept of equilibrium is ever present throughout the development of the doctrines that have been drawn up on this subject.

The summarised presentation of a balance sheet, through its patrimonial masses, allows for relating the sources of financing with the use that is made of the same. But the role of these relations should not finish either with the analysis of events that have occurred or with a study of the current situation, but their objective should be extended to providing for future situations by means of an estimate of these patrimonial masses for successive periods and to being able to select in this way the path to follow in order to attain said objectives.

Apart from the balance sheet other financial statements exist, which constitute an ordered layout of the figures that totally or partially represent the economic or financial activity of an institution (business, corporation, etc).

It is considered that the most representative financial statements are:

- the balance sheet;
- the operating account
- the statement of source and application of funds.

Anna Maria Gil-Lafuente: *Fuzzy Logic in Financial Analysis*, StudFuzz **175**, 37–43 (2005)
www.springerlink.com

Let us recall some of the concepts relative to the balance sheet. We leave it up to the interested reader to consult the complete, varied and extensive texts on accounting statements[1].

The balance sheet constitutes an economic-financial representation of the business at a given moment, and is formed by an economic structure (assets) and a financial structure (liabilities) that show on the one hand (the liabilities) the sources from where the financial means proceed and on the other (assets) the applications of these financial means.

The different elements that form the economic structure are grouped in different levels; the largest group giving rise to fixed assets and working capital.

The working capital is formed by those patrimonial elements that are in constant motion throughout the economic accounting period. The elements that comprise it are:

- Liquid assets: part of the working capital with total liquidity.
- Realisable assets: are formed by those elements the conversion of which into liquid assets is independent of the continuity in the activity of the business.
- Stocks: grouping those elements the transformation of which in financial means depends on the continuity in the activity of the business.

Fixed assets is constituted by those elements which are linked to the business for a long period of time and the destiny of which is not the sale, but that through them the production activity of a business is carried out. Although suffering wear and tear or depreciation, they remain fixed during the economic accounting period.

Also, the financial structure includes the various sources of the financial means, the largest group being the item that includes short-term liabilities and permanent capital.

- Short-term liabilities or current liabilities, also called immediate liabilities, are represented by those debts the requirement of which could occur at any time. The elements comprising this item have the following characteristics:
- Credits for supply: elements that remain in the business waiting to make payments and which, in a certain way, constitute the counterpart of activities already carried out (suppliers, labour, other production expenses for immediate payment, etc.).

Other short-term credits: are outside sources of finance with immediate expiry, such as bridging loans or other credits for immediate payment.

[1] Among others, we can mention, Cañibano L (1988) Contabilidad. Análisis contable de la realidad económica (in spanish). (Ed) Pirámide, Madrid; Fernández Pirla J M (1967) Teoría económica de la contabilidad (in spanish). 5th Ed. Madrid; Mathews R (1984) Contabilidad para economistas. (Ed) Aguilar, Madrid; Pifarré Riera M, Curso de introducción a la teoría de la contabilidad (in spanish). (1989) Faculty of Business and Economic Sciences.

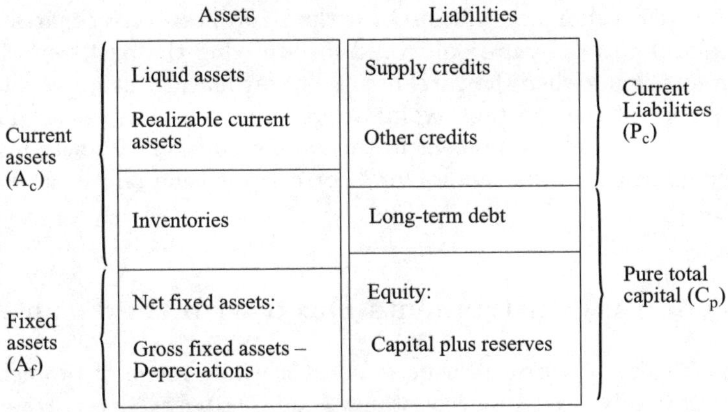

Fig. 3.1.

Permanent capitals include long-term debts, initial contributions and undistributed profits. They can be sub-divided as follows:

- Equity:
  Capital: expresses contributions made by the owners of the business.
  Reserves: is profits that have not been distributed
  Results: differentiates the total amount of income less the total sum of expenses.
- Medium and long-term liabilities: is formed by external financing sources the return of which is only subject to the passing of a certain period of time.

A balance sheet can be represented, among others by the outline shown in Fig. 3.1.

Once we are located within the economic-financial sphere of the business, several questions can be asked about the relation existing between the masses of both assets and liabilities. Among these, and only as an indication, we could mention the difference existing between fixed assets and the sources from where their financing arose. Certain rules exist, accepted with a certain amount of generalisation that relates the sources of financing with their application. It is said then that fixed assets should be financed by means of permanent capitals, with all the relativity that can be given to the word permanent. This is due to the limited capacity that fixed assets have of becoming transformed into monetary means, that is to say, by their limited degree of liquidity.

The use of the financial statements is not only limited to an analysis of the balance sheet, operating account, and source and application of funds from the point of view of the control of the activity carried out, but also its objective should consist in the improvement of business management in the future.

Now, these techniques, anchored in the hypothesis of estimates of data in the field of certainty, are only valid for studying the past, but their use brings up problems when they are directed at estimating future situations, as a consequence of the constant evolution undergone by businesses. The data accumulated at a certain moment in time is not sufficient for arriving at the necessary information that allows for a correct provision relative to events in the future.

3.2 Accounting Instruments Faced with the Future

Economic reality of current business poses a whole range of problems that methods normally used by accountancy cannot resolve, due to the lack of instruments that are adequate for treating the same, in spite of the fact of the introduction over latter years of new techniques that have allowed for widening the limits to which they are confined. The classical elements used that attained their greatest exponent in the application of differential calculus, included in the macro-economic sphere by marginal analysis, has increasingly less importance due to the recent evolution of the economic-accounting approach of business processes.

The appearance of the principles of modern statistics in the study of accountancy opened up new prospects for old problems, contributing solutions that were extraordinarily useful. This has been the cause of the movement away from determinist studies towards probabilistic studies, which describe the evolution of a "system" over time in terms of probability.

Within the probabilistic field discontinuous processes are perhaps, due to their adaptation to real events, those that have captured the attention of micro-economic studies.

But it is in the sphere of uncertainty where the greatest mathematical contribution has occurred over latter years for incorporation into the treatment of business management problems in general, and in accounting forecasts in particular.

The practical use of fuzzy logic allows for the development of the traditional elements of financial management by adapting them, by means of a modifying process, to the new requirements of an uncertain future.

The instruments traditionally used by accountancy in order to report on the patrimonial situation relative to a moment in time, or their historical evolution, are open to transformation in order to convert them into elements of management for the future. The balance sheet, ratios, working capital, cash-flow, etc., can be used for selecting, from among several financial strategies, the one or ones that can provide a good internal financial structure and, therefore, a financial position and image that is optimum in the medium and long term.

Immersed in the business task, a basic aspect is constituted by decision taking. In the majority of cases the businessman will have several strategies to follow in order to comply with the proposed objectives. The follow-up of each one of these strategies will give rise to the attainment of different objectives which will be reflected in the balance sheets, operating accounts, accounting statements, etc., which will all be different depending on the chosen path.

A task of the businessman will be constituted by foreseeing to what degree the objectives are attained, on following each one of the different strategies. Also among the responsibilities is foreseeing which are going to be the respective balance sheets, operating accounts and other financial statements. If we find ourselves in the case where the business will be in the future by following different strategies with sufficient credibility, it could be decided which would be the optimum strategy (or the most suitable) for the business, taking into account the objectives which have been proposed.

Historically, business only had budgeting as an instrument for being aware of its future situation. Budgets were established and based on these, the final situation of the business was determined and a perspective was had of all the financial and economic values (profit and loss statements, cash, client relations, suppliers and financial institutions, etc.).

Advances have occurred in the field of budgeting by arriving at better implementation. The most enterprising authors have manifested the inconveniences of current budgeting techniques. The most important is the difficulty of gathering more data in the sphere of certainty, as well as what is signified by using probabilistic values. In this latter case, voices are being heard with increasing force, voices that are being raised against the use of techniques from the field of randomness, in the field of social sciences, as well as in the sphere of business management, since the phenomena that occur in them do not easily comply with the axioms of Borel-Kolmogrov.

A quantum leap forward has occurred in this field with the incorporation of techniques based on the theory of fuzzy sub-sets and their variations with the scheme of the Fuzzy Zero Based Budget[2], above all taking into account that it is not possible to do the same operations in the field of certainty and in the probabilistic field as it is in the sphere of uncertainty. The structuring of uncertain budgets will permit arriving at those necessary accounting instruments (balance sheet, operating account, etc.) for establishing a financial analysis done ex ante and, therefore, to be able to determine which strategies to follow, which in the most adequate fashion will provide the objectives set by the business.

It is obvious that the budget is a forecast in the short term that must be immersed in certain long-term estimates in accordance with the objectives of the business also for a further removed future. We feel that the budget cannot be an isolated instrument for estimating the prospects of the business, but it has to be coherent with said long-term expectations. In other words,

---

[2] Kaufmann A and Gil Aluja J (1986) Introducción de la teoría de los subconjuntos borrosos a la gestión de las empresas (in spanish). (Ed) Milladoiro, Santiago de Compostela, Chap. 3.

before making a budget for intermediate periods an action framework must be established for the business in the long term; the instruments to be used in the short term cannot coincide with the long-term instruments. Technically there exist methods for establishing long term models of which the best known and the one giving the better results is the Delphi method.

Nevertheless, the difficulties in long-term forecasts are greater than those occurring in short term forecasts. For this adaptations have been made of the Delphi method in order to be able to use them in uncertainty. Fruit of this is the elaboration of the Fuzzy Delphi method[3]. By following this system it is possible to estimate today what the accounting statements will be like within a several accounting periods, which will permit the financial analysis ex ante, as we have stated previously.

The objectives of the business can be many and varied. Attaining these objectives means that determined concepts, such as working capital, statement of source and application of funds, cash-flow, capital cost, profitability for the shareholder, etc, can move within certain intervals, which will be the most suitable, in order for the objectives that have been set to be attained.

What this basically is then is to create certain schemes that are suitable for the treatment of uncertain future situations starting out from those that have been traditionally used by accountancy. The task is not easy because operations are not the same in the sphere of certainty or randomness as in the field of uncertainty. If we can do this, we will have at our disposal the instruments, which will allow us tomorrow to treat accounting statements, projected out into the future just as today we analyse accounting statements relative to the past. This is our objective and our challenge.

3.3 Management in Uncertainty

The fact that economic phenomena are characterised by their constant mutation does permit, in the majority of cases, taking into consideration data from the past in order to be able to establish a sufficiently valid forecast of the future. This tendency is not going to decrease, but the process will accentuate with acceleration giving way to continuously changing situations.

In the attempts to formalise these phenomena preferential attention is paid to the use of crisp data. There is nothing more comfortable than the mechanisms of certainty. Nevertheless, these are inoperative when reality deviates from the elaborated scheme. As a subsidiary solution resort has been increasingly made to stochastic techniques, which will hardly adapt to the situation when the data is impregnated with subjectivity.

Only when these solutions have proven inadequate has the task been tackled to draw up a set of operative techniques that are capable of carry out an adequate treatment of business phenomena when knowledge of the

---

[3] Kaufmann A and Gil Aluja J (1986) op. cit., Chap. 10.

same takes place in an imprecise manner. For this resort is made to the most general of the theories that are capable of describing uncertain environments: the theory of fuzzy sub-sets.

It is very well known that the theory of fuzzy sub-sets is a part of the mathematics that is perfectly adapted to the treatment of both the subjective and the uncertain. It is an attempt to include the phenomena just as they are presented in real life and treat them without attempting to deform them in order to make them more precise and certain. The formalisation of the uncertain, starting out from fuzzy concepts, has given rise to a different manner of thinking that includes the rigor of sequential reasoning and the richness of imagination.

In this way mathematicians and economists have arrived at new schemes that permit a more complete consideration of reality, by avoiding its traditional deformation when resorting to numerical precision. In them, it is assumed that the taking of decisions is done in an environment in which the objectives that it is intended to attain, the limitations to which they are going to be submitted to and even the consequences for each one of the alternatives approached, appear in an imprecise manner.

In order to quantify this imprecision the techniques that are typical of the theory of probabilities are not adequate, since this would mean accepting that the imprecise phenomena are equivalent to random phenomena. It has been seen that on abandoning the requirements of the hypothesis of stochastic models there has been a substantial narrowing of the gap towards the real world.

The possibilities that are offered by fuzzy sub-sets for tackling the problems of decision in the field of business activities are very extensive,[4] and include, among others, long and short term forecasts, investment selection, stock management, equipment renewal, new product investigation, personnel selection and a very large etcetera. Our objective, in this book, is to provide a new way of approaching the problems of financial management by using classical instruments, duly transformed so that they become suitable for treatment in uncertainty.

[4] See Kaufmann A and Gil Aluja J (1986) op. cit.

4 The Estimate of Economic-Financial Values by Means of the Budget

4.1 Budgeting Activity in Business

One of the objectives sought after by businesses as such is their permanence and survival over time. In order to achieve this objective it is necessary that a series of policies and strategies are drawn up, which, in the long term, frame and condition the activity of the business over several periods and, at the same time shape future planning that covers the most general aspects of the same.

In this context of projections towards the future, one of the principal missions of the businessman is to adapt objectives to the medium and short term. In order to achieve this end diverse means are used. From an economic-financial point of view the budget stands out from among those that have been traditionally used the most.

We feel that within the framework of this book it would not be appropriate to make an exhaustive analysis of budgeting techniques. Other specialists with a greater knowledge of the subject have already done this. Nevertheless, we cannot avoid the attraction of making certain considerations on this subject, as a prior step before approaching a new horizon that we feel could be very useful, but above all very much in agreement with the thoughts of the people who direct and continue to direct the financial activities of businesses.

A multitude of definitions exist for the word budget all of which give rise to a varied gamut of positions on the part of their authors. Thus, it can be considered that a budget is a "a co-ordinated set of forecasts by means of which we become aware in advance of some of the results of the business"[1], and also a "formulation of plans for a determined period of time, expressed un quantitative terms".[2] The list could be interminable.

A large portion of experts in the budgeting field, among whom can be found Paul Loeb[3], consider that "the sale is a beginning, which is linked to

[1] Pisón Fernández, I., El control de gestión en las pequeñas y medianas empresas, Universidad de Santiago de Compostela, 1983, p. 36.

[2] Appleby, R.C., Así se dirige una empresa moderna, Salamanca, Anaya, 1971, p. 105.

[3] Loeb, P., El presupuesto de la empresa, Madrid, Aguilar, 1961, p. 22.

Anna Maria Gil-Lafuente: *Fuzzy Logic in Financial Analysis*, StudFuzz **175**, 45–60 (2005)
www.springerlink.com

other management acts, in such a way that the problems of sales cannot be resolved before having resolved those the precede the same".

The possible expanding or depressing tendency and the overall value that invoicing can attain both in the long term as in the short term, affect and at one and the same time are affected by the diversity of activities carried out by the business. Therefore, on designing the plans, both for the immediate future and for a more extended future, it is necessary to take into account this element and grant it the importance that estimating the same warrants.

It is by reason of all the above that the sales estimate appears in the drawing up of budgets as the fundamental basis for arriving at the different budgets relative to the sub-systems that comprise the business.

This leads us to consider the different classification criteria for the budgets. According to Khermarkhen[4], budgets can be classified by taking into account:

- the unit of measurement used;
- the nature of the operations;
- the structure to which they refer.

Taking into account the unit of measurement used, budgets can be expressed in physical units or monetary units. The first are normally used as a consequence of the fact that the greater part of the information required for drawing up the budget is obtained from accounting data, on the one hand and, on the other because the budget values are easier to handle if reduced to a common valuation unit such as monetary units. Nevertheless, in spite of these inconveniences, budgets are usually presented, the unit of measurement used of which are physical units, above all in the budgets for industrial business. An example is the budgets for labour, in which the common value are hours of work. But budgets expressed in monetary units are those that are used the most.

Relative to the nature of the operations to which a budget refers, these are normally classified in three different types:

- operating budgets;
- investment budgets;
- cash budgets.

Operating budgets cover all operations relative to the normal operating cycle of the business, that is, purchasing, supplies, manufacture, warehousing, sales, etc. The principal objective of this budget is the valuation of the monetary and realistic flows linked directly with the normal activities carried out by the business. The budget should include at least:

(a) sales budget;
(b) production budget;

[4] Khermarkhen A (1976). El control de gestión. (Ed) Deusto, Bilbao, p. 86.

(c) manufacturing expenses budget;
(d) personnel expenses budget;
(e) commercial or distribution expenses budget;
(f) purchasing budget;
(g) financial expenses and dividends budget;
(h) administrative expenses budget;
(i) studies and research budget.

In our understanding, this budget includes three spheres, the significance of which in the business is different, and the management of which requires different formulating and performance.

For this we propose separating the elements of this budget into three groups:

- sales budget;
- manufacturing budget;
- management and administration budget.

Investment budgets (equipment) show those operations that the business carries out that bear no relation to the day-to-day operating cycle. These activities can be such as new construction, installations, production machinery, brand names, patents, investment portfolio, etc. They tend to be operations and activities related to the fixed capital of the business. The periods for which these investment budgets are drawn up are normally longer than the accounting period and therefore for longer periods relative to those previously mentioned operating budgets. Investment budgets, therefore, will be based directly on the estimates made for sales.

Cash budgets include sales, manufacturing, management and administration and investment budgets, but are considered from a different perspective. In these budgets are shown the activity of the business on the economic side, while cash budgets have as their objective to show the financial consequences in the sense of determining the times and amounts in which it is expected that collections and payments will take place with the object of avoiding a lack of liquidity. Cash budgets are drawn up from the estimates contained in the manufacturing, investment and management and administration budgets for estimating payments and from the sales budget for collections. Once all this information is gathered on payments that must be made and income from sales, an estimate must be made of the possibilities of obtaining the corresponding financial means from suppliers and financial entities, as well as the credits it is foreseen will be granted to clients.

Due to their special characteristics, on many occasions it is considered necessary to separate the rest of the forecasts from payments arising from the investment process. Thus, along with the ordinary cash budget certain extraordinary requirements for cash may be considered. It is convenient to point out here that, apart from the financial requirements arising from operation, the sources that will be used must be studied (equity, or short, medium or long term credits) in order to cover the requirements for payment means to meet the due dates arising from the investments of the business.

Each one of these budgets is susceptible to being divided taking into account different criteria according to the specific needs of each business. For example, budgets could be drawn up by only taking into account the nature of the operations or the areas of responsibility among others.

Independently of the cash budget, it is possible to draw up other estimating statements that depend on information obtained in the remaining budgets. We are referring to a trial balance sheet and a forecast statement for profit and loss. These two latter documents should allow the businessman to become aware of the economic-financial situation at the end of the accounting period, which is budgeted on the hypothesis that forecasts are achieved.

In Fig. 4.1 the necessary links are shown schematically that are required for obtaining the three basic instruments for efficient financial management: cash budget, trial balance sheet and trial statement of profit and loss.

Finally, and from the point of view of the structure to which they refer, scholars of the subject normally make a separation of the budgets in accordance to whether they are directed towards estimating values relative to the organs of the business or refer to specific projects. This gives rise to mention being made of "organic budgets" or "budgets for projects".

The organic structure of the business is very varied, so much so that it can be said that there are as many organization charts as there are businesses. The organic unit (which normally coincides with a decision centre) may include one or several work centres (sections, services, departments, divisions, submanagements, etc.) and their dimension varies according to the activity of the business and also according to its geographical location.

There are certain businesses, among which stand out those that work to order, that are characterised by the joining together of their production means (personnel, equipment, installations, etc.) in one decision centre in order to produce each individual product. In this event, the budgets refer to each one of the projects, from where stems the name of "budgets for projects".

Obviously we could look at other classifications relative to budgeting activity, but to spend more time on this aspect would mean, we feel, exceeding the object of this work.

We will now move on to describe very briefly determined aspects relative to the drawing up of budgets. (see Fig. 4.1)

4.2 Budgeting Development

The process of drawing up a budget has varied over the years. From the very first strict budgets to the latest variations of the Zero-Base Budget a great deal and very fruitful ground has been covered. Under this heading we are

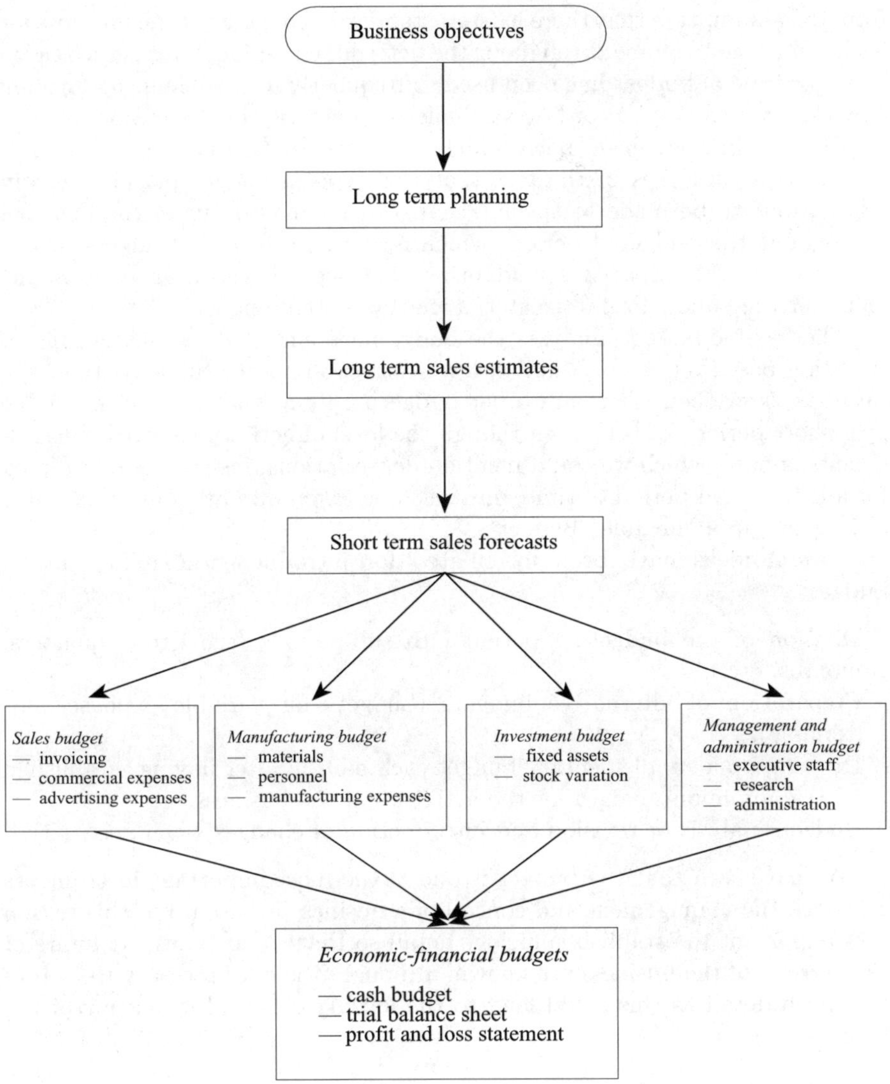

Fig. 4.1. Process of drawing up a budget

going to look at the most significant characteristics of the flexible budget, the budget by programmes, RCB system, and Zero-Base Budget.

4.2.1 Flexible Budgets

Flexible budgets are forecasts that are "established for each sub-period based on the different volume of activity".[5] In other words, flexible budgets start out

[5] Bourdin J (1978) El control de gestión (in Spanish). (Ed) ICE, Madrid, p. 117.

from the assumption that there exists a standard behaviour in the business for each production volume throughout the determined period. This is the reason why this type of budget has been used so frequently for the elements forming expenses, which vary according to the level of activity of the business.[6]

Flexible budgets have been widely accepted due to two basic reasons. The first of these lies in the fact that this type of budget permits certain adaptations to be made to specific changes that may occur in the business throughout the budgeted period, which is contrary to rigid budgets, which do not enjoy this capacity for adapting. The second reason arises from the fact that they allow for far greater efficiency in planning.

The flexible budget analyses the budgeting items that are susceptible to variation based on the production level corresponding to the activity of the business. Now then, there are other budgeting items that remain invariable over short periods of time even though the level of activity changes. Such elements, among which we could mention depreciations, insurances contracted for a determined period of time, must also be taken into account at the time of drawing up a "flexible" budget.

Several ideas have been formulated for introducing flexibility into a budget:[7]

- Division of the budgeting period into sub-periods (semesters, quarters, months, etc.).
- Preparation of alternative budgets (budgets for variable expenses, for example).
- Drawing up a supplementary budget each month, after having established a basic minimum budget for the activities of the business.
- Budget analysis or revision and introduction of changes when necessary.

We have seen that the budget is one of the most important instruments on which the management and control of a business is based. For this reason it is important to establish sufficient detail so that at all times the limits of the activity of the business are known, although it is not necessary to go into too much detail as this could lead to the loosing of overall efficiency of the business.

4.2.2 Budget Planned by Programmes

The budget planned by programmes, budget by programming or *planning programming budgeting system* (PPBS) saw the light of day in the United Sates in 1924 when Du Pont took over the control of General Motors, and had as its aim the establishment of fundamental objectives, define the principal

[6] Daloubeix J (1977) Los presupuestos flexibles (in Spanish). (Ed) ICE, Madrid, p. 89.

[7] Wol H I, Gerber Q N and Porter G A (1988) Management Accounting. Planning and Control. PWS-Kent Publishers, Boston, p. 336.

programmes in order to attain them and to point out the resources available for doing this. Therefore, an attempt is made to resolve in a systematic manner those problems that appear when decisions have to be taken in order to reach certain objectives by using the financial means available in a rational manner.

Later on in 1960, Novick published in his work *New Tools for Planners and Programmers* what constitutes a reformulation of the PPBS system. From this work the North American Department of Defence introduced a new technique in 1961. Later, in 1965, The North American Government applied this process to all Departments of the Federal Government. This use is followed by other countries such as: Sweden, Britain, Belgium, Japan, etc. In 1968, a variation of this procedure, the RCB, was used in France.

As the PPBS was put into practice by Public Administrations, there is no doubt that this has created very wide possibilities for its use in the field of business management.

The budget by programmes constitutes, according to Koontz, O'Donnell and Weihrich,[8] "a means of providing a systematic method of distribution of resources in a business in the most effective way for reaching its objectives". In this definition the relationship between objectives and the means of attaining them is brought to light. On the other hand Schultze[9] considers that the budget by programmes is "and attempt to integrate the formulation of policies with the budgeting assignment of resources and provides a means for systems analysis to be applied regularly to the formulation of policies and the assignment of budget items". We could continue with a whole series of definitions which would lead us to the conclusion that this type of budget is an attempt to include in one and the same scheme all the processes of planning, programming, budgeting and control. Obviously this is not an isolated attempt, because budget schemes, which have appeared over time, attempt to include these elements within a unitary scheme.

In the opinion of Ferreiro Santana[10], the budget by programme has as its aim the covering of failures and to rectify the lack of information inherent in traditional budgets. The most important difference between a classical budget and a budget by programmes is the use, in the latter, of programming criteria that allow for the taking of decisions to select between several alternatives, both relative to the objectives and to the means for attaining them.

Among other characteristics, the PPBS establishes the costs for each one of the projects separately, which permits drawing up a programme with a high

[8] Koontz H O'Donnell C and Weihrich H (1980) Management. (Ed) McGraw Hill 7th Ed. New York, p. 752.

[9] Schultze C (1971) El PBS., política y economía del gasto público (in spanish). Ministerio de Hacienda e Instituto de Estudios Fiscales, Madrid, p. 146.

[10] Ferreio Santana L A (1976) *Introducción a la técnica del presupuesto por programa,* México, Dirección General del Presupuesto por Programas (in Spanish). Series: Study 2, p. 36.

degree of integration. At the same time it allows for reconciling the different programmes with the object of distributing among them the financial means which, in principal, should mean an optimisation relative to the carrying out of each function or sub-function.[11]

Six phases are required for drawing up the PPBS:

- Establishment of objectives
- Determination of the means necessary for reaching the set objectives
- Relationship between objectives and means
- Establishment of solution in the form of programmes
- Valuation, in terms of cost/profit, of the selected programmes
- Taking of rational decisions between the set of alternative solutions.

We will leave this point at this juncture as the objective of this book is not to make an exhaustive description of any particular form of drawing up a budget. What we are only trying to do is to bring to light a range of budgeting techniques that exist at this time and which are fundamentally aimed at arriving at a cash budget, a trial balance sheet and an estimate for the operating account corresponding to a future situation.

4.2.3 The French System RCB

The French system RCB, *rationalisation des choix budgétaires*, can be considered, in a certain manner, as a variation of the PPBS. It first saw the light of day in 1968 as a set of management techniques for application to the ministerial management of the French Government. In February of 1969 a seminary on RCB was held, with the participation of twenty-three general directors from diverse Ministries, among other participants.

The RCB method can be considered as a research method applied to an activity, in this case public, that by "using all the analysis, calculation, forecasting and organisation techniques available, leads towards the efficient and faithful realisation of a policy"[12]. With this criterion an attempt was made to increase efficiency and productivity of the economic activity and eliminate the lack of administrative co-ordination, thus improving the level of services.

The RCB method can be broken down into several phases, as shown in Fig. 4.2.

Thus, the following links appear when carrying out an RCB:

- Formulation of the problem
- Description and classification of objectives and means

[11] Martner G (1967) Planificación y presupuesto por programa (in Spanish). (Ed) Siglo XXI, México, p. 198.

[12] Ferry J (1971) Informes del Consejo Económico y Social francés (in Spanish). Revista de Hacienda Pública Española, No. 11, p. 288.

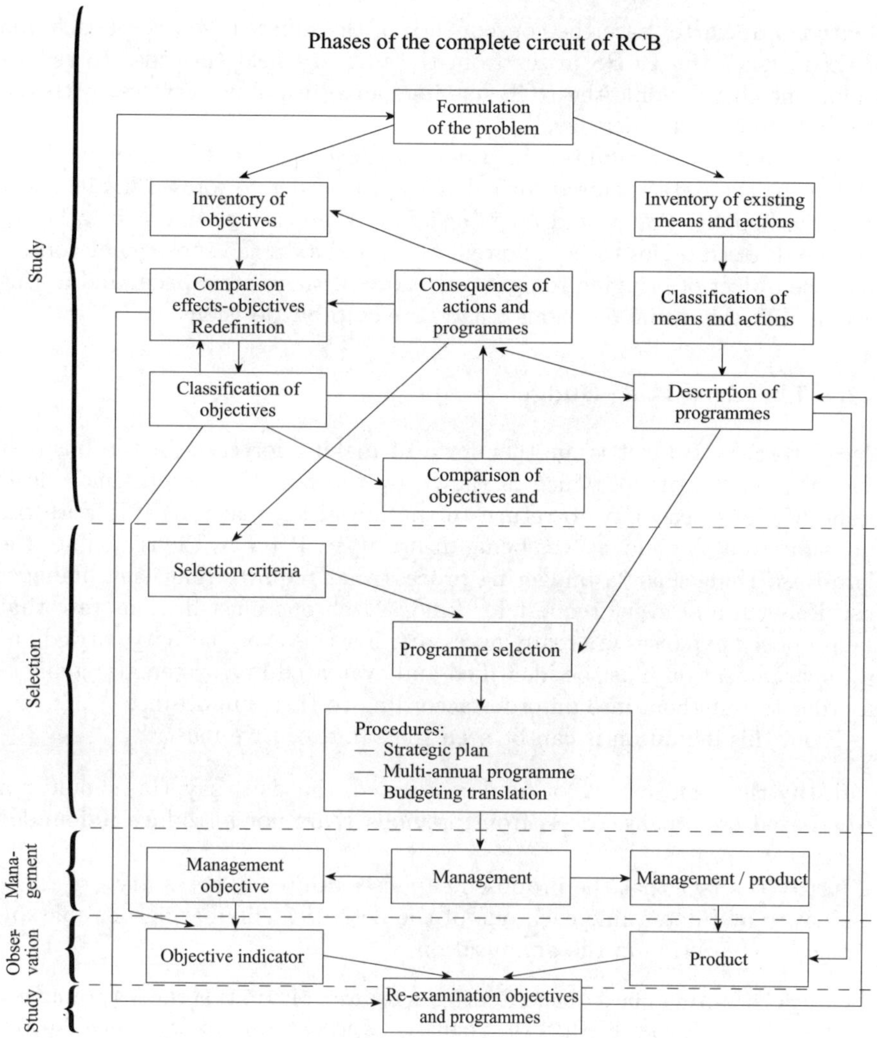

Fig. 4.2. Phases of the complete circuit of RCB

- Comparison between objectives and programmes
- Start up of programmes, overall valuation and decision
- Carrying out and control of decisions.

In fact, both the theoretical and technical aspects used in the RCB method are similar to those drawn up and used in the North American system PPBS. The differences that exist that are pointed out by Ortega Díaz-Ambrona and

Rodríguez Inciarte[13] are the consequence of the different source of each one of them, since the PPBS arises from the military field (in order to reduce armament costs), while the RCB has its source in public services (with the object of improving their level).

The information required in order to draw up the RCB is high, both relative to the data required for calculation in order to foresee in the short term the benefits that would be gained from the decision on one or other of the possible strategies to be followed, and relative to the subsequent control with the object of bringing to light any errors with greater speed, and in this way modify the initial decisions with more suitable decisions.

4.2.4 The Zero-Base Budget

The Zero-Base Budget is another form of making forecasts in the business field, the singularity of which is shown by the fact that, each time one is prepared, it is necessary to return to the initial approach, as if it were the first time that the budget was being drawn up[14]. Peter A. Phyrr, defines the Zero-Base Budget as "a budgeting process that requires that each manager justifies each and every request for funds. Each one must demonstrate that the proposed expenses are really necessary. For this every activity carried out in the organisation must be identified and evaluated in a systematic analysis in order to put them into an order according to their importance"[15].

From this definition it can be seen that an executive must:[16]

- Justify the requirement for funds requested, that is to say, this should not be based on expense levels from previous years nor introduce unfounded security margins.
- Justify the fact that the proposed expenses are really necessary.
- Identify and determine each one of the activities eliminating possible duplications existing in the organisation.

---

[13] Ortega Díaz-Ambrona J M and Rodríguez Inciarte M (1971) Formulas presupuestarias de la eficacia. El PBS (in Spanish). *Hacienda Pública Española* No. 11, pp. 264–265.

[14] Díez de Castro E P (1980) Presupuesto Base Cero: un nuevo instrumento de planificación (in Spanish). Revista de Economía y Empresa, Num. 3 and 4, p. 21; Pereda Millán, M (1980) Presupuesto Base Cero: técnicas para la racionalización y reducción del gasto. Su aplicación práctica a la empresa (in Spanish). (Ed) Cirde, Madrid, p. 23.

[15] Phyrr P A (1977) Zero-Base Budgeting, an unpublished text from a seminary given at the Planning Executive Institute. New York, March 15. Reference taken from Castañeda Ordóñez P (1977) Nuevas técnicas de dirección: el Presupuesto Base Cero para la reducción de gastos (in spanish). Económicas Y Empresariales, No. 8, pp. 119–120.

[16] Martín Ruiz F (1983) El Presupuesto Base Cero. Cuestiones de economía de empresa, Universidad de Málaga, Málaga, p. 127.

- Evaluate, by mans of systematic analysis, activities in order of their justification and need, without limitations of expenses, in order to proceed to placing them in an order based on their importance.

The Zero-Base Budget is supported on an axiom, which is consubstantial to all businesses: "the monetary funds available for covering the expense are a scarce resource"[17].

The Zero-Base Budget is intended to be an instrument that offers the businessman the best possible alternatives for the resources available in the business. This is then, the channelling in the best possible manner of the financial means in order to obtain from them the maximum benefit in the overall operation of the business.

In order to attain this, the Zero-Base Budget is based on the following points:

- Determination of the objectives of the business.
- Evaluation of alternative methods for carrying out each activity.
- Evaluation of the alternative levels of providing resources.
- Evaluation of the labour and execution means.
- Establishment of priorities.

Current economic events both in private businesses as in public institutions, are characterised by the scarcity of financial resources. This fact increasingly makes it necessary to use budgets the operation of which does not give rise to working overloads or additional resources. It is in this sense that the Zero-Base Budget has been conceived.

The Zero-Base Budget is a powerful instrument at the service of management in order for improving its managing capacity, for the rational assignment of financial means and also for expense control. On the other hand, the Zero-Base Budget can be co-ordinated with other processes that may be in use for the determination of the policy of the business, such as those that are used in the PPBS.[18]

This brief resume is not intended to be exhaustive, but it constitutes a limited panorama of the possibilities that business has before it at the time of commencing its budgeting activity, some variations of which, such as the Fuzzy Zero-Base Budget (FZBB), are acquiring notable success in real applications.[19]

---

[17] Seisdedos, Martínez J (1981) Nuevas técnicas presupuestarias. El Presupuesto Base Cero (in Spanish). *Esic-Market*, No. 34, January-April, p. 139.

[18] Mueller K J (1981) Zero-Base Budgeting in Local Government: Attempts to implement Administrative Reform. Univiversity Press of America, Washington, p. 18, p. 32.

[19] We should like to emphasise in this respect Kaufmann A and Gil Aluja J (1987) Técnicas operativas de gestión para el tratamiento de la incertidumbre (in spanish). (Ed) Hispano-Europea, Barcelona, pp. 375–381.

4.3 The Drawing Up of Provisional Statements

As we have often repeated throughout this chapter, one of the objectives of budgeting activity consists in arriving at an estimate for cash needs in order to cover payments which, throughout the budgeted period, will arise as a consequence of production activities and those payments not corresponding to exploitation, as well as the determination of the sources of auto and/or outside financing (cash budget). But it is also necessary to be able to estimate, sufficiently in advance, what will be the structure and contents of the balance sheet and the operating account referring to a moment located in the future (trial balance sheet and operating account).

When we talk about cash estimates it is inevitable that reference be made to the "cash budget". On consulting texts on this subject it will be seen that this concept has several meanings. For instance, Pierre Conso[20] makes a distinction between "cash budget" as such and the "general cash budget". The first refers to estimates made in the short term, that is, within a specific accounting period, its principal objective being the continued control of collections and payments that are hoped will be made by the business throughout the same. On the other hand, in the general cash budget all estimated collections and payments are included, whatsoever their origin and whatsoever the period to which their respective operations refer.

Given that the monetary flows are very sensitive to the operations being carried out throughout the accounting period, cash problems are normally tackled in two stages: the first has as its objective the determination of requirements that the business has for outside financing in order to maintain its level of liquidity in a given period that generally fluctuates between six and twelve months, according to the activity of the business.

The second stage consists in the drawing up of adequate instruments that allows for managing the credits obtained by the business. The duration of this normally fluctuates as a general rule, between one week and one month, also dependent upon the activity of the business.

It should be pointed out that, contrary to other budgets, cash estimates are normally reformulated on a continuous basis throughout time. This is due to the fact that liquidity requirements in the short term require careful control to avoid incurring in a situation of insolvency.

The presentation of this type of budget consists of two parts. The first includes income and expenses relative to exploitation operations and the other part refers to those expenses outside exploitation.

Once all operations to be carried out during an accounting period are taken into account, what will be established in the budget is the existence of an excess of availability, or on the contrary, the need for short term financing. The coverage of these requirements, in any event, is essential for maintaining

[20] Conso, P (1970) La gestión financiera de la empresa, Barcelona (in Spanish). (Ed) Hispano Europea, pp. 345–347.

the solvency of the business. In order to cover this lack of liquidity there are two paths that can be followed: resort to bank credits or make a new budget with the object of limiting the same.

In drawing up cash budgets a safety margin is taken into account "constituted by the difference between total credits granted to the business or authorised facts and the applications of the necessary credits for covering requirements".[21]

In short, therefore, the cash budget is a joint summary of applications assigned to credits.

Now then, in order to become aware with precision of the situation of cash movements the data contained in the budgets is not sufficient, since this only shows the overall situation that is foreseen will occur during the accounting period. In order to ensure at all times the equilibrium of the cash position it will be necessary to resort to what are known as cash estimates. These consist in establishing the movements it is foreseen will occur throughout a short period of time, which is the continuation of the current situation. According to the size of the business, its activity, its financial situation, etc., this period can be ten days, a week, even one day. Obviously, the short period for which these estimates are drawn up will depend on the greater or lesser information, detail and effectiveness of the results.

Generally speaking, these forecasts are independent of the budget, as a consequence of the need to analyse, the monetary funds that affect in a notable manner the movement of funds, in far greater detail. Nevertheless, "the provisional position is an instrument of work for the financial services that are responsible for the cash position, but cannot be used as an analysis and information document on the development of the financial situation".[22]

How should these cash budgets be expressed? If in times in which stability was the characteristic of economic systems, the estimate of values located in the future by means of true values could be accepted, but the current situation, as we have often repeated does not allow us to draw them from the field of certainty. Increasing for this it will be inevitable that forecasts are made based on uncertain estimates, expressed, as the case may be, by means of confidence intervals, or when possible by means of fuzzy numbers.

Through the budgets provisional results and trial balance sheets can also be arrived at. With regard to the first of these statements, in the first place it will be necessary to consider for its estimate, the sales budget, the amount of which will be shown by the result of adding the amounts for foreseen sales in each category of measurement unit, that is to say, labour hours to be invoiced, number of units produced, sales expressed by length, weight, surface area, etc. Obviously the corresponding physical quantities have had to be multiplied by their provisional sales price. From this must be deducted

[21] Conso P, op. cit., p. 354.
[22] Conso P, op. cit., p. 356.

the amount of these foreseen sales calculated at cost price (in the event at standard cost prices), obviously including the operating expenses budget, the budget for material consumption and other supplies, as well as the budget for general services. Normally excluded from this chapter are the budgets for sales and financial expenses, which are added separately under other headings. Therefore, independently of this chapter, sales expenses budget, on the one hand and financial expenses budget on the other also appear as deducted from the sales budget. Therefore:

$$\text{Provisional results} = \text{Sales} - \text{Cost of sales} - \text{Sales expenses} - \text{Financial expenses}$$

Therefore the provisional result would be the consequence of sales forecasts expressed in monetary terms from which three items have been deducted: the total sales valued at cost of sales, the amount for sales expenses and the financial expenses budget. The total will show the provisional result of the business.

If what is desired is to establish a long term financial plan for the business the forecast for profitability and profits will also have to be taken into account in order to determine what portion of the same will be distributed and what portion will remain in the business as a means of future financing (auto-financing).

If the different budgets have been estimated in an uncertain manner, be this by means of confidence intervals, or by fuzzy numbers, in this event we will arrive at an uncertain operating statement.

We will not insist on the fact that faced with an economic system with a degree of evolution, this is the only way of expressing a forecasting statement, if what is desired is that it constitutes a true reflection of what will occur in reality.

Finally, consideration can be given to fact that "the combination of programmes and budgets of budgeting management, investment plans, financial and cash plans will allow for the drawing up of the trial balance sheet for the business".[23]

Professor Depallens[24] established for each of the patrimonial masses that comprise the economic and financial structure of the balance sheet, the sources from which information is gathered in order to be able to establish the trial balance sheet. Thus:

[23] Depallens G (1970) Gestion financière de l'entreprise (in French). (Ed) Sirey, 4th Ed., Paris, p. 574.

[24] Depallens G op. cit., pp. 574–578.

Economic structure:

Fixed assets. Both for the constitution of the company and for the fixed assets the following sources of data can be considered: relative to formation expenses these will obviously be nil when they refer to an already operating business, except in the event that they need to be foreseen for future increases in capital.

Special consideration should be given to those expenses arising from research and studies, which, in certain cases, can be of great importance to the future of the business. In any event, these estimates are tied in to the budget for investments.

The same occurs with reference to fixed assets. For estimating the account for this patrimonial mass in the trial balance sheet it will be necessary to make a summary of the different plans and financing projects, as well as consider the financing plans arising from the same.

Stocks. A differentiation can be made between raw materials and other materials not directly belonging under the heading of exploitation, the estimate for which will be made by the budget for supplies.

Relative to semi-manufactured products, the elaboration process appears in the production budgets.

In order to forecast finished products it will be necessary to resort to a combination of the production budget and the sales budget.

With regard to products in the course of manufacture, as with semi-finished products, these will be seen from the production budget.

Relative to packaging materials and other products that are additional to the products, their amount will be established by means of forecasts included in the supplies budget.

Realisable inventories. This is the consequence of adding items such as suppliers, clients, other debtors, short-term loans, etc. For arriving at this, resort must be made to the supplies budget, cash budget and sales budget.

Liquid assets. The amount for this item will be estimated by means of the periodic cash forecasts.

Financial structure:

Equity. Estimates for this item are made from the long-term financial plan.

Long-term debts. These normally should be included in the trial balance sheet as a consequence of previously existing credits, to which must be added those arising from long term financing.

Short-term debts.	These are comprised of suppliers, clients, other creditors and bills to be paid, among others. The sources of information for forecasting their respective amounts proceeds fundamentally from the provisions of the cash and sales budgets.
Results (profit).	This amount will be given by the estimated operating statement.

As can be seen, both the estimated operating statement and the estimated budget sheet constitute a synthesis of information obtained by means of the different budgets and complementary information. Established in an estimating manner they will provide an orientation relative to the strategy that the business can follow in the foreseeable future. Up to now, these statements, very well known by the specialists, have been expressed in terms of certainty. Our aim consists, as we have already pointed out, in arriving at certain estimates in an "uncertain manner", expression, which should not be confused with "inexact manner". The lack of precision has nothing whatsoever to do with uncertainty. In a situation such as the current one, an estimate in terms of certainty would be more inexact in terms of certainty than an estimate done in the field of uncertainty.

5 Long Term Financial Solvency

5.1 Working Capital

As is often the case with the majority of concepts, the definition of working capital acquires several meanings according to different authors and also according to the circumstances under which said concept has been used. If we follow Conso[1], for example, four definitions can be considered related with other ways of conceiving rotating or working capital:

- Working capital and liquid assets. This meaning has been identified with the so-called "cash funds". This definition has currently fallen into disuse.
- Working capital and current assets. This definition is based on the consideration of all the elements of current assets, that is to say, total assets after deducting the items corresponding to fixed assets.
- Permanent working capital. The contents of this concept is what is currently used with the nature of in general practice and, in the Spanish language, is called indistinctly "fondo de rotación" or "fondo de maniobra", and coincides with what the Anglo-Saxons call working capital. What it is in fact, is that part of the permanent capitals that finances current assets. That is to say, that the amount of permanent working capital is the result of arriving at the difference between permanent capitals and fixed assets.
- Own working capital. This meaning is determined by deducting long-term liabilities from working capital or permanent working capital.

Our opinion is that the term working capital refers to the part of permanent capitals that finance current assets.

Now, as can be easily verified from an arithmetical point of view (within the filed of certainty), its amount coincides with the difference between current assets and current liabilities. In fact, let us assume that the dimensions of the economic and financial structures of a balance sheet are represented by the areas of the rectangles of Fig. 5.1.

We are going to insist on the fact that arithmetical equality occurs as a consequence of our being able to write the following equation:

[1] Conso P (1970) La gestión finaciera de la empresa (in Spanish). (Ed) Hispano-Europea, Barcelona, pp. 56–59.

Anna Maria Gil-Lafuente: *Fuzzy Logic in Financial Analysis*, StudFuzz **175**, 61–76 (2005)
www.springerlink.com

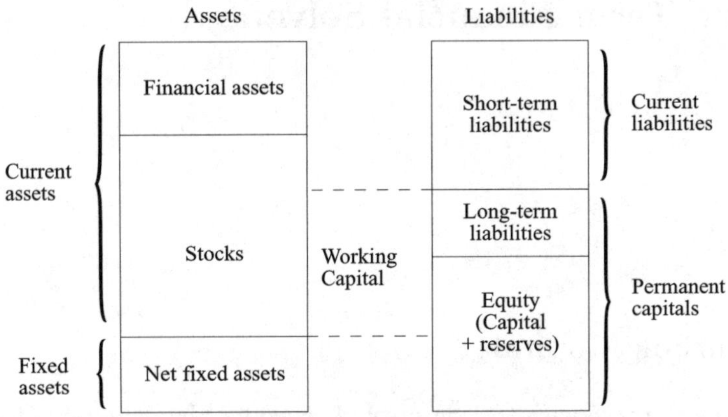

Fig. 5.1. Balance sheet

$$A_c + A_f = P_c + C_P$$

as:

A_c = current assets
A_f = fixed assets
P_c = current liabilities
C_p = permanent capitals

And therefore, the following should be complied with:

$$C_P - A_f = A_c - P_c$$

In the field of certainty there is no problem whatsoever for determining, by means of either of the two members of the equality, the amount of the working capital, but what happens in a sphere of uncertainty?

If a balance sheet for the end of a following accounting period is estimated by means of confidence intervals:

$$\begin{array}{rlrl} \text{Financial assets} = & [a_1, a_2] & \text{Short term liabilities} = & [d_1, d_2] \\ \text{Stocks} = & [b_1, b_2] & \text{Long term liabilities} = & [e_1, e_2] \\ \text{Net fixed assets} = & [c_1, c_2] & \text{Equity} = & [f_1, f_2] \end{array}$$

when using the original definition (first meaning) we arrive at:

$$\text{Working Capital} = FM_1 = C_P - A_f$$

first member of the previous equality, which in this case will be:

$$C_P = [e_1, e_2](+)[f_1, f_2] = [e_1 + f_1, e_2 + f_2]$$

This interval indicates that in the worst circumstances $e_1 + f_1$ monetary units will be available, and the most optimistic situation there will be $e_2 + f_2$ for covering requirements for placing financial means.

It has been stated that fixed assets must be financed by means of permanent capitals and the "remainder" of these used for financing part of current assets (working capital). In this case for financing current assets there will "remain":

$$FM_1 = [e_1 + f_1, e_2 + f_2](-)[c_1, c_2] = [e_1 + f_1 - c_2, e_2 + f_2 - c_1]$$

the significant point of which is as follows. If in the pessimistic situation there are only $e_1 + f_1$ monetary units available and it is necessary to invest the highest possible amount in fixed assets, only $e_1 + f_1 - c_2$ monetary units will remain as working capital. On the other hand, under the most optimistic conditions, if $e_2 + f_2$ monetary units were to be available and it were only necessary to place c_1 monetary units in fixed assets then $e_2 + f_2 - c_1$ could be available for working capital.

Is arithmetical equality maintained in uncertainty? Let us see what happens when we consider the second member of the equality in which:

$$\text{Working Capital} = FM_2 = A_c - P_c$$

In this case it will be:

$$A_c = [a_1, a_2](+)[b_1, b_2] = [a_1 + b_1, a_2 + b_2]$$

which indicates that, in order to cover estimated outlays for current assets, in the best of cases $a_1 + b_1$ monetary units will be required and in the worst circumstances $a_2 + b_2$ monetary units.

Already here a different way can be seen of interpreting the values, since the pessimistic hypothesis does not coincide with the lower extreme nor the optimistic with the upper.

Since current liabilities are not sufficient for financing current assets resort will have to be made to a part of permanent capitals (working capital) in accordance with the second meaning. How much? Given the fact that in the worst of cases (to these effects) d_1 will be available and the best (also to these effects) d_2, what will be missing is:

$$FM_2 = [a_1 + b_1, a_2 + b_2](+)[d_1, d_2] = [a_1 + b_1 - d_2, a_2 + b_2 - d_1]$$

which would give us a different result (except for special cases) as:

$$[e_1 + f_1 - c_2, e_2 + f_2 - c_1] \neq [a_1 + b_1 - d_2, a_2 + b_2 - d_1]$$

Now, if we consider as the best representation of the interval its middle point, we arrive at:

$$\overline{FM_1} = \frac{(e_1 + f_1 - c_2) + (e_2 + f_2 - c_1)}{2} = \frac{e_1 + e_2 + f_1 + f_2 - c_1 - c_2}{2}$$

$$\overline{FM_2} = \frac{(a_1 + b_1 - d_2) + (a_2 + b_2 - d_1)}{2} = \frac{a_1 + a_2 + b_1 + b_2 - d_1 - d_2}{2}$$

In this case $\overline{FM_1} = \overline{FM_2}$ is complied with. In fact what should be complied with is:

$$\frac{e_1 + e_2 + f_1 + f_2 - c_1 - c_2}{2} = \frac{a_1 + a_2 + b_1 + b_2 - d_1 - d_2}{2}$$

since:

$$e_1 + e_2 + f_1 + f_2 - c_1 - c_2 = a_1 + a_2 + b_1 + b_2 - d_1 - d_2$$

and therefore:

$$a_1 + a_2 + b_1 + b_2 + c_1 + c_2 = e_1 + e_2 + f_1 + f_2 + d_1 + d_2$$

because of the requirement that the following is complied with:

$$A_1 = P_1 \text{ and } A_2 = P_2$$

and therefore:

$$A_1 + A_2 = P_1 + P_2$$

which in this case is:

$$\begin{aligned} A_1 &= a_1 + b_1 + c_1 \\ A_2 &= a_2 + b_2 + c_2 \\ P_1 &= d_1 + e_1 + f_1 \\ P_2 &= d_2 + e_2 + f_2 \end{aligned}$$

Thus, by merely making the entropy fall (that is to say, returning to certainty) arithmetical equality[2] is arrived at.

Specialists in fuzzy logic are perfectly aware of the fact that subtracting intervals, as defined, is not suitable for the treatment of equations, therefore resort must be made to a type of ordinary difference, which we will designate by $(\bar{m})$ (in honour of Minkowski).

In this case, as a consequence of the equality between the economic and financial structure of the balance sheet we arrive at:

$$[a_1 + b_1, a_2 + b_2](+)[c_1, c_2] = [d_1, d_2](+)[e_1 + f_1, e_2 + f_2]$$

and finding new $F_m M_1$ and $F_m M_2$ we arrive at the following in which $[d_1, d_2]$ becomes the first member and $[c_1, c_2]$ the second:

$$\begin{aligned} F_m M_1 &= [e_1 + f_1, e_2 + f_2](\bar{m})[c_1, c_2] = [e_1 + f_1 - c_1, e_2 + f_2 - c_2] \\ F_m M_2 &= [a_1 + b_1, a_2 + b_2](\bar{m})[d_1, d_2] = [a_1 + b_1 - d_1, a_2 + b_2 - d_2] \end{aligned}$$

[2] We ought to understand, here, the entropy as a disorder degree, meaning which it is taken under uncertainty (non-probabilistic entropy). Reducing the degree of uncertainty means reducing the disorder.

result which is again different, but which complies with an ordinary rule of subtraction.

Finally if we make the entropy fall we arrive at $\overline{F_m M_1} = \overline{F_m M_2}$. In fact:

$$\overline{F_m M_1} = \frac{(e_1 + f_1 - c_1) + (e_2 + f_2 - c_2)}{2} = \frac{e_1 + e_2 + f_1 + f_2 - c_1 - c_2}{2}$$

$$\overline{F_m M_2} = \frac{(a_1 + b_1 - d_1) + (a_2 + b_2 - d_2)}{2} = \frac{a_1 + a_2 + b_1 + b_2 - d_1 - d_2}{2}$$

results, which coincide with those, obtained previously.

It can be clearly seen that the extension of the interval that represents the working capital is different in every case, indicating a different uncertainty.

5.2 Medium and Long-Term Solvency

The amount of the working capital brings to light a measurement of the financial solvency of a business in the medium and long term.

If the working capital is obtained for any given moment in time, the solvency of the business will be known for that particular moment in time. But more than in a piece of historical data, where our interest really lies is in the evolution of solvency estimated for the future. That is to say we are interested in getting to know how the working capital will move if we adopt one or other of the possible economic strategies that can be used by the business.

In order to study the variation undergone by the working capital it is sufficient it to compare the figures shown at the beginning and at the end of the period under study. Apart from the variation in its absolute value, it is necessary to penetrate deep into its interior and see the modifications that the elements forming it have undergone and the characterisation of the same, as well as the influence its composition exercises on its final value.

It can be said that the analysis of the variation in working capital will only be completed when, apart from considering the quantitative variation, the proportion that the elements that form it have between themselves is taken into account and then the proportion of these with the total value.

In this way the origin and application of working capital arises as an instrument that is suitable for determining the causes for changes in solvency. Let us remember that the application of funds is a use of financial means and the origin of funds is a source of financing. Although the operating account has as its object the calculation of the profit resulting from the productive activity, it is insufficient to show the variations in current assets and current liabilities that determine the dimension and characteristics of the working capital. Now then, any modification of operating conditions, as a consequence

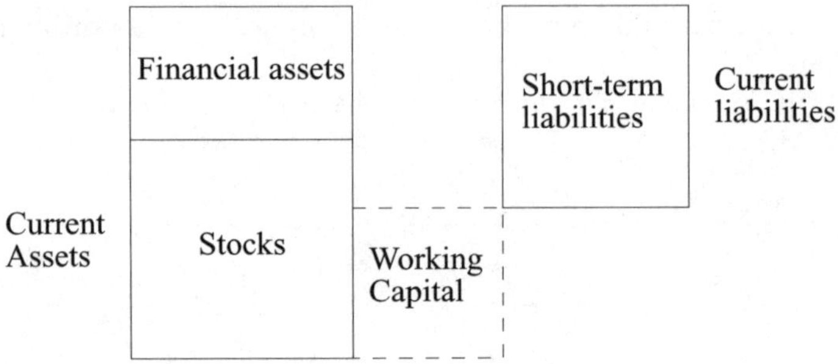

Fig. 5.2.

of changes in purchasing, sales or other policies, will vary the amount of patrimonial elements from which the working capital depends.

In this way, the production process, shown in the operating account, acts on working capital in the following possible manners (see Fig. 5.2):

(A) Expenditure from cash in payments in cash due to purchases for operation: they reduce financial assets → decreasing working capital.
(B) Purchases for operation used in the same and not paid in cash: increase current liabilities → decreasing working capital.
(C) Existing stocks are used for operation; stocks are reduced → decreasing working capital.
(D) Cash sales of manufactured products: increase financial assets → increasing working capital.
(E) Credit sales of goods and/or services: increase financial assets → increase working capital.
(F) Unsold finished products and products in the course of manufacture: increase stocks → increase working capital.
(L) Payment or accounting entry of taxes and dividends: constitutes an application of working capital resulting from operations → decreases working capital.

The only element forming costs that does not have an influence on working capital is depreciation, therefore:

$$\text{Operating profit} = \text{Working capital} - \text{Depreciation}$$

But also working capital can be modified by variations in fixed assets and permanent capitals (see Fig. 5.3):

(G) Sale of fixed assets in cash or for short-term credit: reduces fixed assets, increasing current assets → increasing working capital.
(H) An increase in equity or medium and long-term debts (not arising from an increase in fixed assets): increases permanent capitals → increasing working capital.

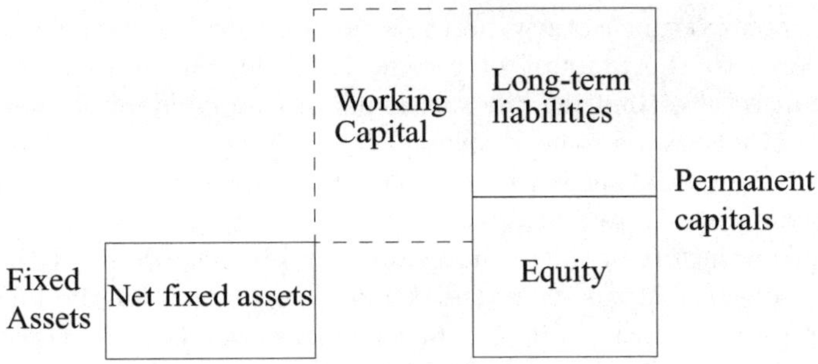

Fig. 5.3.

(I) Purchase of fixed assets in cash or with short-term credits: increases fixed assets without varying permanent capitals → decreasing working capital.

(J) Repayment of medium and long-term credits: reduces permanent capitals without modifying fixed assets → decreasing working capital.

According to which economic structure is followed by the business for future periods, its economic-financial structure will be different. Estimation of the possible working capitals will permit for taking a decision on the strategy to be followed, if the objective is to attain a determined degree of solvency.

5.3 Variations in Working Capital

Below we are going to take a look at some didactic examples of modification in working capital.

At the 31st of December a business presents the following balance sheet:

Assets		Liabilities	
Financial assets	100	Short term liabilities	50
Stocks	250	Long term liabilities	100
Fixed assets	400	Equity	600
Total	750	Total	750

The products manufactured by this business are later sold through several points of sale. We are considering then a hypothesis of manufacture that is not for prior orders, therefore their sale is subject to the acceptance they may have from potential clients.

In this context uncertainty pertains not only to the purchase price and sale, but also to the amount of production that the market may absorb. The difficulties of estimating the variations in the patrimonial masses of the balance sheet foreseen for the following accounting period, as well as a certain estimate of the operating account, makes it necessary to do a valuation by means of confidence intervals.

Thus after a study of the available data and in accordance with the experience and sensations relative to the economic evolution for the following accounting period, the possibility can be considered of following a finite number of different strategies, each one of which will give rise to different variations in the patrimonial masses of the balance sheet.

In this example[3] we are going to consider one of these that principally includes the renewal of a piece of equipment for another, which will give rise to the following operations for the different concepts:

Estimated variations in the operating account:

(A)	Cash payments for operating purchases	[600, 675]
(B)	Operating purchases pending payment	[75, 75]
(C)	Operating use of in house stocks (stock reduction)	[50, 100]
(D)	Collected sales in cash for operation	[650, 700]
(E)	Sales pending payment	[125, 150]
(F_1)	Unsold finished products	[125, 150]
(F_2)	Products in course of manufacture	[25, 50]

Estimated variations in fixed assets and permanent capitals:

(G)	Possible cash or short term sale of equipment at its residual price	No
(H_1)	Increase in share capital	No
(H_2)	Undistributed profits	[0, 100]
(I_1)	Cash purchase industrial equipment	No
(I_2)	Work in workshop paid in the short term	No

Other estimates:

(K)	Taxes to pay	Not considered
(L)	Dividends to distribute	[75, 225]
(M)	Other income	Do not exist
(N)	Corporate income tax	Not considered

[3] This example has been taken from Gil Aluja J (1989) La instrumentación financiera en la incertidumbre. Real Academia de Ciencias Económicas y Financieras, Barcelona, pp. 19–26.

Estimates that have no influence on working capital:

(G′)	Delivery of old equipment the value of which will serve as a down payment on the purchase of new equipment	[250, 250]
(I'_1)	Purchase of new equipment, with the following payments:	
	Old equipment	[250, 250]
	Long-term credit	[0, 50]
	Deferred	[75, 100]
(I'_2)	Works done with payment deferred over the long term	[25, 50]
(O)	Obtaining long term credit for immediate payment of equipment	[0, 50]
(P)	Payment deferred over the long term for equipment and works	[100, 150]
(Q)	Accounting period depreciation	Will not be done

The patrimonial masses will be modified as follows due to the consequences of the activity of the business:

Financial assets:			
Cash sales	[650, 700]		
Sales pending payment	[125, 150]		
Cash purchases		[650, 765]	
	[775, 850] (−)	[650, 765]	= [100, 250]
Stocks:			
Unsold production	[125, 150]		
Increase in works in course	[25, 50]		
Use of stocks		[50, 100]	
	[150, 200] (−)	[50, 100]	= [50, 150]
Fixed assets:			
Purchase equipment	[325, 400]		
Carrying out works	[25, 50]		
Sales old equipment		[250, 250]	
	[350, 450] (−)	[250, 250]	= [100, 200]
Short-term liabilities:			
Purchases not paid for		[75, 75]	
Dividends to be paid		[75, 225]	
			= [150, 300]
Long-term liabilities:			
Machine and pending works		[100, 150]	
Long-term credit		[0, 50]	
			= [100, 200]
Equity:			
Increase in reserves		[0, 100]	
			= [0, 100]

These estimated variations will give rise to values relative to the balance sheet foreseen for the end of the accounting period and the estimated operating statement, which for this alternative will be:

Financial assets	[200, 350]
Stocks	[300, 400]
Net fixed assets	[500, 600]
Short-term liabilities	[200, 350]
Long-term debts	[200, 300]
Equity	[600, 700]
Sales	[775, 850]
Cost of sales	[525, 700]
Stock increase	[50, 150]
Depreciation	[0, 0]

In this way the foreseen balance sheet and estimated operating statement will be those shown in table below.

Working capital in the initial balance sheet is

$$FM_1 = 700 - 400 = 300$$
$$FM_2 = (100 + 250) - 50 = 300$$

and after variables established in the foreseen working capital will be:

Foreseen Balance Sheet			
Assets		Liabilities	
Financial assets	[200, 300]	Short-term liabilities	[200, 350]
Stocks	[300, 400]	Long-term liabilities	[200, 300]
Net fixed assets	[500, 600]	Equity	[600, 700]
Total	[1.000, 1.350]	Total	[1.000, 1.350]

Foreseen Operating Statement			
Sales		[775, 850]	
Cost of sales*		[525, 700]	
	Gross profit		[75, 325]
Depreciation		[0, 0]	
	Net profit		[75, 325]
Profit distribution:			
	Dividends		[75, 225]
	Reserves		[0, 100]

* From payments for purchases we have deducted the part corresponding to those products that have moved into stock:

$$[600, 675](+)[75, 75] = [675, 750]$$

$$[600, 675](-)[50, 50] = [525, 700]$$

$$FM_1 = ([200, 300](+)[600, 700])(-)[500, 600] = [200, 500]$$
$$FM_2 = ([200, 350](+)[300, 400])(-)[200, 350] = [150, 550]$$

where it can be seen that FM_1 does not coincide with FM_2.[4]

If we use the Minkowsky difference this will be:

$$F_m M = [800, 1000](\bar{m})[500, 600] = [500, 750](\bar{m})[200, 300] = [300, 400]$$

It is easy to see that, by making the entropy fall, the same result is arrived at. In fact:

$$\overline{FM_1} = \frac{200 + 500}{2} = 350$$
$$\overline{F_m M} = \frac{150 + 550}{2} = 350$$

For comparative effects, it would appear to be adequate to use the working capital found from the Minkowsky difference. In this way we will have:

$$DFM = F_m M^{(2)} - FM^{(1)} = [300, 400](-)300 = [0, 100]$$

It can be seen, then, that the increase in working capital will be found in the interval $[0, 100]$

Finally, when we study the increases in the monetary masses of the balance sheet between these two periods we arrive at:

In Assets		In Liabilities	
Financial assets	[100, 250]	Short-term liabilities	[150 300]
Stocks	[50, 150]	Long-term liabilities	[100, 200]
Net fixed assets	[100, 200]	Equity	[0, 100]
Total	[250, 600]	Total	[250, 600]

[4] It is possible sometimes, a particular case where both calculations of the working capital have the same result. Then we will be in front of an exception and the normal substraction between intervals cannot be used in the resolution of such equations. That can be seen through the following example:

$$[20, 50](+)[40, 60](+)[70, 90] = [30, 40](+)[20, 30](+)[80, 130]$$

Moving one part of the equality to the other part, we observe that:

$$[20, 50](+)[40, 60](-)[30, 40] \neq ([20, 30](+)[80, 130])(-)[70, 90]$$

been $[20, 80] \neq [10, 90]$. Only when we make fall the entropy we arise the equality:

$$(20 + 80)/2 = (10 + 90)/2 = 50$$

Therefore:

$$FM_1 = [100, 300](\bar{m})[100, 200] = [0, 100]$$
$$FM_2 = [150, 400](\bar{m})[150, 300] = [0, 100]$$

5.4 Statement of Source and Application of Working Capital in Uncertainty

The drawing up of the statement of source and application of working capital is not, in uncertainty, as immediate as it is in the sphere of certainty as a consequence of the need to use alternatively the subtraction of intervals and the Minkowsky difference.

And this is so because in order to arrive at the net increments or decreases of a patrimonial mass it is necessary to resort to normal subtraction of intervals, since the result must include all possible solutions, from the most pessimistic to the most optimistic. On the other hand, on arriving at the difference between one patrimonial mass of the economic (or financial) structure and one from the financial (or economic) structure implies an operation within an equation (assets = liabilities), which requires the use of the Minkowsky difference.

In the case under study, the statement of source and application of working capital can be drawn up in the manner shown in Fig. 5.2.

It is not difficult to imagine that this new approach to the concepts of working capital and statement of source and applications of the working capital can be made extensive to other elements of financial analysis, such as the cash-flow and even to studies for the determination of the optimum financial structure of the business.

5.5 The Intervention of Counter-Experts

We now move on to the study of the results arrived at for working capital. Independently of the chosen concept and technique, it can be seen that the intervals found are characterised by their extension, that is, by their high degree of uncertainty. This fact, jointly with a very elevated charge of subjectivity that these initial estimates include, makes it in this case, as it does in many others, very difficult to take an adequate decision. With the object of avoiding these difficulties, we propose resorting to the opinion of counter-experts.

Let us assume, in the first place, that the concept and technique for allowing us to find FM_1 is accepted and therefore:

$$FM_1 = [200, 500]$$

and that a group of experts (who now act as counter-experts) is requested to give their opinion relative to this confidence interval, expressing their positioning, that is to say, expressing "in some way or other" if they feel they are

Statement of source and application of working capital			
Application (use of financial means)		Origin (sources of financing)	
Operating			
A Decrease	$A_c = [600, 675]$	D Increase	$A_c = [600, 700]$
C Decrease	$A_c = [50, 100]$	E Increase	$A_c = [125, 150]$
		F_1 Increase	$A_c = [125, 150]$
	$A(+)C = [650, 775]$	F_2 Increase	$A_c = [25, 50]$
		$D(+)E(+)F1(+)$	$F_2 = [925, 1.050]$
B Increase	$P_c = [75, 75]$	B Increase	$P_c = [75, 75]$
C Increase	$P_c = [75, 225]$	C Increase	$P_c = [75, 225]$
	$B(+)L = [150, 300]$		$B(+)L = [150, 300]$
Non operating			
		H_2 Increase	$C_p = [0,100]$
I'_1 Decrease	$A_f = [325, 400]$	G' Decrease	$A_f = [250, 250]$
I'_2 Decrease	$A_f = [25, 50]$		
	$I'_1\ (+)\ I'_2 = [350, 450]$	P Increase	$C_p = [100, 150]$
		O Increase	$C_p = [0, 50]$
		C Increase	$P_c = [100, 200]$

Increase in current assets = [925, 1.050] (−) [650, 775] = [150, 400]
Increase in current liabilities = [50, 300]

Increase FM_2 = [150, 400] (m) [150, 300] = [0, 100]

Increase in fixed assets = [350, 450] (−) [250, 250] = [100, 200]
Increase in permanent capitals = [100, 200] (+) [0, 100] = [100, 300]

Increase FM_1 = [100, 300] (m) [100, 200] = [0, 100]

close to 200, more closer to 200 than to 500, etc., or even they feel that the interval $[200, 500]$ is too narrow. With the object of rationalising the process of gathering information, certain rules are going to be established which the counter-experts must follow:

1. Their opinion will be expresses by means of an interval [0,1], that is $[\alpha_1, \alpha_2] \in [0,1]$.
2. They should take into account a semantic correspondence expressed by means of the hendecagonal system:

$a = 0$	for 200
$a = 0.1$	for practically 200
$a = 0.2$	for nearly 200
$a = 0.3$	for near to 200
$a = 0.4$	for closer to 200 than to 500

$a = 0.5$	for as close too 200 as to 500
$a = 0.6$	for closer to 500 than to 200
$a = 0.7$	for near to 500
$a = 0.8$	for nearly 500
$a = 0.9$	for practically 500
$a = 1$	for 500

3. They will accept that the selection of an interval will give rise to a correspondence in overall values, estimated by linear transformation. Thus, if the initial working capital is $FM_1 = [A_*, A^*]$, the estimate of an expert positioned by means of $[\alpha_1, \alpha_2] \in [0,1]$ will be:

$$[A_1, A_2] = A_*(+)((A^* - A_*)(\cdot)[\alpha_1, \alpha_2])$$

Before continuing with our example, we should point out that if one or several experts consider that the interval of reference $[A_*, A^*]$ is not sufficiently wide and they were to propose $[A'_*, A^{*\prime}] \in [A_*, A^*]$ then A'_*, $A^{*\prime}$ would substitute A_*, A^* in the above semantic correspondence.

Under these conditions, the experts consulted, which in this particular case we will assume was 13, express their opinion by means of the following intervals:

Expert 1:	[0,5; 0,7]	Expert 8:	[0,8; 0,9]
Expert 2:	0,6	Expert 9:	[0,4; 0,5]
Expert 3:	[0,4; 0,8]	Expert 10:	[0,3; 0,6]
Expert 4:	[0,6; 0,9]	Expert 11:	[0,5; 0,8]
Expert 5:	0,4	Expert 12:	[0,2; 0,6]
Expert 6:	[0,7; 0,8]	Expert 13:	[0,6; 0,7]
Expert 7:	[0,3; 0,6]		

The use of the linear transformation will give rise to the following opinions of the counter-experts:

$$\begin{array}{ll}
\text{Expert1}: & 200 + (500 - 200)(\cdot)[0{,}5; 0{,}7] = [350; 410] \\
\text{Expert2}: & 200 + (500 - 200)(\cdot)[0{,}6; 0{,}6] = [380; 380] = 380 \\
\text{Expert3}: & [320; 440] \\
\text{Expert4}: & [380; 470] \\
\text{Expert5}: & [320; 320] = 320 \\
\text{Expert6}: & [410; 440] \\
\text{Expert7}: & [290; 380] \\
\text{Expert8}: & [440; 470] \\
\text{Expert9}: & [320; 350] \\
\text{Expert10}: & [290; 380] \\
\text{Expert11}: & [350; 440] \\
\text{Expert12}: & [260; 380] \\
\text{Expert13}: & [380; 410]
\end{array}$$

In view of these results it could be thought that the counter-experts could have provided their opinions directly by means of the resulting intervals. Indeed, the direct process is always possible, although in this case we feel that the expert would be deprived of a recourse that, in our opinion, allows for expressing the subtlety of the nuances of the hendecagonal scale far better, and this is so useful when the giving of an opinion is imbued with a very high charge of subjectivity.

Be it then by the procedure we are proposing, or by direct determination, we are now faced with a set of confidence intervals, each one of which represents the estimate made by each counter-expert on the working capital. These estimates considered separately one from the others, offer valuable information provided the number of counter-experts consulted is sufficiently small. But when the number of experts begins to increase, it will be difficult for an executive taking decisions to get an "overall" idea of the opinions provided. It then becomes necessary to find a procedure that allows for making an aggregate.

For this several paths can be followed. From a very simplistic point of view, which at one and the same time provides the greatest security that all the opinions of the counter-experts has been included, with the thirteen opinions a single interval can be constructed which has as its lower extreme the lowest of all the lower extremes of the set, and as its upper extreme the highest of all the upper extremes of the set, In this case what we have is:

$$FM_1^* = [290, 470]$$

Taking into account the wide cross section of the opinions given we have arrived at an interval which, although narrower than the initial interval, still includes a great deal of uncertainty. What is more it would have been sufficient for just one counter-expert to have given a 0 as a lower extreme in the hendecagonal scale, and that another (or the same) a 1, for the initial estimate [200, 500] to have been maintained. That is to say that we would have arrived at the singular conclusion that the opinion of the expert coincides with the opinion of "all" the counter-experts. Also, if only a single counter-expert were to have gone beyond the initial interval, because of considering it too small, at one or other of the extremes, the interval of the aggregate could well have been more uncertain than the original interval.

Let us move on then to another path, the principal advantage of which apart from its operativeness, is also its simplicity. What is attempted is to arrive at an interval the lower extreme of which is the arithmetical average of the lower extremes, and the upper extreme is the arithmetical average of the upper extremes[5]. In this way we would have:

[5] In that process underlay the principle that all opinions have an equal importance. The weight as a consequence of valuing each opinion differently doesn't implies additional difficulties.

$$A_1 = \frac{\sum_{j=1}^{n} A_{1j}}{n} = \frac{4.490}{13} = 345{,}38$$

$$A_2 = \frac{\sum_{j=1}^{n} A_{2j}}{n} = \frac{5.270}{13} = 405{,}38$$

With this the following confidence interval is formed representing the aggregated estimate of the working capital from all the experts:

$$FM_1^{**} = [345{,}38; 405{,}38]$$

It goes without saying that many other aggregating procedures can be imagined[6].

We end here as we consider that the budget has sufficient guarantees for being used for the problems that may arise in current reality.

Whatever the aggregating procedure used, it will be seen that the counter-expertise can become converted into an efficient aid for taking decisions, since, in a certain way, what is arrived at is the mitigation of subjectivity, and in most cases, the reduction of uncertainty.

[6] Along this book we will use some of them, as the main the one which takes as a support the R^+ Expertons. It can be demonstrated that the mathematic expectation of an R^+ Experton obtained from the same information is equal to the interval found through simple arithmetic means.

6 Short Term Financial Solvency

6.1 The Cash-Flow

In order to be able to carry on with its activity, a business requires making certain immediate payments; for this it must be able to have available financial means in the form of short-term liquidity. These highly liquid assets are included in the cash accounts and allow for financing the economic structure of the business. This is cash in the safe, bank accounts and other immediately available cash. These liquid assets and quasi-liquid assets will be closely linked with those payments that the business must make in the immediate future and, therefore posses a high degree of liability.

In this way appears the need to arrive at a balance between the liquid nature of certain assets and the requirement of certain liabilities. On the other hand there also arises the need for arriving at a new equilibrium between two opposing forces: on the one had the business must ensure that it has liquid financial means in order for it to cover its commitments and avoid a serious problem of lack of liquidity; while on the other hand it must maximise profitability and the profitability of liquid financial means or quasi liquid financial means is practically nil. Liquidity and profitability are aspects that are juxtaposed with regard to the optimum amount of cash. From here stems the need to arrive at an equilibrium of counter-position.

The business must know, of course, what its cash movements were in past accounting periods. This historical over-view is interesting in itself but, above all, it is since it constitutes information that will permit having a feeling for how its solvency will evolve in the future. Naturally, the business looks to the future and extracts from the past all the experience it can incorporate into its estimates.

Cash-flow is a term formed by two English words, cash (which corresponds, as we have seen, to numerical concepts, availability, ready money, etc.) and flow (proceed, issue from, circulate, etc). Cash-flow is a semantic approximation of what the specific term signifies. This cash flow can be seen as the conversion of the resources of the business into money, to be understood as money cash available in the safe, current account balances, and other immediately available cash.

We can refer to cash-flow as a current of income and payments in money during a determined period (understanding money–cash), which has become

Anna Maria Gil-Lafuente: *Fuzzy Logic in Financial Analysis*, StudFuzz **175**, 77–93 (2005)
www.springerlink.com

to be called cash-flow statement (as against a commonly accepted meaning, although incorrect, which has been called cash-flow generated).

Among the meanings of cash-flow (cash-flow statement) mention can be made of the definition by Professor Fernández Peña for whom cash-flow constitutes certain "reduced information of all collections and payments specified in an overall manner by the different items of the balance sheet and operating account and corresponds to a determined period of time with the initial and final cash situation"[1].

Having defined the concept and specified the contents of cash-flow, this now appears as an interesting instrument for the business since it refers to being aware of and treating its solvency in the short term.

In this sense, analysis of the cash-flow constitutes a formalisation that permits the study of the currents of payment means and the determination of their difference. This occurs by means of an analysis of comparative statics and a pseudo dynamic analysis:

- Analysis of comparative statics. The cash-flow is compared at two times, at the commencement and at the end of the period, and the difference will show the movement of the cash-flow during the period:

$$C_{T_2-T_1} = C_{T_2} - C_{T_1}$$

- Pseudo dynamic analysis. By means of the balance sheet and operating account the movements that form a part of and do not form a part of operations are studied in order to determine what movements give rise to variations in the cash situation of the business. What this means is a study of the currents done by means of the statement of source and application of the cash-flow for the study and treatment of the solvency of the business:
  1. The historical aspect of the cash-flow, which stems from the figures that the business has gathered in the past and which bring to light the capacity of the business for financing itself internally (historical cash analysis).
  2. The prospective aspect of the cash-flow, where this becomes a complement to the cash budget at the time of establishing forecasts on the solvency of the business, co-operating in the detection of imperfections that could cause financial problems. The importance of the cash-flow relative to how it affects financial management arises from this prospective and estimating approach.

A brief mention at this point should be made of the difference between the concept and contents of the cash-flow and of profit, based on the financial nature (difference between collections and payments) of the first and economic nature of the second (difference between income and costs).

An alternative meaning of cash-flow, which has already been referred to, is that of cash-flow generated. In fact, the custom of including at the end of

[1] Riebold G (1971) The cash-flow (spanish version Ed. ICE, Madrid) p. 18.

the balance sheet and operating account what is generally known as cash-flow generated has become quite extended in business. This is arrived at by:

$$\text{Cash-flow generated} = \text{Profits} + \text{Depreciation} + \text{Supplies}$$

or also

$$= \text{Profits} + \text{Depreciation} + \text{Supplies} - \text{Tax/Profits} - \text{Dividends}$$

this latter concept having also received the name of net cash-flow generated.

Therefore, it is understood that the cash-flow generated (or net-cash-flow generated) means the excess that the business generates through its activity. This is conceptually incorrect and is not a proper meaning for the cash-flow since it confuses what is economic profit with financial excess (and in this respect we refer to the difference between profit and cash-flow).

That the cash-flow generated does not respond entirely to the concept of cash-flow does not mean, however, that it is useless, but all to the contrary. By means of this element it is possible to know the potential for generating resources by the business, as well as establish comparisons between businesses within the same sector (it should be pointed out that the cash-flow generated is not disguised by depreciation or dividend distribution policies).

A frequent use of the net cash-flow generated takes place in the sphere of the stock exchange, because as it refers to the resources of the accounting period that are available to the business, taking into account its policies on depreciations and dividend distribution, it allows for getting to know the position of the business at any determined moment, as a complement to the information that could represent the distribution of dividends, which, in itself, does not take into account the depreciation process.

6.2 Statement of Source and Cash-Flow Application

As has been mentioned when referring to working capital, the statement of source and application of funds brings to light the variations in that working capital. Similarly, the statement of source and application of the cash-flow (or cash) shows the variations in available funds, by means of the origins and applications of the cash, that is to say of short term cash and other accounts.

With the object of studying the EOAT (statement of source and application of cash-flow) it is sufficient to use the patrimonial masses of the balance sheet used up to this point, but making a separation in current assets (A_c) between the patrimonial masses of liquid assets (D) and inventories (obviating the difference between certain and conditioned inventories). In this way we arrive at Fig. 6.1.

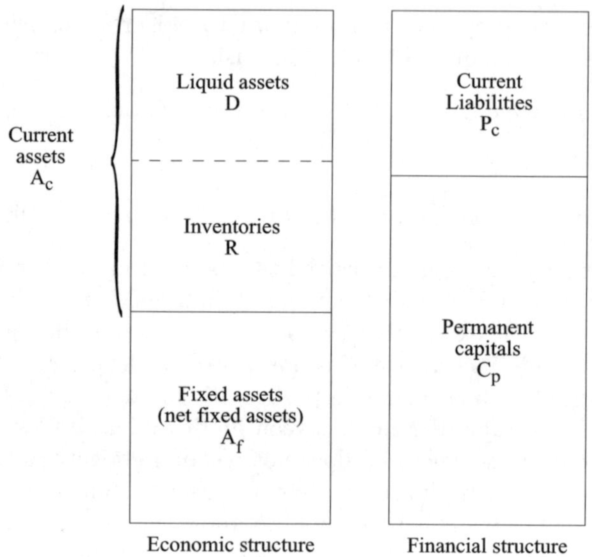

Fig. 6.1.

The variation in liquid assets, considering the economic and financial structures formed by the already mentioned patrimonial masses will be shown by:

- Sources. Cash entries through increases in operating credits (P_c), or permanent capitals (C_p) or decreases in inventories ($-R$) or in fixed assets ($-A_f$).
- Applications. Cash payments due to increases in inventories (R) or of fixed assets (A_f), or by decreases in operating credits ($-P_c$) or permanent capitals ($-C_p$).

The statement of source and application of cash-flow brings to light how cash currents can proceed, on the one hand, from typical operations, that is payments and collections for purchasing and sales operations, and from investment/disinvestments, or permanent capitals.

We should ultimately point out that, for greater simplicity, current assets (P_c) have been considered as only formed by short term credits included within current liabilities, basically bank credits, the increases and decreases of which will constitute cash sources and applications, respectively, which can be as a consequence of financing-investment operations (and not exploitation operations). The extension of this case is simple, since it is sufficient to consider current liabilities (P_c) subdivided into credits for supplies (C_s) and bank credits (C_b) and follow the same process as already described.

6.3 Estimating the Cash-Flow Faced with an Uncertain Future

We do not feel it is necessary to insist once again on the importance represented for the business by the possibility of working with instruments that are suitable for establishing forecasts for the future, above all when this is immersed in the sphere of uncertainty. In order to comply with our objective, that is to say, provide the business with a series of techniques that are adequate for making forecasts relative to the future in a context of uncertainty, it is sufficient to start out from classical instruments. This is simply the application to these classical models certain new techniques, the mathematical calculations of which take uncertainty into consideration. In this way the businessman at the time of taking a decision on the strategy to be followed in the future sees the margin of possibility for success or failure limited.

Focusing our attention on the forecast cash-flow and statement of cash runs, the attempt to adapt known instruments to the new situations of uncertainty could become interesting for estimating an interval in which the cash variations will move. Attaining this objective will allow us to get to know, although in an uncertain manner, the evolution of liquidity and solvency in the short term, with which it will be possible to take correcting action.

In order to make a study of the cash flows in the field of uncertainty confidence intervals will be used for estimating the values of the future balance sheet, since the business does not know the exact numbers that will be shown by the different patrimonial masses although it will be able to estimate between what minimum and maximum these will be found. The numbers of the starting out balance sheet, on the other hand, will be expressed in crisp numbers as this is data from facts that have already occurred.

Let us consider as the initial balance sheet the one we used in the previous example, making a separation in financial assets, with an amount of 100, between "cash and banks" for 50 and "clients and bills receivable" also for 50. Below is the resulting balance sheet:

Assets		Liabilities	
Financial assets		Short term liabilities	50
– Cash and Banks	50		
– Clients and bills receivable	50		
Stocks	250	Long term liabilities	100
Fixed assets	400	Equity	600
Total	750	Total	750

The operations that are estimated will be carried out during the following period and which affect financial assets and short-term liabilities, in this case:

(D)	Operating sales collected in cash	[650; 700]
(A)	Payments in cash for operating purchases	[650; 675]
	Variation "Cash and Banks"	[-25; 50]
(E)	Sales pending payment	[125; 150]
	Variation "Clients and bills receivable"	[125; 150]
(B)	Operating purchases pending payment	[75; 75]
(L)	Dividends to be distributed	[75; 175]
	Variation "Short-term liabilities"	[150; 250]

With these operations, the corresponding patrimonial masses would be estimated, at the end of the period as follows:

Cash and Banks	50 (+) $[-25; 50] = [25; 100]$
Clients and bills receivable	50 (+) $[125; 150] = [175; 120]$
Financial Assets	[200; 300]
Short term liabilities	50 (+) $[150; 250] = [200; 300]$

These are the movements that affect, on the one hand, the cash position and on the other short term debts. In order to arrive at the statement of source and application of the cash flow it is sufficient to group these movements either under a source or cash application. In this didactic example we have limited ourselves to operations that only affect production activity "strictu sensu" (operations) with the object of reducing numerical operations to the maximum; but it is obvious that in a real case as well as the "operating cash-flow" the "non operating cash-flow[2]" would also be included.

In Table 6.1, the statement of source and application of cash-flow is shown including some of the possible causes of cash movements due to operations that are not strictly exploitation, obviously without having been assigned any estimate whatsoever, for the sole effect of pointing out their possible presence in other cases which may be brought up.

The result of Table 6.1 indicates that the cash situation is estimated will have a variation, which will be from a decrease of 25 in the most unfavourable event to an increase of 50 in the optimum situation (from this point of view).

6.4 Estimating the Cash-Flow by Means of the Opinion of Experts

Up to this point we have started out from the fact according to which the forecasts for the patrimonial masses of the balance sheet were uncertain and

[2] As it is well-known, the "non operating cash-flow" include the cash and banks variations whom are direct consequence of changes made in fixed assets and permanent capitals and the ones due to other operations alien to exploitation.

Table 6.1.

Income, origins or sources (inflow)	
Operational sales for cash ($-\Delta$ inventories)	[650; 700]
Long term credits obtained (ΔC_p)	[0; 0]
Collection of client debts and bills receivable ($-\Delta$ inventories)	[;]
Materialisation of capital increases ($-\Delta C_p$)	[0; 0]
Sales of fixed assets for cash ($-\Delta A_f$)	[0; 0]
Total sources	[650; 700]
Payments, applications or destination (outflow)	
Purchases in cash (Δ inventories)	[650; 675]
Payment of bills to be paid and suppliers ($-\Delta E_c$)	[0; 0]
Payment short term credits ($-\Delta E_c$)	[0; 0]
Purchase of fixed assets for cash (ΔA_f)	[0; 0]
Total applications	[650; 675]
Net cash variation = [650; 700] (−) [650; 675] = [−25; 50]	

this uncertainty was not structured, in the sense that between the extremes of the interval there existed total ignorance on what might occur.

At this juncture, let us assume the existence of an expert whose mission in the business consists of giving opinions on future prospects of cash movements. If we asked this professional for a forecast on any accounting value for the following period, it is almost sure that the answer would not be expressed with an exact number. In other words, if we ask our expert to tell us what cash purchases will be for operations, the answer will not be a specific number. The expert will not tell us that purchases for cash operations will be for example 670 or 650. In the best of cases the expert will provide three figures, of which the first will be a value below which, in the opinion of the expert, it is impossible that the numbers representing purchases for cash operations will be.

The second value, which we are sure will be provided by our expert, will be the value above which, in his professional opinion it will be impossible to find the figure representing cash purchases for exploitation.

And lastly, the expert, in accordance with his way of seeing things, will give us a figure that he sees as having the greatest possibility of complying for the next period relative to cash purchases for operations.

It is possible to take this opinion, which we have been given by an expert, to the field of fuzzy calculations by transforming the same into a triangular fuzzy number in the ternary form. Conversion is very simple. The minimum estimate of the expert becomes the lower extreme of the triangular fuzzy number; the maximum estimate, becomes the upper extreme; and the estimate given by the expert with the most possibilities of complying is the number for the maximum presumption in the triangular fuzzy number expressed in the ternary form.

For example, our expert might say:

- Sales for cash operations will be no less than 650
- Neither will they be greater than 675
- What I feel is the most possible is that they will be 670

The resulting triangular fuzzy number in the ternary form will be: (650, 670, 675).

In this sense, it could be thought that our hypothetical expert when establishing the estimate on "cash sales from operations" says that they will not be less than 650, nor greater than 700 but feels (with the maximum intensity) that they will be 660. He has expressed in this way his opinion by mean of a T.F.N. $V = (650, 660, 700)$. As the extremes have a level of presumption 0 and the intermediary number a level of presumption of 1, this fuzzy number can be represented as in Fig. 6.2.

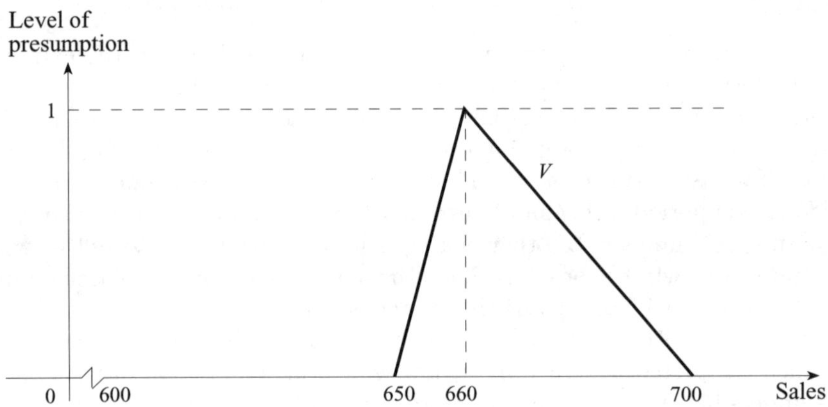

Fig. 6.2.

As is well known[3], a triangular fuzzy number can also be expressed in the form of α-cuts and in the form of four equations. Therefore if we know a triangular number expressed in any of the above forms (that is to say, in the ternary form, in the α-cut form or in the four equations form) we can arrive at its expression in any one of the others that we have mentioned.

Taking into account the graphical scheme of what constitutes a triangular fuzzy number, we can easily move from one form of expression to another. For example, knowing the triangular fuzzy number in the ternary form, as is $\underset{\sim}{\mathbf{V}} = (650; 660; 700)$ we can make some simple calculations in order to convert this into a triangular fuzzy number in the form of α-cuts. If we look at the

[3] Kaufmann A and Gil Aluja J (1987) Técnicas operativas de gestión para el tratamiento de la incertidumbre (in Spanish). (Ed) HISPANO-EUROPEA, Barcelona, p. 65.

graphical representation in Fig. 6.2, we will see that, for level of presumption $\alpha = 0$, the level of sales is 650, as the first piece of information and as the second that for level of presumption $\alpha = 1$, the sales level is 660. Then to find the lower extreme (for any level of presumption) from our T.F.N. in the form of α-cuts, we must find the equation of the straight line that passes through the previously mentioned two points:

$$\begin{aligned}\alpha_1 &= 0 \quad \text{if } v_1 = 650 \\ \alpha_2 &= 1 \quad \text{if } v_2 = 660 \\ \alpha - 0 &= \frac{1-0}{660-650}(v-650) \\ 10\alpha &= v - 650\end{aligned}$$

$10\alpha = v - 650$ is our lower extreme of the T.F.N. in the form of α-cuts.

In order to find the upper extreme it will be sufficient to find the straight line that joins the point of maximum presumption and the point representing the maximum value relative to the T.F.N. in the ternary form. To do this we act in the same way as before. We find the points in question and then apply the formula of a straight line:

$$\begin{aligned}\alpha_1 &= 1 \quad \text{if } v_1 = 660 \\ \alpha_2 &= 0 \quad \text{if } v_2 = 700 \\ \alpha - 1 &= \frac{0-1}{700-660}(v-660) \\ 40 - 40\alpha &= v - 660\end{aligned}$$

$v = 700 - 40\alpha$ is our upper extreme of the T.F.N. in the form of α-cuts.

In this way we could pass over from the expression of a T.F.N. in the ternary form such as (650; 660; 700) to another expression of the same T.F.N. but in the form of α-cuts such as $[650 + 10\alpha; 700 - 40\alpha]$.

In order to do the reverse operation, that is to say to pass over from a T.F.N in the form of α-cuts to a T.F.N. in the ternary form, it is sufficient to give the α the value of zero in order to respectively arrive at the lower extreme and the upper extreme in the ternary form and give a unit value to any of the α in order to arrive at the maximum presumption. $\alpha = 1$ can be done for either of the extremes of the confidence interval because both coincide when the level of presumption is the unit.

Another form of expression of a T.F.N. is via four equations. Its determination is quite easy. For example it is enough to start out from the expression of a T.F.N. in the form of α-cuts or confidence intervals and express it based not on the α, but on the variable we are analysing. For example, taking the T.F.N we found before $[650 + 10\alpha; 700 - 40\alpha]$, we arrive at:

$$\begin{aligned}\alpha &= 0 && \text{if } x < 650 \\ \alpha &= \frac{x-650}{10} && \text{if } 650 \leq x \leq 660\end{aligned}$$

$$\alpha = \frac{700 - x}{40} \qquad \text{if } 660 \leq x \leq 700$$
$$\alpha = 0 \qquad \text{if } 700 < x$$

In order to invert the process, it is sufficient to clear the variable in question and express the function in α, arriving in this way at the new expression of the T.F.N. in the form of α-cuts from which we started out originally.

Returning to the object of our analysis, in the same way as we have expressed the forecast on "operating sales made for cash" as was done by an expert, we can also proceed in the same way so that this same expert gives an opinion on "cash purchases" by means of the T.F.N. $\underset{\sim}{\mathbf{C}} = (650; 660; 675)$. This could be designated as follows:

$$C_\alpha = (650 + 20\alpha; 675 - 5\alpha)$$

The net cash variation would be expressed, in this case, also by means of a T.F.N., which in the ternary form would be:

$$\underset{\sim}{\mathbf{T}} = (650; 660; 700)(-)(650; 670; 675) = (-25; -10; 50)$$

and expressed as α-cuts:

$$\begin{aligned} T_\alpha = V_\alpha(-)C_\alpha &= [650 + 10\alpha; 700 - 40\alpha](-)[650 + 20\alpha; 675 - 5\alpha] \\ &= [(650 + 10\alpha) - (675 - 5\alpha); (700 - 40\alpha) - (650 + 20\alpha)] \\ &= [-25 + 15\alpha; 50 - 60\alpha] \end{aligned}$$

As can be easily seen, the result arrived at with the T.F.N. coincides with the result that was found by using confidence intervals, when considering the extremes (that is to say, between which maximum and minimum figures will the cash position vary). In fact, on doing $\alpha = 0$ in T_α the result is $T_0 = [-25; 50]$, which indicates that the uncertainty is present throughout $50 - (-25) = 75$ units. The level of presumption is, in this case, equal to 0. But as presumption increases, uncertainty is reduced. Thus, when $\alpha = 0{,}4$:

$$T_0 = [-25 + 15 \times 0{,}4; 50 - 60 \times 0{,}4] = [-19; 26]$$

When we arrive at the maximum presumption $\alpha = 1$, the interval is reduced, obviously, to a crisp number; in this case:

$$T_1 = [-10; -10] = -10$$

We are now going to generalise this scheme, providing it with a higher degree of objectivity, assuming that resort is made to the opinion of several experts with the object of their giving their valuations[4] relative to the values which define the net cash variation.

[4] We mean by valuation a subjective given value.

Table 6.2.

Expert	Inflow	Outflow	Net cash variation
1	(650; 660; 700)	(650; 670; 675)	(−25; −10; 50)
2	(660; 690; 720)	(640; 660; 670)	(−10; 30; 80)
3	(600; 640; 660)	(620; 630; 655)	(−55; 10; 40)
4	(630; 640; 680)	(580; 600; 620)	(10; 40; 100)
5	(640; 650; 710)	(600; 620; 630)	(10; 30; 110)
6	(590; 660; 690)	(540; 570; 590)	(0; 90; 150)
7	(610; 650; 670)	(560; 580; 620)	(−10; 70; 110)
8	(670; 690; 720)	(600; 640; 650)	(20; 50; 120)
9	(620; 640; 680)	(570; 600; 620)	(0; 40; 110)
10	(640; 670; 690)	(600; 620; 630)	(10; 50; 90)

In the didactic example which we developed both the total of the "sources" and that of the "applications" are formed, each one of them, by a single component. Nothing varies in the scheme when these totals are the sum of several sources and different destinations. Continuing with our example and let us assume here, for greater ease of calculation, that 10 experts give their valuations, relative to the source and with the application of cash runs by means of T.F.N. (see Table 6.2).

This table allows us to observe the existence of a first "group of fuzzy numbers" relative to the inflow of cash foreseen by the experts, a second group corresponding to estimates for cash outflows and finally a third group that constitutes the net cash variations which the 10 experts feel will occur.

The valuation of the "inflow" made by expert i is designated by $\underset{\sim}{\mathbf{V}}_i$ and the "outflow" given by the same expert by $\underset{\sim}{\mathbf{C}}_i$. The difference between these T.F.N is represented by $\underset{\sim}{\mathbf{T}}_i$, $i = 1,2,\ldots,10$, and brings to light the estimates that expert i gives for the net cash variations.

In order to arrive at the best representation of each group by means of a fuzzy number, we are going to find the corresponding "average fuzzy numbers[5]".

As we know, if we have a bundle of fuzzy numbers:

$$A_\alpha^{(i)} = \left[a_1^{(i)}(\alpha); a_2^{(i)}(\alpha)\right] \quad i = 1,2,\ldots,n$$

that express Ai in the form of α-cuts, the average fuzzy number $\overset{m}{A}_\alpha$ will be:

$$\overset{m}{A}_\alpha = \overset{m}{a}_1(\alpha); \overset{m}{a}_2(\alpha)$$

where:

[5] For a better knowledge of the "average fuzzy number" see Kaufmann A and Gil Aluja J (1986) Introducción a la teoría de los subconjuntos borrosos a la gestión de las empresas (Spanish version). (Ed) Mlladoiro, Santiago de Compostela. pp. 223–230.

$$\overset{m}{a}_1(\alpha) = \frac{1}{n}\sum_{i=1}^{n} a_1^{(i)}(\alpha)$$

$$\overset{m}{a}_2(\alpha) = \frac{1}{n}\sum_{i=1}^{n} a_2^{(i)}(\alpha)$$

In the case of T.F.N. for greater simplicity, the ternary form can be used. In our example we would have:

$$\overset{m}{\underset{\sim}{\mathbf{V}}} = \frac{1}{n}\sum_{i=1}^{n} (a_i,b_i,c_i) = \frac{1}{n}(\cdot)\left(\sum_{i=1}^{n} a_i, \sum_{i=1}^{n} b_i, \sum_{i=1}^{n} c_i\right)$$
$$= \frac{1}{10}(\cdot)(6.310; 6.590; 6.920) = (631; 659; 692)$$

Likewise we can arrive at $\overset{m}{\underset{\sim}{\mathbf{C}}}$.

$$\overset{m}{\underset{\sim}{\mathbf{C}}} = \frac{1}{10}(\cdot)(5.960; 6.190; 6.360) = (596; 619; 636)$$

And finally, $\overset{m}{\underset{\sim}{\mathbf{T}}}$ will be:

$$\overset{m}{\underset{\sim}{\mathbf{T}}} = \frac{1}{10}(\cdot)(-50; 400; 960) = (-5; 40; 96)$$

Therefore, if it is considered that the group of fuzzy numbers is a good aggregate formalisation of the opinion of the experts and the average fuzzy number is accepted as its best representation, the result arrived at indicates that, in the period under consideration, the cash variation will move between a reduction of 5 and an increase of 96. And the maximum presumption is that it will be 40.

6.5 The Difference of Opinion Between Experts

The result arrived at, however, must be submitted to a certain analysis which situates it in the position that it really is within the schemes for treating uncertainty.

The first consideration to be made refers to the extension of the differences of opinion between all the experts. If it was desired to operate within the strict limits of security provided by the schemes of uncertainty and we were to accept as possible "all" the opinions of the experts, what should really be taken into account as the lower extreme of the T.F.N. representing the aggregate, is the smallest of the lower extremes of the fuzzy numbers, and as the upper extreme the highest of the upper extremes. In this event the uncertainty would be much higher, since for the "cash variation" −55 should

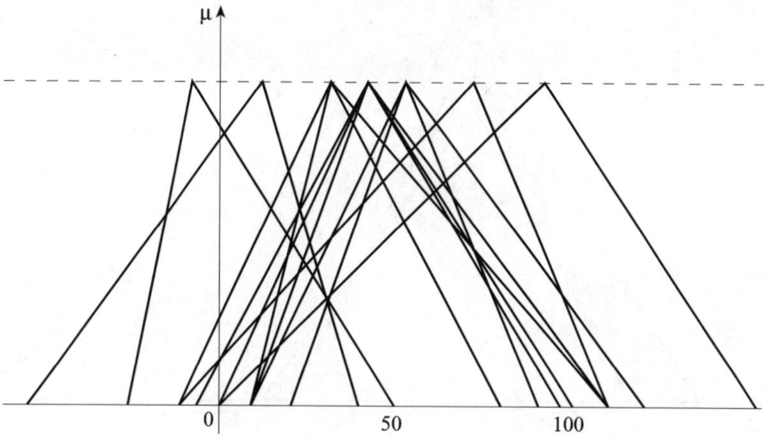

Fig. 6.3.

be taken as the lower extreme and 150 as the upper extreme. The range of the uncertainty would make this scheme inoperable.

We are now going to graphically represent (see Fig. 6.3) the T.F.N. corresponding to the opinion of each one of the experts relative to the "net cash variations" and we have included the average fuzzy number, so as to make this aspect more visible.

The fact that there are opinions of experts that are notably removed from what we have considered representative of the aggregate, the need arises for attempting wherever possible to draw these opinions somewhat closer. For this we propose using a path, previously tried with success in the Delphi[6] techniques, which consists of informing each expert of the distance that their opinions and those of the aggregated opinion of all the experts, with the object of their reconsidering.

This leads to the use of a concept of distance in the T.F.N., the so-called "linear distance" as the average value between the "linear distance to the left" and the "linear distance to the right". Let us look at a specific case of distance, always in relation to the "net cash variation", between the opinion of the aggregate $T = (-5; 40; 96)$ and the opinion of expert 10, $T^{(10)} = (10; 50; 90)$, for example. If we consider the graphical representation we arrive at the Fig. 6.4.

It will be seen that the linear distance to the left is given by the area of the trapezoid lined diagonally, while the distance to the right by the sum of the areas of the two triangles lined horizontally. This example brings to light the fact that the calculation will be different according to whether the sides (either of the left or the right) of the triangles cross over (distance to

[6] See, for example, Kaufmann A and Gil Aluja J (1986) Introduccion de la teoría de los subconjuntos borrosos a la gestion de las empresas. Op. cit., pp. 131–137.

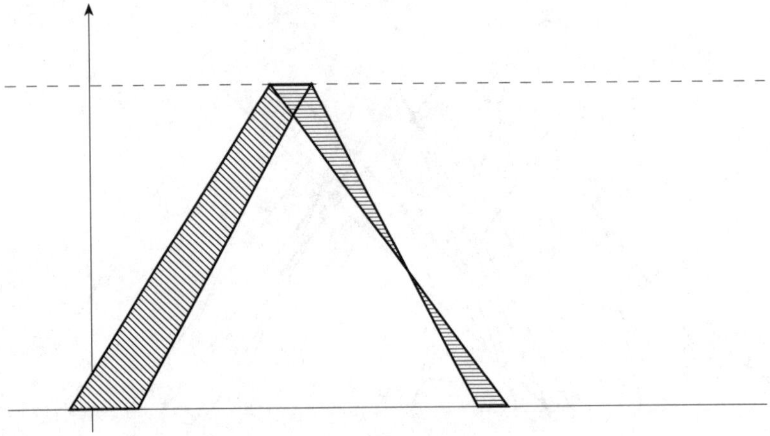

Fig. 6.4.

the right in this example) or they do not cross over (distance to the left in this example). The immediate question that arises is how to know, without the need to resort to a graphical representation, if the sides cross over or not. This will be a prior step in order to be able to do one or other calculation of distance.

In order to know if a cross over exists or not between the sides of two triangles, the sign of the product of the differences should be arrived at between the two bases (either of the trapezoid or the two triangles). If this is positive or nil then there is no cross over, and there will be a cross over of the result is negative. In fact, if with a general character the aggregate is given by:

$$\overset{m}{\underset{\sim}{\mathbf{T}}} = \left(\overset{m}{a_1}; \overset{m}{a_2}; \overset{m}{a_3} \right)$$

and the opinion of the expert by:

$$\underset{\sim}{\mathbf{T}}^{(i)} = \left(a_1^{(i)}, a_2^{(i)}, a_3^{(i)} \right)$$

there will be no cross over to the left (neither to the right) if:

$$\left(a_1^{(i)} - \overset{m}{a_1} \right) \cdot \left(a_2^{(i)} - \overset{m}{a_2} \right) \geq 0$$

and on the other hand there will be a cross over, for example to the right, if:

$$\left(a_3^{(i)} - \overset{m}{a_3} \right) \cdot \left(a_2^{(i)} - \overset{m}{a_2} \right) < 0$$

In this specific case, since $a_1^{(10)} = 10$, $\overset{m}{a_1} = -5$, $a_2^{(10)} = 50$, $\overset{m}{a_2} = 40$, $a_3^{(10)} = 90$, $\overset{m}{a_3} = 96$ we will have:

$$[10-(-5)]\cdot(50-40)=150>0 \quad \text{no cross over}$$
$$[90-96]\cdot(50-40)=-60<0 \quad \text{cross over exists}$$

Arriving at the distance in absolute terms, when no cross over exists, is limited to the calculation of the area of the trapezoid:

$$D_1=\frac{\left|a_1^{(i)}-\overset{m}{a}_1\right|+\left|a_2^{(i)}-\overset{m}{a}_2\right|}{2}=\frac{|10+5|+|50-40|}{2}=\frac{25}{2}$$

In order to arrive at a number included between 0 and 1 it is normal to take a referential:

$$P\geq\bigvee_i a_3^{(i)}-\bigwedge_i a_1^{(i)}$$

which in the case of our example with 10 experts would be:

$$P\geq 150-(-55)=205$$

Using 205 we will arrive at the relative distance to the left as follows:

$$d_1=\frac{D_1}{P}=\frac{25}{2}\quad\frac{1}{205}=\frac{5}{82}=0{,}06$$

In order to arrive at the distance in absolute terms when a cross over occurs between the sides it is necessary to find the sum of the areas of the two triangles and consequently find the height of the same (the bases are known).

If we call the height of a triangle α, the height of the other will be $1-\alpha$. By similarity the following will be complied with:

$$\frac{\alpha}{\left|a_3^{(i)}-\overset{m}{a}_3\right|}=\frac{1-\alpha}{\left|a_2^{(i)}-\overset{m}{a}_2\right|}$$

and also

$$\left|a_3^{(i)}-\overset{m}{a}_3\right|-\left|a_3^{(i)}-\overset{m}{a}_3\right|\cdot\alpha=\alpha\cdot\left|a_2^{(i)}-\overset{m}{a}_2\right|\cdot\left|a_2^{(i)}-\overset{m}{a}_2\right|+\left|a_3^{(i)}-\overset{m}{a}_3\right|$$
$$=\left|a_3^{(i)}-\overset{m}{a}_3\right|$$

therefore:

$$\alpha=\frac{\left|a_3^{(i)}-\overset{m}{a}_3\right|}{\left|a_2^{(i)}-\overset{m}{a}_2\right|-\left|a_3^{(i)}-\overset{m}{a}_3\right|}$$

And, then:

$$1-\alpha=\frac{\left|a_2^{(i)}-\overset{m}{a}_2\right|}{\left|a_2^{(i)}-\overset{m}{a}_2\right|-\left|a_3^{(i)}-\overset{m}{a}_3\right|}$$

The sum of the areas of the triangles will be:

$$D_D = \frac{\left|a_3^{(i)} - \overset{m}{a}_3\right| \cdot \alpha}{2} + \frac{\left|a_2^{(i)} - \overset{m}{a}_2\right| \cdot (1-\alpha)}{2}$$

By substituting α and $1-\alpha$, we arrive at:

$$D_D = \frac{1}{2}\left[\frac{\left(a_3^{(i)} - \overset{m}{a}_3\right)^2}{\left|a_2^{(i)} - \overset{m}{a}_2\right| + \left|a_3^{(i)} - \overset{m}{a}_3\right|} + \frac{\left(a_2^{(i)} - \overset{m}{a}_2\right)^2}{\left|a_2^{(i)} - \overset{m}{a}_2\right| + \left|a_3^{(i)} - \overset{m}{a}_3\right|}\right]$$

$$= \frac{1}{2} \cdot \frac{\left(a_3^{(i)} - \overset{m}{a}_3\right)^2 + \left(a_2^{(i)} - \overset{m}{a}_2\right)^2}{\left|a_2^{(i)} - \overset{m}{a}_2\right| + \left|a_3^{(i)} - \overset{m}{a}_3\right|}{}_7$$

The relative distance to the right would be:

$$d_P = \frac{D_D}{P} = \frac{1}{2P} \cdot \frac{\left(a_3^{(i)} - \overset{m}{a}_3\right)^2 + \left(a_2^{(i)} - \overset{m}{a}_2\right)^2}{\left|a_2^{(i)} - \overset{m}{a}_2\right| + \left|a_3^{(i)} - \overset{m}{a}_3\right|} = \frac{1}{410}\,\frac{100+36}{10+6} = \frac{136}{6.560} = 0{,}02$$

And finally the relative linear distance would be:

$$d = \frac{d_1 + d_1}{2} = \frac{0{,}06 + 0{,}02}{2} = 0{,}04$$

The information received by each expert will allow for the reconsideration or not of the opinion given initially. As can be seen in this case, the opinion of the experts does not coincide, since while some of them, as is the case with the opinion of expert 10 the distance of which we have just arrived at, are close to the aggregate opinion, other experts, for example, expert 3 or expert 6, are notably removed from what has become considered as the opinion of the group.

In this specific case, as in many others, it would be convenient to remit to all and each one of the experts consulted the result of the aggregated opinions, which in the event is given by the average T.F.N. $(-5; 40; 96)$, jointly with the distance between the T.F.N. showing his particular opinion and the aggregated opinion (which in the case of expert 10 would be 0,04). A consideration relative to the proximity of the average T.F.N. compared to the other experts could be useful for him for giving or not a new opinion, that is to say to eventually reconsider.

Normally, each new phase of information/reconsideration that is established implies a closing up of positions of the experts that are further removed from the aggregate opinion, although this does not necessarily have to occur.

When one or several experts exist who insist on opinions that are separated from the group, it is then convenient to analyse the cause why this disparity exists and which can be motivated by reasons of training, profession, geographical location, among others.

Finally it is convenient to establish the number of phases that will form the process for requesting information and providing an opinion. It is quite obvious that this process cannot be indefinite.

It is usually convenient to establishing ex ante the number of times that an opinion is going to be asked for: 3, 4 or 5 times in the majority of cases. Also it can be established to stop the process either when the distance of the greater part of the experts is no higher than a determined number, or when the greater part of the opinions that are given in a determined phase does not differ substantially from the previous phase.

In this chapter we have attempted to introduce certain new techniques that are suitable for treating uncertainty in a classical problem such as are cash movements. With this we do not intend to say that the techniques are new. The novelty, in any event, consists in having incorporated the same to certain schemes that with them, we feel, are more suitable for the treatment of current and future problems.

Our modest contribution could become a reality in the establishment of some general formulae for arriving at the distances to the left and to the right both in the event of the existence of a cross over or not of the sides of the triangles representing the T.F.N.

This, in conjunction to the clear establishment of the existence of any cross over between the sides of the triangles representing the T.F.N. without the need to resort to a graphical representation, can be the addition of our grain of sand for the improvement of this management instrument called cash-flow.

7 Financial Analysis by Means of Ratios

7.1 Preliminary Considerations

The profound changes that are taking place in the economic systems of countries that intervene in the economic-financial concert of our day and age, make it increasingly necessary to analyse the situation of businesses by following not only their internal evolution but also their position compared to other businesses within the same market.

Economic-financial analysis basically concentrates its attention on the movements taking place in economic and financial structures, on the one hand, and in the study of the variations that take place in financial currents within the business. It is necessary for a certain correspondence to exist between liquidity of the economic structure and the liabilities of the financial structure.

Among the elements that are used traditionally for economic-financial analysis, the one standing out due to its interest is the ratio method. To be understood by "ratio" is a quotient between patrimonial elements or masses that brings to light a particularity in the economic or financial situation. An attempt is made to express the representation as some aspect of the economic-financial activity by means of a number.

From a traditional point of view, ratios offer a wide range of possibilities for determining both the situation of the business at any given moment, such as the evolution of the same over time and also its positioning compared to other similar businesses. Therefore, four types of ratios can be considered based on:

(A) Static analysis. A balance sheet and operating account is considered from which data is extracted informing on the situation of the business at a determined moment in time.
(B) Comparative statistics. The situation of the business is compared at different times. Its history is not studied in a continuous way, but this is done between two or more situations. Normally these are used for studying the evolution of the history of the business relative to predetermined elements.

Anna Maria Gil-Lafuente: *Fuzzy Logic in Financial Analysis*, StudFuzz **175**, 95–116 (2005)
www.springerlink.com

(C) Positioning of the business. By means of ratios a comparison of the current situation of the business with the current situation of other businesses in the same economic sector or sub-sector.
(D) Comparison with standard ratios. A comparison is made between the currents of the business with certain standard ratios, which allow for setting limits for desirable situations or simply to compare it with an ideal ratio (objective to which the business should be inclined).

Without prejudice to the fact of insisting on this, further on we will be advancing an idea that presides over this work: this is that an analysis of past or present situation contributes important information, obviously partial, that is necessary for efficient management of the business. Nevertheless, only prospective analysis will allow for the taking of decisions that are suitable relative to possible future strategies. From here stems the need for creating certain ratios that are apt for treatment in the field of uncertainty.

Thus, in classical studies a start is made from data offered by the accounts, mainly but not exclusively, obtained from the flowing summarising documents:

- Balance sheet
- Operating account
- Profit and loss account.

From these documents data is traditionally extracted that is treated by the ratio method.

With regard to the first of these documents, the balance sheet, its accounts are usually grouped taking into account a degree of liquidity, relative to the accounts of the assets side of the balance sheet and to the degree of liability, when referring to the accounts of the liabilities side of the balance sheet. Therefore, by following this scheme what could be arrived at, among other things, is the structure shown by Table 7.1 (ordered from high to low liquidity and liability). Another presentation of the balance sheet, somewhat more summarised, could be that as shown in Table 7.2.

Table 7.1. Balance Sheet

Assets	Liabilities
Liquid assets	Short term liabilities
Certain inventories	Long term liabilities
Stocks	
Fixed assets	Capital and reserves

Table 7.2. Balance Sheet

Assets	Liabilities
Current assets	Short term liabilities
Fixed assets	Permanent capitals

By considering a balance sheet the relation between two of its patrimonial masses can be arrived at from two different perspectives:

- Vertical analysis. Determination of the relation over total assets or liabilities represented by liquid assets, inventories or fixed assets or short-term liabilities, long-term liabilities and capital and reserves respectively. Generally speaking these ratios will provide a certain idea about the business in question. For example, in an industrial business, the mass of fixed assets represents, over total assets, a greater percentage than if we were to be talking with reference to a commercial business. For example if we see that a business keeps a great deal of liquid assets it may be due to the fact that it cannot obtain credits, etc.
- Horizontal analysis. The masses of assets are compared to those of liabilities. It is generally accepted that the masses of less liquidity have a certain correspondence with those of greater liabilities. We will return to this subject at a later stage.

It appears obvious then that the situation and prospects of a business cannot be defined by a single ratio, but it is necessary to take into account several of these in order for them to be used as indicators of the multiple aspects of the economic-financial situation in order to attain an overall view.

The first alternative that arises consists in deciding if to use a few ratios or on the contrary a high number of ratios should be used.

For the second alternative the decision has to be taken on setting, which should be the ratios to be taken into account. Obviously, there cannot be just a single answer, as the selection has a lot to do with the type of business, its dimension and the aspect of this business which is need to be investigated. The positions adopted by the specialists are multiple. Below we mention the classification followed by Suárez[1]:

(A) Situation ratios
(B) Activity or management ratios (rotation)
(C) Profitability ratios

[1] Suárez Suárez A (1985) Decisiones óptimas de inversión y financiación en la empresa, (in Spanish). (Ed) Pirámide, Madrid, p. 138.

(D) Productivity ratios
(E) Stock exchange ratios

We have no intention of making a description, even a brief one, of the meaning and characteristics of each one of these groups of ratios. These introductory aspects should only serve as a support for demonstrating the efficiency of the new techniques that we are proposing.

Merely as an indication, and in accordance with the subject we are developing, we are going to make a special mention of the ratios of financial summary or equilibrium.

In order to analyse the capacity that a business has to meet its financial commitments it is customary to compare certain elements of assets with others of liabilities, reaching the conclusion that there is sufficient solvency when asset liquidity permits covering liability payments.

7.2 Forecasting Profitability and Solvency by Means of Ratios

We are going to elaborate somewhat on this point and will consider, on the one hand, solvency in the medium and long-term and, on the other, solvency in the short term.

With regard to medium and long-term solvency it is necessary to point out that there must exist a certain amount of correspondence between investments that have a low degree of liquidity and permanent capitals with a low degree of liability (capital, reserves, medium and long-term debts). Financing investments in fixed assets with short-term debts on a continuous basis may lead the business to a situation of suspension of payments. For this reason an attempt should always be made to ensure that the average duration of debt should be approximately equal to the average life of all the investments which are financed from outside sources. If the average duration of the debts is less than the average life of the investments, a moment may come in which we find ourselves without being able to attend to the normal operation of the business (the case of suspension of payments). If the contrary occurs, that is to say, that the average duration of the debts is very much longer than the average life of the investments, we will be wasting economic resources not achieving profitability of the economic resources we have available.

The above principle (correspondence between the average duration of debts and average life of investments) is accompanied by a second principle: permanent capital should also finance, as well as fixed assets, a part of current assets in order to be able to count on a cushion against possible temporary imbalances. As we saw in a previous chapter, this difference between permanent capitals and fixed assets constitutes the so called working capital.

Among the most used medium and long-term solvency ratios[2] we could mention the fixed assets financing ratio. This indicates the relationship existing between permanent capitals and net fixed assets.

By equity (own capital) we understand not only equity in the strictest sense (that is capital plus reserves), but we also include under this the amount for supplies, since these can also be considered as a type of reserve.

Net fixed assets is constituted by the total value of all fixed assets once the respective depreciation has been discounted:

$$\text{Fixed assets financing ratio} = \frac{\text{Permanent capitals}}{\text{Net fixed assets}}$$

Nevertheless, it should be taken into account that on not including the value for depreciations in net fixed assets, under the heading permanent capitals amounts relative to depreciation funds are similarly excluded.

As has been mentioned before, working capital, can never be negative, that is to say, that the value of fixed assets can never be higher than the value of permanent capitals. In order for this to be complied with the ratio of fixed asset financing cannot be less than the unit. Nevertheless, this does not imply that the most adequate situation is that this ratio be much higher than the unit, since, in this way, we would be faced with a situation in which a large part of permanent capitals would be financing current assets; in which case a situation would be incurred in of decrease of the general profitability of the business. To conclude, we can see that the desirable situation for a business would be one in which, on analysing its ratio of fixed asset financing, we were to find a value close to the unit, but always higher than one.

In order in a way to complete the information offered by this ratio, which we have just mentioned, we can use others, such as those below:

$$\text{Ratio of indebtedness} = \frac{\text{Total debts}}{\text{Equity}}$$

$$\text{Long term of indebtedness} = \frac{\text{Long term debts}}{\text{Equity}}$$

$$\text{Medium and long term of indebtedness} = \frac{\text{Medium and long term debts}}{\text{Equity}}$$

If for example we analyse the medium and long-term indebtedness, we can easily deduce that the solvency of the business is not good if this ratio is greater than the unit, that is, when medium and long-term debts are very much higher than equity, which makes it very difficult for the business to increase its medium and long-term debts. On the other hand, the fact that the ratio of medium and long-term indebtedness were to be equal to the unit, would mean that the business had a financial policy that was excessively conservative, not taking advantage in this way with the opportunity for obtaining, by using external financial means, higher profitability on its equity.

[2] Suárez Suárez A. Op. cit., pp. 141–142.

Experience shows us that overall medium and long-term debts should not be any more than 100% of the value of equity.

We now move on to the analysis of short-term solvency and it should be stated that with this the intention is to study the possibilities the business has for making the payments, which normally appear as a consequence of the daily activity of the business. In this way, independently of the instruments we have already studied (like working capital, cash-flow etc.), which constitute some good indicators for short-term solvency, it would be convenient to consider other ratios that would allow us to take into account the composition of current assets and, in the same way, that for short-term liabilities. Therefore, analysis of short-term solvency of a business can be determined, as well as by the values studied before, by cash ratios, among others. To these effects it can be taken into account that, for example, the so-called liquidity ratio, which, when working in the field of certainty, provides us with a relative measurement of what constitutes working capital. The liquidity ratio is expressed as a relation between current assets and short-term debts:

$$\text{Liquidity ratio} = \frac{\text{Current assets}}{\text{Short term debts}}$$

Given that working capital must always be positive, that is to say, current assets must always be higher than short-term debts, the liquidity, the liquidity ratio must always be equal to or higher than the unit. This ratio acquires special interest since it allows us, as it is a relative value, to make both inter-business and inter-period comparisons.

Another ratio that is particularly important at the time of determining short-term solvency of the business is what is known as the ordinary cash ratio, the expression of which relates the values of liquid assets and true inventories with short-term liabilities:

$$\text{Ordinary cash ratio} = \frac{\text{Liquid assets and true inventories}}{\text{Short term liabilities}}$$

For a better representation of what constitutes the reality of the business in this case it is convenient not to include stock or inventories in the numerator of the ordinary cash ratio, due both to the problems of valuation and the difficulty of determining their degree of liquidity. Taking into account the fact that the stocks can act as a buffer for possible fluctuations that the ordinary cash ratio may suffer caused by liquid assets and true inventories, a value for this ratio would be acceptable that varied between 0,5 and 1 according to the type of business and activity it carries out.

Finally we should mention relative to short-term solvency the immediate cash ratio, which is expressed as a relation between the values of liquid assets and short-term liabilities:

$$\text{Inmediate cash ratio} = \frac{\text{Liquid assets}}{\text{Short term liabilities}}$$

With regard to this ratio, a ratio of between 0,1 and 0,3, once again clearly stating that this will be according to the type of activity carried on by the business, since a ratio higher than 0.3 normally indicates an excess of liquidity, which can have an effect, which may result in a lesser degree of profitability. The interest in this ratio, however, lies in its momentary nature for special situations.

It would not be correct to finish this rapid description of ratios without making a brief mention of profitability ratios, which relate different notions of the profit with diverse aspects representing the life of the business. Therefore, if the sales figure is considered, we have:

$$\text{Net margin ratio} = \frac{\text{Anual net profit}}{\text{Annual sales figure (less sales tax)}}$$

Another aspect of profit is shown by the remuneration of equity. The financial profitability ratio supplies this value in percentage terms. A business with future prospects must posses a certain hope that will allow it to assume that this ratio, in the future, will be higher than the interest rate that holders of funds can obtain in the financial market, taking into account the fact that profit must not only cover distribution of dividends, but also auto-financing. The ratio is defined as:

$$\begin{aligned}\text{Financial profitability ratio} &= \frac{\text{Net profit before taxes}}{\text{Equity}} \\ &= \frac{\text{Net profit before taxes}}{\text{Capital + Reserves}}\end{aligned}$$

This same ratio could also be established with the sole inclusion of the share capital and, therefore, eliminating reserves from the denominator. We feel that there is no sense in acting in this way since these are also the property of the shareholders.

When it is desired to establish the efficiency of all the capitals available to a business (own and outside) the economic profitability ratio is used:

$$\begin{aligned}&\text{Economic profitability} \\ &= \frac{\text{Net profit before taxes + interest on long, medium and short term debts}}{\text{Equity + long, medium and short term debts}}\end{aligned}$$

To round this off we should take a look at an interesting scheme[3] which is normally called "the Du Pont ratio cascade", in which, starting out from certain data gathered from the balance sheet and operating account, successive ratios are arrived at which end up in the ratio that supplies the profitability on equity expressed as a quotient between the ratio of profitability on

[3] This scheme has been reproduced in many works on the subject. We mention: Bachiller, Lafuente y Salas (1982) Gestión económico-financiera del circulante. (Ed) Pirámide, Madrid,, Chap. 6.

financial resources used and the proportion between equity and permanent capitals.

This scheme can be represented as is shown in Table 7.3.

The summary we have just made is obviously incomplete, since we have only attempted to bring to light the interest that the ratios method has for financial analysis ex post. What we certainly have mentioned are the ratios that are most related to solvency and profitability of the business in order to show the effective possibilities of their use. But we must not loose sight of the principal objective of this book, which is the drawing up of certain techniques that are suitable for the management of the business of tomorrow, that is, those which can face the challenge of a future that we consider uncertain. For this we are going enter the spheres of uncertainty by means of confidence intervals.

7.3 Representation of Ratios by Confidence Intervals

The incorporation of confidence intervals as forecasts for ratios in the future is facilitated by the fact that a ratio is always a quotient of positive numbers. For this reason the problem is limited to the solution of the following questions:

1. How do we arrive at the quotient of confidence intervals?
2. How do we compare confidence intervals?

7.3.1 Arriving at the Quotient Between Confidence Intervals

Given two confidence intervals in $R^+ : \underset{\sim}{\mathbf{N}} = [N_1, N_2]$ and $\underset{\sim}{\mathbf{D}} = [D_1, D_2]$, we can obtain the quotient as follows:

$$\underset{\sim}{\mathbf{R}} = \underset{\sim}{\mathbf{N}} : \underset{\sim}{\mathbf{D}} = [N_1, N_2] : [D_1, D_2] = \left[\frac{N_1}{D_2}; \frac{N_2}{D_1}\right] = [R_1, R_2]$$

This interval we have obtained is also a member of the set of positive rational numbers. Let us take a look at an example. If:

$$N - [500; 600]$$
$$D = [800; 1.000]$$
$$\underset{\sim}{\mathbf{R}} = \underset{\sim}{\mathbf{N}} : \underset{\sim}{\mathbf{D}} = [500; 600] : [800; 1.000] = \left[\frac{500}{1.000}; \frac{600}{800}\right] = [0{,}5; 0{,}75]$$

Then:

$$R = [R_1, R_2] = [0{,}5; 0{,}75]$$

Table 7.3. Du Pont scheme for financial control

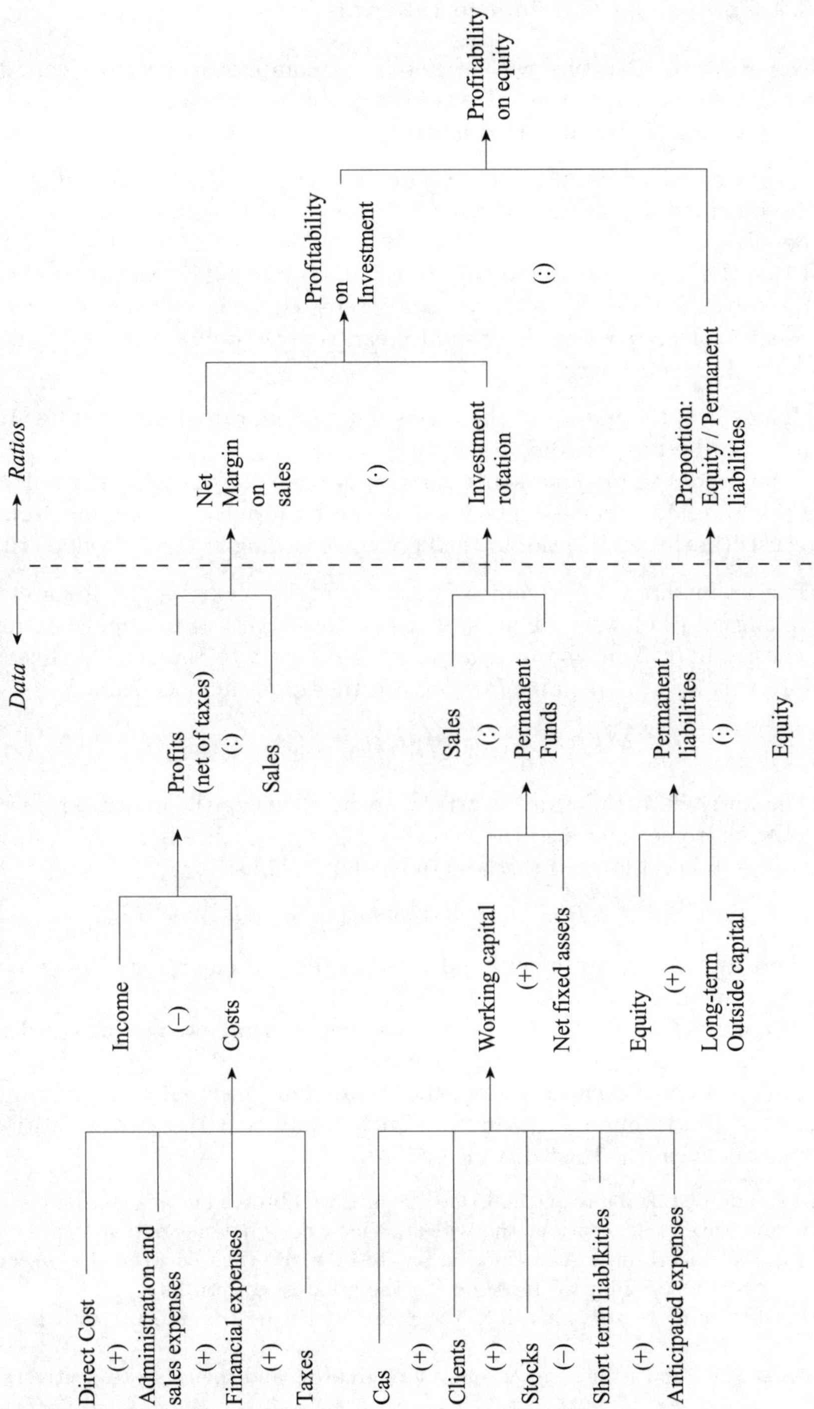

7.3.2 Comparing Confidence Intervals

There is more than one way of making a comparison between confidence intervals. According to the nature of the problem, comparison can principally be established by two different means:

- Place the intervals in increasing or decreasing order. Profitability ratios are included in this case as what interests us here is to choose the highest possible.
- Place the interval between two determined limits. This would be the case for solvency. In fact: what we are interested in is solvency but not excessive solvency, since this would mean that we would be holding idle or unproductive resources.

Therefore, the nature of the ratios must be known in order to be able to determine the type of comparison to be used.

The first case (comparison by order) is based on arriving at the highest (or the lowest) confidence interval that can be formed starting out from the intervals (in this application ratios) given, according to the following criteria:

- The upper limit (supremum) is arrived at by choosing the highest from among all the lower extremes, and also the largest of the upper extremes. The result will be the interval located furthest to the right[4]. Given two intervals $A = [a_1; a_2]$ and $B = [b_1; b_2]$, the upper limit is defined as:

$$A \vee B = [a_1; a_2](\vee)[b_1; b_2] = [a_1 \vee b_1; a_2 \vee b_2]$$

- The lower limit (inferum) is arrived at by choosing the lowest between the lower extremes and also the lowest of the upper extremes. The result will be the interval located furthest to the left[5]. Thus:

$$A \wedge B = [a_1; a_2](\wedge)[b_1; b_2] = [a_1 \wedge b_1; a_2 \wedge b_2]^6$$

For both cases, supremum and inferum the chosen extremes may or may not be a member of the same confidence interval.

Having arrived at the supremum (or the inferum) we can now order the ratios from highest to lowest (or else from lowest to highest). For this we resort to the distance of each ratio from the supremum (eventually the inferum) by means of the Hamming distance, in such a way that the greatest ratio will be the one with the least distance:

---

[4] It should be taken into account that the interval obtained in this way (supremum) is not, save for exception, the widest, which does not mean that it posses the greatest uncertainty. A distinction has to be made then between the supremum interval and the interval representing the greatest uncertainty.

[5] For the same reason, this does not mean it will be the interval with the least uncertainty.

[6] As is well known the signs $\vee$ and $\wedge$ the greatest and the least, respectively.

$$d(A, A \vee B) = |a_1 - a_1 \vee b_1| + |a_2 - a_2 \vee b_2|$$
$$d(B, A \vee B) = |b_1 - a_1 \vee b_1| + |b_2 - a_2 \vee b_2|$$

The relative distance will be obtained by dividing the overall distance by two.

For ordering the ratios from least to greatest the same process is followed, but arriving at the distance relative to the inferum. The lowest ratio will be the one to be found with the least distance.

Let us look at an example. We can assume the following two intervals $\underset{\sim}{\mathbf{A}}$ and $\underset{\sim}{\mathbf{B}}$):

$$\underset{\sim}{\mathbf{A}} = [70; 110]$$
$$\underset{\sim}{\mathbf{B}} = [90; 100]$$

The supremum and the inferum will be:

$$\text{Supremum} = [70; 110](\vee)[90; 100] = [90; 110]$$
$$\text{Inferum} = [70; 110](\wedge)[90; 100] = [70; 100]$$

If what is wanted is to order then from greater to least do:

$$d(A, A \vee B) = |70 - 90| + |110 - 110| = 20 + 0 = 20$$

$$d(B, A \vee B) = |90 - 90| + |100 - 110| = 0 + 10 = 10$$

so that $B \succ A$.

So as to order them from least to greatest:

$$d(A, A(\wedge)B) = |70 - 70| + |110 - 100| = 0 + 10 = 10$$

$$d(B, A(\wedge)B) = |90 - 70| + |100 - 100| = 20 + 0 = 20$$

so that $A \prec B$.

This comparison process is easy and immediately usable for the case of n confidence intervals[7].

For the second case (comparison of an ideal ratio which need not necessarily be the supremum or the inferum) resort can also be made to the notion of distance[8].

In fact, let us assume that in order to reach a determined objective, a ratio should not be located below e_1 nor above e_2, that is, that ratio $E = [e_1; e_2]$ is established as the ideal in the field of the R^+. It should be pointed out

[7] For this generalisation see Kaufmann A and Gil Aluja J (1986) Introducción de la teoría de los subconjuntos borrosos a la gestión de las empresas (in Spanish). (Ed) Milladoiro, Santiago de Compostela, Chap. VIII.

[8] This case is not mentioned in the work by Kaufmann A and Gil Aluja J mentioned previously, although in a certain way, it could be considered a variation of the one studied there.

that underlying in this reasoning is an implicit hypothesis that consists in assuming that a deviation below the lower limit is as bad as a deviation above the upper limit. If the possibilities for action of a business is given, for example, by two different strategies, leading to ratios $A = [a_1; a_2]$ and $B = [b_1; b_2]$ for deciding which path to follow, resort can be made to a distance, for example the Hamming distance[9], selecting that strategy which provides a greater distance. This criterion warrants certain considerations, which we will do later on. In this case, let:

$$d(A; E) = |a_1 - e_1| + |a_2 - e_2|$$

$$d(B; E) = |b_1 - e_1| + |b_2 - e_2|$$

If the relative distances are considered:

$$\delta(A; E) = \frac{d(A; E)}{2} = \frac{|a_1 - e_1| + |a_2 - e_2|}{2}$$

$$\delta(B; E) = \frac{d(B; E)}{2} = \frac{|b_1 - e_1| + |b_2 - e_2|}{2}$$

We now take a look at an example. Let us assume that the ideal ratio is.

$$\underset{\sim}{\mathbf{E}} = [0{,}50; 0{,}75]$$

and the ratios representing each one of the strategies are:

$$\underset{\sim}{\mathbf{A}} = [0{,}60; 0{,}95]$$
$$\underset{\sim}{\mathbf{B}} = [0{,}40; 0{,}70]$$

In this way we will arrive at the relative distances:

$$\delta(A; E) = \frac{|0{,}60 - 0{,}50| + |0{,}95 - 0{,}75|}{2} = \frac{0{,}10 + 0{,}20}{2} = 0{,}150$$

$$\delta(B; E) = \frac{|0{,}40 - 0{,}50| + |0{,}70 - 0{,}75|}{2} = \frac{0{,}10 + 0{,}05}{2} = 0{,}075$$

It is obvious then in this example, that the strategy represented by ratio B is the preferred one in the event of having to select between the two strategies represented above.

Now, what would happen in the event that the distances of the two strategies was to be the same? Could it be stated, in this event, that the selection of one of the two strategies was immaterial? From a perspective such as we have followed up to this point, the answer, in any event, would be positive, that is to say, that the selection would be immaterial between one of the two

[9] It is also possible to use other notions of distance, such as the Euclidean or Minkowski distances which generalise them.

strategies when the relative distances were to coincide. Nevertheless, a consideration accompanied by an example will allow us to adopt a complementary criterion.

Let us return to the previously described example and take a new ideal ratio $\underset{\sim}{\mathbf{E}} = [0{,}50; 0{,}75]$, and let us assume two new ratios for effects of comparison:

$$\underset{\sim}{\mathbf{C}} = [0{,}45; 0{,}85]$$

$$\underset{\sim}{\mathbf{D}} = [0{,}55; 0{,}65]$$

In this way we arrive at:

$$\delta(\underset{\sim}{\mathbf{C}}, \underset{\sim}{\mathbf{E}}) = 0{,}075$$

$$\delta(\underset{\sim}{\mathbf{D}}, \underset{\sim}{\mathbf{E}}) = 0{,}075$$

If in the first place the uncertainty of $\underset{\sim}{\mathbf{C}}$ and $\underset{\sim}{\mathbf{D}}$ is reduced and consider the average point as a better representation of the interval, we arrive at:

$$\bar{\underset{\sim}{\mathbf{C}}} = \frac{0{,}45 + 0{,}85}{2} = 0{,}65$$

$$\bar{\underset{\sim}{\mathbf{D}}} = \frac{0{,}55 + 0{,}65}{2} = 0{,}60$$

On doing the same operation with our $\underset{\sim}{\mathbf{E}}$ ratio, we find:

$$\bar{\underset{\sim}{\mathbf{E}}} = \frac{0{,}50 + 0{,}75}{2} = 0{,}625$$

It will be seen that the distance to the ideal ratio both for one and the other ratios continues to be the same, yet, nevertheless there is something that differentiates ratio $\underset{\sim}{\mathbf{C}}$ from ratio $\underset{\sim}{\mathbf{D}}$. We are inclined to think that the basic difference can be found in the degree of uncertainty, that is, in the extension found in the confidence interval. For this we propose establishing a complementary criterion according to which when an equality of distances occurs in relation to the ideal ratio, the ratio which is the least uncertain will be selected, that is to say, the difference of which between the upper extreme and the lower extreme is the least. In this case ratio $\underset{\sim}{\mathbf{D}}$ would be chosen in place of ratio $\underset{\sim}{\mathbf{C}}$, since:

$$0{,}85 - 0{,}45 = 0{,}40 \quad \text{for ratio } \underset{\sim}{\mathbf{C}}$$

$$0{,}65 - 0{,}55 = 0{,}10 \quad \text{for ratio } \underset{\sim}{\mathbf{D}}$$

7.4 Use of Intervals in the Event of Several Strategies

Now let us consider a case in which a business has the possibility of carrying out three alternative strategies estimating the dimensions (in confidence intervals) that it is felt that the different patrimonial masses will take in each

Table 7.4.

Expert	Alternative I	Alternative II	Alternative III
Financial assets	[200; 300]	[150; 250]	[250; 400]
Stocks	[300; 400]	[400; 600]	[250; 300]
Net fixed assets	[500; 600]	[1.000; 1.300]	[600; 700]
Short-term liabilities	[200; 300]	[300; 450]	[300; 450]
Long-term debts	[200; 300]	[600; 800]	[200; 300]
Equity	[600; 700]	[650; 900]	[600; 650]
Sales	[775; 850]	[2.500; 3.000]	[1.900; 2.000]
Sales expenses	[575; 700]	[1.400; 1.800]	[1100; 1.200]
Increase in stocks	[50; 150]	[150; 350]	[0; 50]
Depreciation	[0; 0]	[200; 300]	[150; 200]

one of these alternative strategies (see Table 7.4). The first strategy we have made coincides with the example taken in the previous subject.

The foreseen balance sheet and provisional operating statement for each strategy are those shown in Table 7.5.

Now let us consider, for each one of the alternatives, some of the ratios that show the capacity of the business to cover its long-term commitments. Among others, the following may be considered:

$$(1) = \frac{\text{Total debts}}{\text{Equity}}$$

$$(2) = \frac{\text{Long-term debts}}{\text{Equity}}$$

$$(3) = \frac{\text{Permanent capitals}}{\text{Equity}}$$

With these it is possible to estimate the best alternative relative to long-term solvency. On applying the ratios we have described to the data of the three strategies we are going to consider we will arrive at the following results:

$$(1) = \frac{[400; 600]}{[600; 700]} = \left[\frac{4}{7}; 1\right] = [0{,}75; 1]$$

$$(2) = \frac{[200; 300]}{[600; 700]} = \left[\frac{2}{7}; \frac{30}{60}\right] = [0{,}28; 0{,}50]$$

$$(3) = \frac{[800; 1.000]}{[600; 700]} = \left[\frac{8}{6}; \frac{100}{50}\right] = [1{,}33; 2]$$

As has been pointed out before, in this case there is no reason to arrive at the upper limits, which would be, respectively:

$$(1) = [1; 1{,}92]$$

$$(2) = [0{,}66; 1{,}23]$$

$$(3) = [1{,}33; 2]$$

Table 7.5.

Estimated balance sheet (Strategy I)			
Assets		Liabilities	
Financial assets	[200, 300]	Short-term liabilities	[200, 350]
Stocks	[300, 400]	Long-term liabilities	[200, 300]
Net fixed assets	[500, 600]	Equity	[600, 700]
Total	[1.000, 1.350]	Total	[1.000, 1.350]
Provisional operating statement (Strategy I)			
Sales		[775, 850]	
Cost of sales*		[575, 700]	
	Gross profit		[75,275]
Depreciation			[0, 0]
	Net profit		[75,275]
	Taxes*		[0, 0]
Profit distribution:			
	Dividends	[75, 175]	
	Reserves	[0, 100]	

* For purposes of identification with the previous example it is assumed in this case that no taxes are paid. On the other hand, tax payment will be included in Strategies II and III.

Estimated balance sheet (Strategy II)			
Assets		Liabilities	
Financial assets	[150; 250]	Short-term liabilities	[300, 450]
Stocks	[400; 600]	Long-term liabilities	[600; 800]
Net fixed assets	[1.000; 3.000]	Equity	[650; 900]
Total	[1.550, 2.150]	Total	[1.550, 2.150]
Provisional operating statement (Strategy I)			
Sales		[2.500; 3.000]	
Cost of sales*		[1.400; 1.800]	
	Gross profit		[700; 1.600]
Depreciation			[200; 300]
	Net profit		[400; 1.400]
	Taxes*		[140; 490]
Profit distribution:			
	Dividends	[210; 610]	
	Reserves	[50; 300]	

Estimated balance sheet (Strategy III)			
Assets		Liabilities	
Financial assets	[250; 400]	Short-term liabilities	[300; 450]
Stocks	[250; 300]	Long-term liabilities	[200, 300]
Net fixed assets	[600; 700]	Equity	[600, 650]
Total	[1.100; 1.400]	Total	[1.100; 1.400]

Provisional operating statement (Strategy III)			
Sales		[1.900; 2.000]	
Cost of sales*		[1.100; 1.200]	
	Gross profit		[700; 900]
Depreciation			[150; 200]
	Net profit		[500; 750]
	Taxes*		[175; 260]
Profit distribution:			
	Dividends	[325; 1.440]	
	Reserves	[0; 50]	

since the comparison would provide little information, given that the problem does not consist of arriving at the smallest possible ratio (in the event that this were to be an ordering from lowest to highest) but what it is, is an attempt to find the ratio closest to the one we consider as the ideal ratio.

It is quite obvious that the lower the value of the ratio, the higher solvency of the business will be over the long-term, but it is also true to say that when we are faced with a business with profitability higher than the market interest rate, an excess of solvency would mean the infra-use of available financial means. In this way, it is convenient to establish for each of the indicators, an interval that brings to light the lower limits and the upper limits between which the solvency of a business is considered good.

With regard to the other ratio (1), a measurement that is acceptable is the one the value of which does not exceed 1,2, without solvency being affected in any way. On the other hand, a figure less than 0,8 is not acceptable, in this case, as it is an excessively conservative position. Therefore the ideal ratio would be established as $E_1 = [0{,}8; 1{,}2]$.

It should be taken into account that ratio (2) should not, in principle, be any higher than 1, although the upper limit is acceptable. If 0,65 is accepted as the lower limit we would have as the ideal ratio $E_2 = [0{,}65; 1]$.

Finally, with regard to ratio (3), that it be higher than 1 is considered adequate in order for a positive "working capital" to exist, although it should not be very much higher than 1. We accept as the ideal ratio $E_3 = [1{,}2; 1{,}4]$.

Now we must find the Hamming distance for each of the ideal ratios and for those arrived at by the three strategies:

Ratio (1):

$$\delta(E_1;\mathrm{I}) = \frac{|0{,}57 - 0{,}80| + |1 - 1{,}2|}{2} = \frac{0{,}43}{2} = 0{,}215$$
$$\delta(E_1;\mathrm{II}) = \frac{|1 - 0{,}80| + |1{,}92 - 1{,}2|}{2} = \frac{0{,}92}{2} = 0{,}460$$
$$\delta(E_1;\mathrm{III}) = \frac{|0{,}76 - 0{,}80| + |1{,}25 - 1{,}2|}{2} = \frac{0{,}09}{2} = 0{,}045$$

For this ratio the order of preference would be: III, I and II.
Ratio (2):

$$\delta(E_2;\mathrm{I}) = \frac{|0{,}28 - 0{,}65| + |0{,}50 - 1|}{2} = \frac{0{,}87}{2} = 0{,}435$$
$$\delta(E_2;\mathrm{II}) = \frac{|0{,}66 - 0{,}65| + |1{,}23 - 1|}{2} = \frac{0{,}24}{2} = 0{,}120$$
$$\delta(E_2;\mathrm{III}) = \frac{|0{,}30 - 0{,}65| + |0{,}50 - 1|}{2} = \frac{0{,}85}{2} = 0{,}425$$

For this ratio the order of preference would be: II, III and I.
Ratio (3):

$$\delta(E_2;\mathrm{I}) = \frac{|0{,}28 - 0{,}65| + |0{,}50 - 1|}{2} = \frac{0{,}87}{2} = 0{,}435$$
$$\delta(E_2;\mathrm{II}) = \frac{|0{,}66 - 0{,}65| + |1{,}23 - 1|}{2} = \frac{0{,}24}{2} = 0{,}120$$
$$\delta(E_2;\mathrm{III}) = \frac{|0{,}30 - 0{,}65| + |0{,}50 - 1|}{2} = \frac{0{,}85}{2} = 0{,}425$$

For this ratio the order of preference would be III, II, and I.

As occurs on many occasions, the order of preference of each of the strategies taken into consideration is not the same for the three ratios, which, on the other hand show differentiated aspects of the solvency of a business over the long-term. When this happens it is necessary to take a decision to decide which of them is the most representative of the solvency, and consider the others as complementary. If we consider ratio (3) as the main one, the conclusion will immediately be reached that the best strategy to carry out is III, which on the other hand is well placed in the remaining ratios.

To summarise, we could conclude by saying that strategy III allows us to foresee better long-term solvency.

Another of the objectives to be considered could be that of obtaining the best possible economic result, which can be expressed, among others, by the following ratios:

$$\text{(1) Net margin} = \frac{\text{Net profit (before taxes)}}{\text{Sales}}$$

as also:

$$\text{(2) Financial profitability} = \frac{\text{Net profit (before taxes)}}{\text{Equity}}$$

In this case the ratios acquire, for each of the strategies, the following numbers:

Strategy I

$$(1) = \frac{[75; 275]}{[775; 850]} = \left[\frac{3}{34}; \frac{11}{31}\right] = [0{,}088; 0{,}354]$$

$$(2) = \frac{[75; 275]}{[600; 700]} = \left[\frac{3}{28}; \frac{11}{24}\right] = [0{,}107; 0{,}458]$$

Strategy II:

$$(1) = \frac{[400; 1.400]}{[2.500; 3.000]} = \left[\frac{2}{15}; \frac{14}{25}\right] = [0{,}133; 0{,}560]$$

$$(2) = \frac{[400; 1.400]}{[650; 900]} = \left[\frac{4}{9}; \frac{28}{13}\right] = [0{,}444; 2{,}153]$$

Strategy III:

$$(1) = \frac{[500; 750]}{[1.900; 2.000]} = \left[\frac{1}{4}; \frac{15}{38}\right] = [0{,}250; 0{,}394]$$

$$(2) = \frac{[500; 750]}{[600; 650]} = \left[\frac{10}{13}; \frac{5}{4}\right] = [0{,}769; 1{,}250]$$

In this case ordering from greatest to lowest is of great interest. For this a confidence interval can be arrived at that constitutes the upper limit for (1) and (2), which will be:

$$L_{(1)} = [0{,}250; 0{,}560]$$
$$L_{(2)} = [0{,}769; 2{,}153]$$

We immediately proceed to look for the corresponding distances. In this case, consideration will be given to the absolute distances, with the object of covering all the possibilities offered by this technique. We then arrive at:

$$d(\mathrm{I}; L_1) = |0{,}088 - 0{,}250| + |0{,}354 - 0{,}560| = 0{,}368$$
$$d(\mathrm{II}; L_1) = |0{,}133 - 0{,}250| + |0{,}560 - 0{,}560| = 0{,}117$$
$$d(\mathrm{III}; L_1) = |0{,}250 - 0{,}250| + |0{,}394 - 0{,}560| = 0{,}166$$

$$d(\mathrm{I}; L_2) = |0{,}107 - 0{,}769| + |0{,}458 - 2{,}153| = 2{,}357$$
$$d(\mathrm{II}; L_2) = |0{,}444 - 0{,}769| + |2{,}153 - 2{,}153| = 0{,}325$$
$$d(\mathrm{III}; L_2) = |0{,}769 - 0{,}769| + |1{,}250 - 2{,}153| = 0{,}903$$

Obviously if we are looking for the maximisation of profits as our objective, either one of the ratios indicate the order of preference, which will be given by the least possible distances with the upper limit, that is, with the best result possible considering the highest of the lower limits and also that of the upper limits, whichever the interval one or the other may be a member of.

Thus the order of preference will be:

$$\text{II} \succ \text{III} \succ \text{I}$$

The analysis could obviously continue with the use of other ratios representing other aspects of the economic-financial situation of the business.

7.5 The Du Pont Scheme in Uncertainty

Lastly let us take a look at what aspect is acquired, in uncertainty, by the Du Pont scheme. For this we will use as our didactic example the data relative to strategy I. But before applying the confidence intervals to the ratio cascade it will be necessary to make certain considerations, which will be important taking into account that it is not possible to make a direct and immediate use of the scheme that we used in the sphere of certainty.

The Du Pont scheme is based on the interposing of determined arithmetical operations (products, divisions) between certain elemental data and a final ratio (profitability of equity) in order to find other ratios and concepts. This way of acting, which is valid for common arithmetic, that is to say, when we find ourselves within the sphere of certainty, is not exactly adequate for the arithmetic of confidence intervals, that is, when we are working in the sphere of uncertainty. The essence itself of uncertainty is the fact that as we increase the number of operations, uncertainty also increases, that is to say, the confidence intervals become wider.

In fact, let us place ourselves within $R+$ (field of the positive rationales) and start out from the estimates of determined values by means of intervals: Therefore if:

$$\begin{aligned} \text{Profits} &= [b_1; b_2] \\ \text{Permanent capitals} &= [c_1; c_2] \\ \text{Sales} &= [v_1; v_2] \end{aligned}$$

Economic profitability would be:

$$R = \frac{\text{Profits}}{\text{Permanent capitals}} = [b_1; b_2](:)[c_1; c_2] = \left[\frac{b_1}{c_2}; \frac{b_2}{c_1}\right]$$

In the "field of certainty" the following would be complied with:

$$R = \frac{\text{Profits}}{\text{Sales}} \cdot \frac{\text{Sales}}{\text{Permanent capitals}} = \frac{\text{Profits}}{\text{Permanent capitals}}$$

But on the other hand, if we move into the sphere of uncertainty by means of confidence intervals, we will arrive at:

$$[b_1; b_2](:)[v_1; v_2] = \left[\frac{b_1}{v_2}; \frac{b_2}{v_1}\right]$$

and also:

$$[v_1; v_2](:)[c_1; c_2] = \left[\frac{v_1}{c_2}; \frac{v_2}{c_1}\right]$$

Therefore:

$$\left[\frac{b_1}{v_2}; \frac{b_2}{v_1}\right](\cdot)\left[\frac{v_1}{c_2}; \frac{v_2}{c_1}\right] = \left[\frac{b_1}{v_2} \cdot \frac{v_1}{c_2}; \frac{b_2}{v_1} \cdot \frac{v_2}{c_1}\right] = \left[\frac{b_1}{c_2} \cdot \frac{v_1}{v_2}; \frac{b_2}{c_1} \cdot \frac{v_2}{v_1}\right]$$

Now then, $v_1 < v_2$ is always complied with except when the confidence interval is reduced to a common number, in which case $v_1 = v_2$ will be complied with. Therefore:

$$v_1 < v_2 \Rightarrow \left(\frac{v_1}{v_2}\left\langle 1; \frac{v_2}{v_1}\right\rangle 1\right)$$

$$(v_1 = v_2(\text{exact number})) \Rightarrow \left(\frac{v_1}{v_2} = \frac{v_2}{v_1} = 1\right)$$

from there the final result is:

$$\frac{b_1}{c_2} \cdot \frac{v_1}{v_2} < \frac{b_1}{c_2}$$

and also:

$$\frac{b_2}{c_1} \cdot \frac{v_2}{v_1} > \frac{b_2}{c_1}$$

As the lower extreme of the interval, arrived at with more operations, is less than the corresponding interval of the direct ratio and also the upper extreme of the interval with more operations than the direct ratio is also greater, the former will be wider than the latter and therefore will represent greater uncertainty.

For both results to coincide it will be necessary to affect each of the extremes with a quotient:

For the lower extreme $k_1 = \frac{v_2}{v_1}$
For the upper extreme $\frac{1}{k_1} = \frac{v_1}{v_2}$

That is, the quotient or inverse of the quotient between the lower extreme and the upper extreme of the value has acted as an intermediary (in this case sales). In this way we will arrive at:

$$\left[\frac{b_1}{c_2}; \frac{v_1}{v_2}\right] k = \frac{b_1}{c_2} \cdot \frac{v_1}{v_2} \cdot \frac{v_2}{v_1} = \frac{b_1}{c_2}$$

$$\left[\frac{b_2}{c_1}; \frac{v_2}{v_1}\right] \frac{1}{k} = \frac{b_2}{c_1} \cdot \frac{v_2}{v_1} \cdot \frac{v_1}{v_2} = \frac{b_2}{c_1}$$

Table 7.6. Du Pont scheme in uncertainty

Data *Ratios*

Cost of sales [575, 700]
(+)
Depreciation [0, 0]
(+)
Taxes [0, 0]
} Costs [575, 700]

Sales [775, 850]
(−)
Costs [575, 700]
} Profits (net of taxes) [075, 275]

Profits (net of taxes) [075, 275]
(:)
Sales [775, 850]
} Net margin [0,088, 0,354]

Financial assets [200, 300]
(+)
Stocks [300, 400]
(+)
Anticipated Expenses [0, 0]
($\overline{m}$)
Short-term liabilities [200, 300]
} Working capital* [300, 400]

Working capital* [300, 400]
(+)
Net Fixed assets [500, 600]
} Permanent Capitals [800, 1.000]

Sales [775, 950]
(:)
Permanent Capitals [800, 1.000]
} Capital rotation [0,775, 1,062]

Net margin [0,088, 0,354]
Capital rotation [0,775, 1,062]
} Pseudo-profitability on investments [0,068, 0,375]

↓

Adjustment:
$\left[0{,}068\ (\cdot)\ \frac{850}{775},\ 0{,}375\ (\cdot)\ \frac{775}{850}\right]$

↓

Economic profitability [0,075, 0,343]

Equity [600, 700]
(+)
Long-term liabilities [200, 300]
} Permanent Capitals [800, 1.000]

Equity [600, 700]
(:)
Permanent Capitals [800, 1.000]
} Proportion equity / permanent capitals [0,600, 0,875]

Economic profitability [0,075, 0,343]
(:)
Proportion equity / permanent capitals [0,600, 0,875]
} Pseudo-profitability on equity [0,085, 0,571]

↓

Adjustment:
$\left[0{,}085\ (\cdot)\ \frac{1.000}{800},\ 0{,}571\ (\cdot)\ \frac{800}{850}\right]$

↓

Profitability on equity [0,107, 0,458]

*It should be recalled that this is the working capital, which we have denominated F_mM, which does not coincide with FM_1, nor withl FM_2.
In our opinion it would be more correct, in uncertainty, to use the following calculation:

Long-term liabilities [200, 300]
(+)
Equity [600, 700]
(−)
Net fixed assets [500, 600]
} Working capital [200, 500]

See in this respect the chapter on working capital

which are nothing further than the lower and upper extremes arrived at by direct quotient.

It was necessary to make this adjustment both for arriving at the "economic profitability" and for "profitability on equity". With this the Du Pont scheme for strategy 1 would be as shown in Table 7.6.

This application brings to light the fact that it is impossible to use the Du Pont scheme directly in uncertainty, unless an increase in the uncertainty is accepted, which is reflected in the increase of the difference separating the upper and lower extremes of the interval in question. The adjustment made, on the one hand, and the need to accept an inadequate working capital, on the other, makes it necessary to revise this mechanism if what is wanted is that it should also be useful for an overall view of future estimates.

Let us leave this subject at this point, which as has been seen has not been fully covered. Further studies may well shed more light on the subject.

8 Risk Analysis

8.1 Risk in Management Studies

The notion of risk is subject to several meanings all referring to different aspects. On the one hand, we talk about studies in the sphere of risk when basing ourselves on hypotheses relative to the nature of probability and, on the other it is said that risk exists when any danger appears due to the instability of certain economic variables. In this second sense, it is said that the economic risk is that which arises from the instability of net profit. Gross profit (or operating profit) represents the excess generated by the assets of the business after having made the corresponding contributions to depreciation, estimates and provisions, but before deducting interest for debts or financial expenses. Net profit is obtained by deducting from gross profit interest for debts and, therefore, not only do asset elements take part but also elements from liabilities.

In economic risk all those eventualities that may affect gross profit take part, among which can be mentioned market fluctuations, labour incidents, strikes, etc., that may modify the amount of the same, and therefore, in a certain manner may be considered overall as constituting an internal risk for the business.

In financial risk debts are made to intervene, above all those relative to the medium and long-term, which constitute the basis of the true financial risk. Nevertheless, there are authors who also include short-term debts because they consider that within the risk it should be taken into account that there is a possibility of difficulties arising both in the short and long-term.

The existence of an economic risk and a financial risk make the use of determined ratios, such as economic profitability and the index of financial profitability, among others, convenient, and these can be expressed as follows:

$$\text{Economic profitability index} = \frac{\text{Gross profit}}{\text{Total assets}}$$

$$\text{Financial profitability index} = \frac{\text{Net profit}}{\text{Equity}}$$

These ratios allow for the taking of decisions on the convenience of resorting in the future to external financing. In fact, if the economic profitability is

Anna Maria Gil-Lafuente: *Fuzzy Logic in Financial Analysis*, StudFuzz **175**, 117–133 (2005)
www.springerlink.com

greater than the cost of the debts, on increasing indebtedness financial profitability is increased (shareholders profitability), but also the financial risk is greater as a consequence of increase in the possibilities of insolvency.

We can specify this relation in a schematic form as follows:

When economic profitability is > the cost of the debts:

1. The greater the indebtedness, the greater the financial profitability.
2. The greater the indebtedness, the higher the risk.

8.2 The Relationship Between Indebtedness and Financial Profitability in Uncertainty

Let us now move on to a practical case that will allow us to compare, in uncertainty, the relation existing between the variation of indebtedness and financial profitability. For this we are going to place a business in a context of uncertainty, so that some of its variables estimated for the future will be expressed by fuzzy numbers that, initially and for greater simplification we will consider as triangular.

We commence with an estimated balance sheet as follows:

Assets		Liabilities	
Financial assets	(20; 22; 30)		
Inventories	(30; 36; 40)		
Net fixed assets	(50; 58; 60)	Equity	(100; 116; 130)
Total	(100; 116; 130)	Total	(100; 116; 130)

It can be seen then that the business has no debts, an eminently exceptional case. If the estimated profit is:

$$\underset{\sim}{\mathbf{B}} = (14; 16; 20)$$

we arrive at:

$$\text{Economic profitability} = \text{Financial profitability} = \frac{\underset{\sim}{\mathbf{B}}}{\underset{\sim}{\mathbf{A}}} = \frac{(14; 16; 20)}{(100; 116; 130)}$$

In order to arrive at this quotient[1] we resort to expressing these T.F.N. in the form of α-cuts:

$$B_\alpha = [14 + 2\alpha; 20 - 4\alpha]$$
$$A_\alpha = [100 + 16\alpha; 130 - 14\alpha]$$

[1] It is sufficiently well known that the quotient of triangular fuzzy numbers (likewise the product) is not normally a triangular fuzzy number.

Table 8.1.

α	$14+2\alpha$	$20-4\alpha$	$100+16\alpha$	$130-14\alpha$	$\frac{14+2\alpha}{130-14\alpha}$	$\frac{20-4\alpha}{100+16\alpha}$
0	14,0	20,0	100,0	130,0	0,1077	0,2000
0,1	14,2	19,6	101,6	128,6	0,1104	0,1992
0,2	14,4	19,2	103,2	127,2	0,1132	0,1983
0,3	14,6	18,8	104,8	125,8	0,1161	0,1975
0,4	14,8	18,4	106,4	124,4	0,1190	0,1966
0,5	15,0	18,0	108,0	123,0	0,1220	0,1957
0,6	15,2	17,6	109,6	121,6	0,1250	0,1947
0,7	15,4	17,2	111,2	120,2	0,1281	0,1937
0,8	15,6	16,8	112,8	118,8	0,1313	0,1927
0,9	15,8	16,4	114,4	117,4	0,1346	0,1916
1	16,0	16,0	116,0	116,0	0,1379	0,1905

The economic and financial profitability will have the following form:

$$\underset{\sim}{\mathbf{R}}_E = \underset{\sim}{\mathbf{R}}_F = \frac{[14+2\alpha; 20-4\alpha]}{[100+16\alpha; 130-14\alpha]} = \left[\frac{14+2\alpha}{130-14\alpha}; \frac{20-4\alpha}{100+16\alpha}\right]$$

In order to obtain the resulting fuzzy number we will use the hendecagonal system by taking $\alpha \in 0; 0{,}1; 0{,}2; \ldots; 0{,}9; 1$ (see Table 8.1):

It can be seen that the profitability will be no less than 10,76% nor higher than 20%. The greatest probability in this case coincides with a profitability of 13,79%, that is to say that this latter figure will be the value with the greatest possibility of profitability.

If for example, a level of presumption were to be accepted of $\alpha = 0{,}6$, we would arrive at the fact that profitability would be no less than 12,50% nor higher than 16,05%.

Again if instead of taking the resulting fuzzy number, its triangular approximation were to be considered sufficient, we would have:

$$\underset{\sim}{\mathbf{R}}_E = \underset{\sim}{\mathbf{R}}_F = (0{,}1076; 0{,}1379; 0{,}2000)$$

If presented in the form of confidence intervals we will have:

$$[0{,}1076 + 0{,}0303\alpha; 0{,}2000 - 0{,}0621\alpha)$$

and then making the entropy fall:

$$\overline{\underset{\sim}{\mathbf{R}}_E} = \overline{\underset{\sim}{\mathbf{R}}_F} = \frac{0{,}1076 + 2 \times 0{,}1379 + 0{,}2000}{4} = 0{,}1458$$

It should be noted that if we consider the triangular approximation instead of the fuzzy number resulting from the application of confidence intervals, this implies no difference either in the extremes or in the point of maximum presumption, but does so in the rest of the intermediary points.

In fact, for a determined intermediary level, for example $\alpha = 0{,}6$, the fuzzy number is the interval $[0{,}1250; 0{,}1605]$, while for its triangular approximation we have $[0{,}1257; 0{,}1627]$. In this case as in many others, the differences are not significant, above all if taking into account that we are faced with uncertainty and that the estimates are charged with a very high degree of subjectivity.

8.3 The Approach with Confidence Triplets as a Triangular Approximation

If the triangular approximation is accepted for simplicity reasons, it would be sufficient to operate with the so-called confidence triplets.

As is known, a confidence triplet $(a_1; a_2; a_3)$ is formed by three values, of which a_1 represents the lower extreme, a_3 the upper extreme, and a_2 the value of maximum probability and is located between a_1 and a_3 in such a way that the following should be complied with:

$$a_1 \leq a_2 \leq a_3$$

The difference existing between a confidence triplet and a triangular fuzzy number is that, as for the triplet only values a_i are considered providing that $i = 1; 2; 3$ and giving to a_1 and a_3 a level of presumption zero and to a_2 a maximum presumption, the T.F.N. will be assigned values corresponding to the presumption relative to all the dominion of the rationales. That is to say, values are given to the function characteristic of membership.

The fact of having the possibility of developing the calculations from confidence triplets instead of doing so with T.F.N. will allow us greater flexibility from an operative point of view.

Given the fact that the technique that should be used in the event of working with T.F.N. has already been mentioned, we will now continue the process in the case of using confidence triplets.

Let us now assume that the business has several alternatives.

8.3.1 First Alternative

This consists of a decrease in capital in the future of 15 monetary units by acquiring a long-term external financial contribution for the same amount with a financial cost expressed by the following confidence triplet $I = (0{,}10; 0{,}12; 0{,}15)$. If we consider the previous balance sheet and we assume it is expressed in confidence triplets, this operation will give rise to the following variables in the financial structure:

$$\begin{aligned} \text{Equity} &= C_1 = (100; 116; 130)(-)15 = (85; 101; 115) \\ \text{Liabilities} &= E_1 = 15 \\ \text{Net profit} &= B_1 = (14; 16; 20)(-)15(\cdot)(0{,}10; 0{,}12; 0{,}15) = (11{,}75; 14{,}2; 18{,}5) \end{aligned}$$

The economic profitability obviously remains unvaried. But the financial profitability is modified in the numerator as a consequence of the payment of interest for the credit obtained. The denominator, on the other hand, also varies in this case due to the decrease in capital, also affecting, on the other hand, the financial profitability. In this way we arrive at:

$$\begin{aligned}\text{Financial profitability} = R_F^{(1)} &= \frac{(11{,}75; 14{,}2; 18{,}5)}{(85; 101; 115)} \\ &= (0{,}1021; 0{,}1405; 0{,}2176)\end{aligned}$$

It is not difficult to imagine a procedure that allows for bringing to light the fact that, in this case, financial profitability $R_F^{(1)}$ is higher than R_F that was obtained previously. But for this it would be sufficient to make the entropy fall, or to carry out an ordering such as has been done in previous chapters.

What has been brought to light then is the fact that when a business without debts commences a process of indebtedness it obtains an increase in financial profitability, as a consequence of the fact that the economic profitability is greater than the cost of the debts.

8.3.2 Second Alternative

A possible future decrease in capital is considered in an amount of 30 monetary units with an increase in long-term credits for the same amount in order to obtain the same level of financing.

We now arrive at:

$$\begin{aligned}C_2 &= (100; 116; 130)(-)30 = (70; 86; 100) \\ E_2 &= 30 \\ B_2 &= (14; 16; 20)(-)(3; 3{,}6; 4{,}5) = (9{,}5; 12{,}4; 17)\end{aligned}$$

And, therefore, the new financial profitability will be:

$$\mathbf{R}_F^{(2)} = \frac{(9{,}5; 12{,}4; 17)}{(70; 86; 100)} = (0{,}0950; 0{,}1441; 0{,}2428)$$

It will be seen that, given that the economic profitability remains constant the higher indebtedness produces an increase in the financial profitability as a consequence of the fact that the economic profitability is higher than the cost of the debts.

Now we are going to see what happens when the indebtedness is increased even further.

8.3.3 Third Alternative

We now take into consideration a reduction in a greater amount of the capital, for example 60 monetary units, taking on long-term debts for a similar amount.

The variables will be as follows:

$$C_3 = (100; 116; 130)(-)60 = (40; 56; 70)$$
$$E_3 = 60$$
$$B_2 = (14; 16; 20)(-)(6; 7{,}2; 9) = (5; 8{,}8; 14)$$

In this case, we can see that the financial profitability becomes:

$$\mathbf{R}_F^{(3)} = \frac{(5; 8{,}8; 14)}{(40; 56; 70)} = (0{,}0714; 0{,}1571; 0{,}35)$$

where it can be seen that the financial profitability continues to increase.

We now move on to take the confidence triplets as triangular approximations, converting them into triangular fuzzy numbers. In this way we arrive at the following T.F.N. expressed as percentages.

$$\underset{\sim}{\mathbf{i}} = (10{,}00; 12{,}00; 15{,}00)$$
$$\underset{\sim}{\mathbf{R}}_F = (10{,}76; 13{,}79; 20{,}00)$$
$$\underset{\sim}{\mathbf{R}}_F^{(1)} = (10{,}21; 14{,}05; 21{,}76)$$
$$\underset{\sim}{\mathbf{R}}_F^{(2)} = (9{,}50; 14{,}41; 24{,}28)$$
$$\underset{\sim}{\mathbf{R}}_F^{(3)} = (7{,}14; 15{,}71; 35{,}00)$$

Now we continue by drawing the graphical representation shown in the Fig. 8.1:

By looking at the T.F.N. representing the cost of the debts and of the successive financial profitability, as well as of their corresponding graphical representation, the following conclusions can be made:

1. As equity is substituted by outside capital, the uncertainty of the financial profitability increases. It can be seen on passing from $\mathbf{R}_F$ to $\mathbf{R}_F^{(1)}$, from $\mathbf{R}_F^{(2)}$ to $\mathbf{R}_F^{(3)}$ and so on successively, that the lower extreme gets gradually smaller and that the upper extreme gets gradually larger, consequently the bases of the successive triangles representing the T.F.N. become gradually larger.

There are several different ways of estimating uncertainty. On this occasion we are going to propose a process that consists in determining for each triangular approximation its greatest representation by means of a crisp

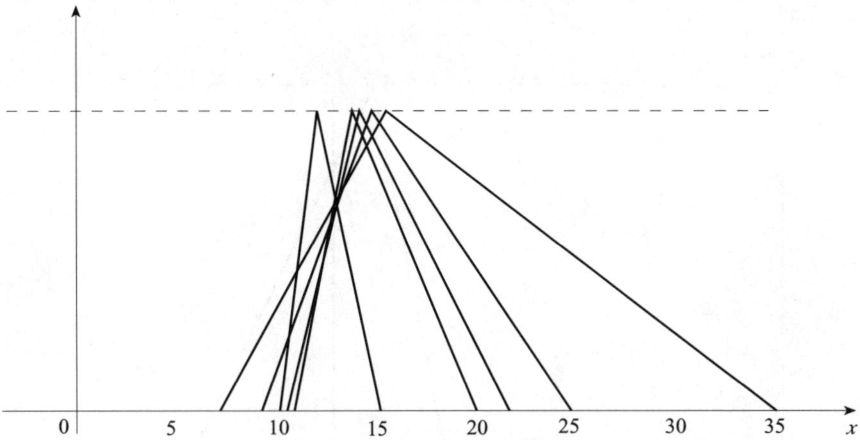

Fig. 8.1.

number, so that later on we can look for the corresponding distance between the crisp number and the respective triangular[2] approximation.

Let us see what happens in the case of our example. The greatest representation for each one of the cases of financial profitability which have been analysed previously, can be as follows:

$$\overline{\underset{\sim}{\mathbf{R}}_F} = \frac{10{,}76 + 2 \times 13{,}79 + 20{,}00}{4} = 14{,}58$$

$$\overline{\underset{\sim}{\mathbf{R}}_F^{(1)}} = \frac{10{,}21 + 2 \times 14{,}05 + 21{,}76}{4} = 15{,}01$$

$$\overline{\underset{\sim}{\mathbf{R}}_F^{(2)}} = \frac{9{,}50 + 2 \times 14{,}41 + 24{,}286}{4} = 15{,}65$$

$$\overline{\underset{\sim}{\mathbf{R}}_F^{(3)}} = \frac{7{,}14 + 2 \times 15{,}71 + 35{,}00}{4} = 18{,}39$$

Now then, a crisp number can be considered as a particular case of a T.F.N. In order to clarify this idea, we could resort to the graphical image according to which a crisp number (a straight line parallel to the axis of and height equal to the unit) is none other than a triangle the sides of which close (that is, the base of which has become reduced) until becoming a single straight line on which the lower extreme, maximum probability and upper extreme were to coincide relative to the respective values of the abscissa (in this case the triangular base would be reduced to a single point).

If we accept this approach, there would be no problem whatsoever for obtaining the distance between a common number (crisp) and an T.F.N., using, for example, the linear distance as the sum of the distance to the left and the distance to the right.

[2] As is particularly well known, uncertainty, this type of disorder, can be valued by means of the distance relative to order.

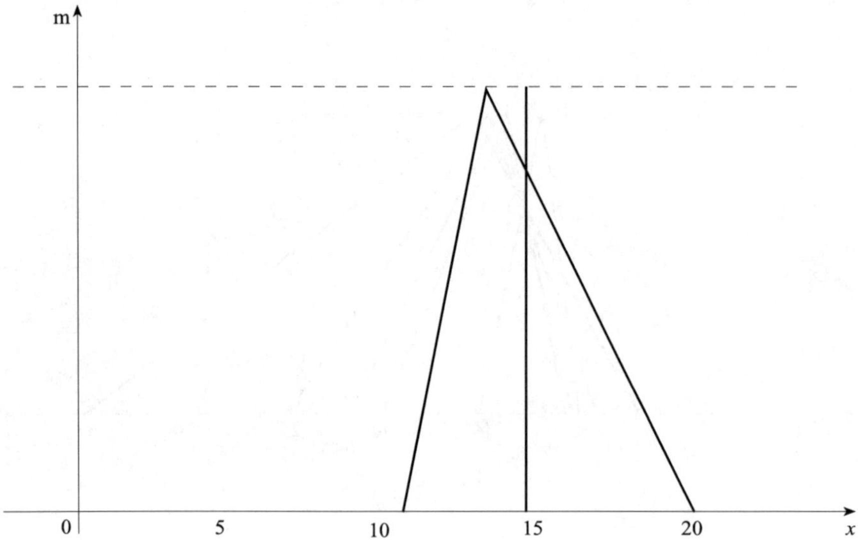

Fig. 8.2.

What we are doing then, is looking for the distance between $\overline{\mathbf{R}_F}$ and $\underset{\sim}{\mathbf{R}}_F$, that is to say, $d(\overline{\mathbf{R}_F}, \underset{\sim}{\mathbf{R}}_F)$, as well as $d(\overline{\mathbf{R}_F^{(1)}}, \underset{\sim}{\mathbf{R}}_F^{(1)})$, $d(\overline{\mathbf{R}_F^{(2)}}, \underset{\sim}{\mathbf{R}}_F^{(2)})$ and also $d(\overline{\mathbf{R}_F^{(3)}}, \underset{\sim}{\mathbf{R}}_F^{(3)})$ as representative of the uncertainty. We commence by arriving at $d(\overline{\mathbf{R}_F}, \underset{\sim}{\mathbf{R}}_F)$, Graphically this can be represented as shown in Fig. 8.2.

The distance to the left would be:

$$d_I(\overline{\mathbf{R}_F}, \underset{\sim}{\mathbf{R}}_F) = \frac{(14{,}58 - 10{,}76)+(14{,}58 - 13{,}79)}{2} = \frac{3{,}82 + 0{,}79}{2} = 2{,}305$$

The distance to the right:

$$d_D(\overline{\mathbf{R}_F}, \underset{\sim}{\mathbf{R}}_F) = \frac{0{,}87 \times (20 - 14{,}58)}{2} + \frac{0{,}13 \times (14{,}58 - 13{,}79)}{2}$$

$$= 2{,}35 + 0{,}05 = 2{,}40$$

And the total distance:

$$d(\overline{\mathbf{R}_F}, \underset{\sim}{\mathbf{R}}_F) = 2{,}305 + 2{,}40 = 4{,}705$$

We now move on to arrive at $d(\overline{R_F^{(1)}, \underset{\sim}{\mathbf{R}}_F^{(1)}})$:

$$d_I\left(\overline{\mathbf{R}_F^{(1)}}, \underset{\sim}{\mathbf{R}}_F^{(1)}\right) = \frac{(15{,}01 - 10{,}21) + (15{,}01 - 14{,}05)}{2} = \frac{4{,}8 + 0{,}96}{2} = 2{,}88$$

$$d_R\left(\overline{\mathbf{R}_F^{(1)}}, \underset{\sim}{\mathbf{R}}_F^{(1)}\right) = \frac{0{,}87 \times (21{,}76 - 15{,}01)}{2} + \frac{0{,}13 \times (15{,}01 - 14{,}05)}{2}$$

$$= 2{,}93 + 0{,}06 = 2{,}99$$

$$d\left(\overline{\mathbf{R}_F^{(1)}}, \underset{\sim}{\mathbf{R}}_F^{(1)}\right) = 2{,}88 + 2{,}99 = 5{,}87$$

And so on successively it will be seen that:

$$d\left(\overline{\mathbf{R}_F}, \underset{\sim}{\mathbf{R}}_F\right) \prec d\left(\overline{\mathbf{R}_F^{(1)}}, \underset{\sim}{\mathbf{R}}_F^{(1)}\right) \prec d\left(\overline{\mathbf{R}_F^{(2)}}, \underset{\sim}{\mathbf{R}}_F^{(2)}\right) \prec d\left(\overline{\underset{\sim}{\mathbf{R}}_F^{(3)}}, R_F^{(3)}\right)$$

2. When we pay attention to the "maximum presumption", it so happens that as the degree of indebtedness increases so does the financial profitability. In this case we pass from 13.79% to 14.05% and then to 14.41% and finally, to 15.71%.

These results should come as no surprise provided that it is taken into account that the arithmetical operations that affect the maximum presumption are identical to those that would be done if we were using crisp numbers in any of the four basic operations.

8.4 Considerations on Estimating the Uncertainty of Profitability

In the previous section we have brought to light the fact that different ways exist for estimating uncertainty and for finding a solution to the problem that we treated, the concept of distance was used as the basis for calculation on establishing the idea that uncertainty can be considered as "disorder" and, therefore, there will be more uncertainty the greater the distance existing relative to "order".

Now let us reflect on the meaning, characteristics and relations existing between the words uncertainty and disorder. For this, as we usually do, we will resort to the use of numerical examples.

In the simplest of cases when a value is estimated by means of a confidence interval, the uncertainty is limited, that is, it is to be found between the two extremes of the interval but not outside it, where in all surety the estimated value will not be found. The "disorder" acquires the form of a doubt on assigning a value between the extremes of the interval, without there being any exact information on the position it may take. As the extremes of the interval come closer together this doubt gets increasingly less and the disorder in our minds is much more reduced. From here it can be said that the narrower the interval the lower is the uncertainty. If the extremes get so close together that they become confused in one point, uncertainty disappears, and there is no doubt in our mind, that disorder gives way to order and we are faced with a "crisp number".

Let us now move on to a more structured uncertainty and refer to the T.F.N. that are dealt with in this chapter.

In the previous section it was accepted that the uncertainty contained in a T.F.N. could be expressed by the distance between this T.F.N and its best crisp (certain) representation. This conception on the forecasting of uncertainty does not vary then with regard to what has been expressed relative

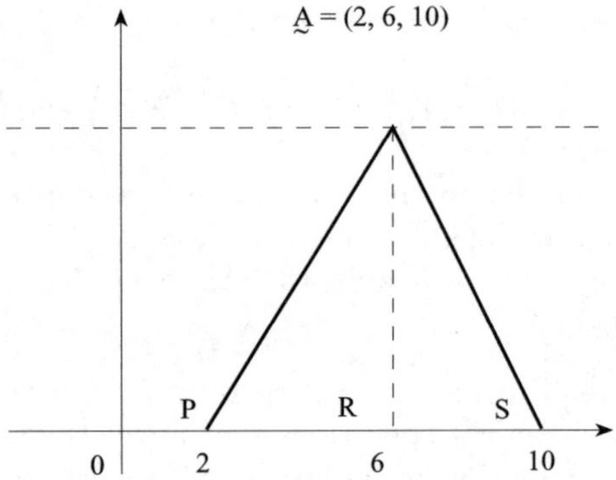

Fig. 8.3.

to confidence intervals. From a geometric point of view it will be seen that the result will "normally" be the sum of the area of a trapezoid and of two triangles. And we say normally because in the event that the graphical representation of a T.F.N. is an isosceles triangle, the distance with its crisp representation will be equal to the sum of the area of the two triangles, which also is equivalent to saying the area of the T.F.N. We will now see this by means of some examples. These are the T.F.N represented in Figs. 8.3 and 8.4.

The best crisp representations will be:

$$\bar{\underset{\sim}{\mathbf{A}}} = \frac{2 + 2 \times 6 + 10}{4} = 6 \quad \bar{\underset{\sim}{\mathbf{B}}} = \frac{2 + 2 \times 4 + 12}{4} = 5{,}5$$

and their respective distances:

$$d\left(\underset{\sim}{A};\bar{\underset{\sim}{A}}\right) = \overset{\triangle}{PQR} + \overset{\triangle}{RSQ} = \overset{\triangle}{PQS} = \frac{4 \times 1}{2} + \frac{4 \times 1}{2} = 4$$

$$\begin{aligned} d\left(\underset{\sim}{B};\bar{\underset{\sim}{B}}\right) &= \overset{\frown}{PQUT} + \left(\overset{\triangle}{TVS} + \overset{\triangle}{QVU}\right) \\ &= \frac{3{,}5 + 1{,}5}{2} + \left(\frac{6{,}5 + 0{,}8125}{2} + \frac{1{,}5 + 0{,}1875}{2}\right) \\ &= 2{,}5 + (2{,}640625 + 0{,}140625) = 5{,}28125 \end{aligned}$$

since the cross over point V is 0,8125.

We now ask ourselves if it is possible to establish a relation between the area of the triangle representing the T.F.N. and the distance of this to its best crisp representation. Let us take a look at the more general case of a scalene triangle.

In the case represented in Fig. 8.4 the distance $d(\underset{\sim}{\mathbf{B}}; \bar{\underset{\sim}{\mathbf{B}}}) = 5{,}28125$ is the result of adding the areas of the trapezoid and the two triangles. It can also

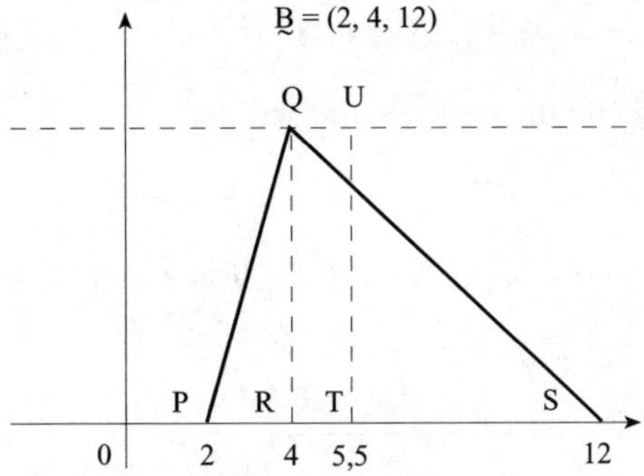

Fig. 8.4.

be arrived at by adding to the area of the triangle representing the T.F.N. (equal to $\widehat{PQS}$) double the area of triangle $\widehat{QVU}$. This is

$$\widehat{PQS} + 2 \cdot \widehat{QVU} = 5 + 2 \times 0{,}140625 = 5{,}28125$$

Generally speaking, it can be seen that the area of the trapezoid is also:

$$\widehat{PQUT} = \widehat{QVU} - \widehat{TVS} + \widehat{QVU}$$

and the distance:

$$d(\underset{\sim}{B},\bar{\underset{\sim}{B}}) = \widehat{PQUT} + \widehat{TVS} + \widehat{QVU} = \widehat{PQS} - \widehat{TVS} + \widehat{QVU} + \widehat{TVS} + \widehat{QVU}$$
$$= \widehat{PQS} + 2 \cdot \widehat{QVU}$$

In this way the following proposal can be stated: "The distance between a T.F.N. and its best crisp representation will always be greater or equal to the area of the triangle that represents it."

$$\boxed{d(\underset{\sim}{B},\bar{\underset{\sim}{B}}) \geq \widehat{PQS}}$$

Equality is only complied with when the maximum presumption is equidistant from the extremes. In other words when the maximum presumption coincides with the best crisps representation of the T.F.N.

The inequality between the distance and the area of the triangle requires, in our opinion, determined comments relative to the estimate of the degree of uncertainty of a T.F.N. For this lest us assume that a determined value has been estimated by means of the following T.F.N.:

$$\underset{\sim}{A} = (4; 6; 18)$$

which, expressed in the form of α-cuts, will be:

$$A_\alpha = [4 + 2\alpha; 18 - 12\alpha]$$

If the hendecagonal system is used Table 8.2 will be arrived at:

Table 8.2.

α	$4 + 2\alpha$	$18 - 12\alpha$	Average Point of the Interval
0	4,0	18,0	11,0
0,1	4,2	16,8	10,5
0,2	4,4	15,6	10,0
0,3	4,6	14,4	9,5
0,4	4,8	13,2	9,0
0,5	5,0	12,0	8,5 $\leftarrow \bar{\underset{\sim}{A}}$
0,6	5,2	10,8	8,0
0,7	5,4	9,6	7,5
0,8	5,6	8,4	7,0
0,9	5,8	7,2	6,5
1	6,0	6,0	6,0

As is all too well known, in order to find a representation of the certainty of the T.F.N. a weighted average is used of the extremes and the maximum presumption. In the greater majority of cases the following is accepted:

$$\bar{\underset{\sim}{A}} = \frac{4 + 2 \times 6 + 18}{4} = 8{,}5$$

The same result can be arrived at by doing:

$$\frac{A_{(\alpha=0)} + A_{(\alpha=1)}}{2} = \frac{11 + 6}{2} = 8{,}5$$

or, what adds up to the same, by taking the median point of the interval at level $\alpha = 0{,}5$. Therefore:

$$A_{(\alpha=0{,}5)} = \frac{5 + 12}{2} = 8{,}5$$

In Table 8.2, the median point has been calculated for each interval corresponding to each of the levels of α, with which the best crisp representation of each one of the levels has been arrived at. Therefore, bearing this in mind, a measurement of the uncertainty at each level could be expressed by the sum of the distances of each extreme to its crisp representation (median point)

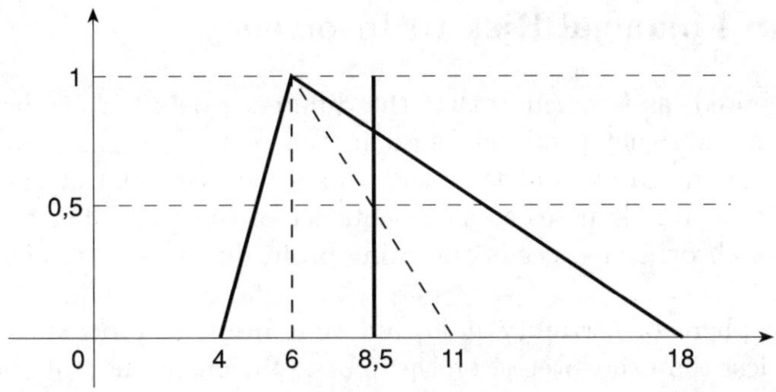

Fig. 8.5.

which, obviously, gives rise to the difference between the extremes of the interval. If we pass over from the hendecagonal system (11 cuts) to an infinite number of cuts (hypothesis of continuity), the sum of the distances for all of the levels will give rise to the area of the triangle that graphically represents the T.F.N. as can be seen from Fig. 8.5.

The numerical estimate of the uncertainty in this particular case does not generally coincide (if normally excepting starting out from the distance of the T.F.N.) with its representation by means of weighting the extremes by the unit and by two the maximum presumption, since, with the oft repeated exception, the latter will always be higher than the area of the triangle.

This disparity in the quantitative estimate of the uncertainty can give rise to an apparent contradiction relative to the concept itself of the degree of uncertainty.

In our opinion, said contradiction is purely numerical and is caused by the imprecision of the notion of "crisp representation of a T.F.N". Indeed, throughout this book we have been particularly careful in writing a representation and not the representation. With this we have attempted to point out that it is possible for several and even many possible representations can exist for a T.F.N in certainty. Throughout these financial applications we have used one of these which, in our opinion, has in conjunction with certain inconveniences obvious advantages, on the understanding that although it is a good representation it does not necessarily have to be the best in every case.

On the other hand, it can be easily seen that there exists perfect parallelism in conceiving the uncertainty by means of the area of the triangle or else through the distance $d(\underset{\sim}{\mathbf{A}}; \bar{\underset{\sim}{\mathbf{A}}})$ in the sense that although the figures that constitute the estimate in one or the other case do not coincide, what is always complied with is the fact that as the area of the triangle increases so does the distance. In short, once more the fact is brought to light, that in uncertainty a numerical estimate is not as important in itself as the comparison of estimates arrived at homogeneously.

8.5 From Financial Risk to Insolvency

If it is accepted, as is normal, that the difference between net profit or operating profit and net profit is found in the cost of outside capitals (or if preferred, outside capitals at the medium and short-terms), it can be stated that the financial risk arises as a consequence of the possibility that the cost of the debts absorbs or exceeds operating profit during a determined number of periods.

In the sphere of certainty it is said that insolvency occurs when gross profits are less than the interest on the debts. Obviously this statement should be conditioned in the sense that it is not sufficient for the inequality to take place in a single accounting period for insolvency to appear strictu sensu. These limiting statements have little meaning in the financial activity of our day and age.

What is also said, from a technical point of view of "probability of insolvency", defined as the probability that a business cannot cover payment of the interest on its debts due to the lack of sufficient gross profit. We will not insist again on the difficulties that are practically insurmountable, of arriving at valid probabilities for values located in the future, in this case profits and the cost of outside capital. Therefore the use of any law of probability (including the law represented by normal distribution) in the majority of cases becomes a merely formal exercise with no contact with reality.

The uncertainty that fills all circumstances of our day and age requires totally different handling. Future profits are uncertain, as is the future cost of outside capital. Therefore their representation should be done by means of those elements that underline this circumstance. We are referring specifically to confidence intervals and fuzzy numbers, among others.

Up to this point we have limited ourselves to working with confidence intervals, confidence triplets and triangular fuzzy numbers. Here we propose a greater generalisation by using fuzzy numbers that we will place in Z (that is, within the field of integer numbers), and in this way we will treat the problem in the sphere of caution which, on the other hand, is very useful in solving real problems.

Thus, in stead of expressing a value by means of a lower extreme (below which there is no possibility of finding said value), and an upper extreme (above which there is no possibility of finding the same value), as occurs with confidence intervals; or else by adding a figure, between that selfsame lower and upper extreme, which represents the maximum probability, and at the same time linearity may or may not be required when moving to the left (lower extreme) or to the right (upper extreme), as we are dealing in this case with T.F.N. or confidence triplets, respectively, we are going to structure the interior of the interval by estimating levels of probability that follow no predetermined law (linear, quadratic, etc.) but that are the result of the sensations of an expert. In this way, we can assume that an expert estimates the future profit of a company by means of the following fuzzy number[3]:

$\underset{\sim}{\mathrm{B}} =$

14	15	16	17	18	19	20
0	0,7	1	0,9	0,5	0,1	0

and the future cost of outside capital:

$\underset{\sim}{\mathrm{C}} =$

10	11	12	13	14	15	16	17
0	0,2	0,8	1	0,9	0,8	0,4	0

With this is shown the fact that the expert considers that the profit will not be any lower than 14, which has a probability of 0,7 (quite high) that it will be 15, his maximum supposition is that it will be 16, and he also seriously thinks 0,9 that it may be 17, ..., and finally he is convinced that it cannot be any higher than 20. The same argument can be applied to the "impressions" of the expert relative to the cost of outside capitals.

In the event of wishing to find the net profit, it will be sufficient to carry out the subtraction: $\underset{\sim}{\mathbf{N}} = \underset{\sim}{\mathbf{B}}(-)\underset{\sim}{\mathbf{C}}$. For this resort can be made to the max-min convolution for the subtraction. From the following expression we arrive at the values of the characteristic function for each one of the elements of the subtraction (net profit)[4]:

$$\mu_{\underset{\sim}{\mathrm{N}}}(Z) = \bigvee_{z=x-y} (\mu_{\underset{\sim}{\mathrm{B}}}(X) \wedge \mu_{\underset{\sim}{\mathrm{C}}}(Y))$$

For our example we arrive at:

$$\begin{aligned}
\mu_{\underset{\sim}{\mathrm{N}}}(-3) &= \mu_{\underset{\sim}{\mathrm{B}}}(14) \wedge \mu_{\underset{\sim}{\mathrm{C}}}(17) = 0 \wedge 0 = 0 \\
\mu_{\underset{\sim}{\mathrm{N}}}(-2) &= (\mu_{\underset{\sim}{\mathrm{B}}}(14) \wedge \mu_{\underset{\sim}{\mathrm{C}}}(16)) \vee (\mu_{\underset{\sim}{\mathrm{B}}}(15) \wedge \mu_{\underset{\sim}{\mathrm{C}}}(17)) \\
&= (0 \wedge 0{,}4) \vee (0{,}7 \wedge 0) = 0 \\
\mu_{\underset{\sim}{\mathrm{N}}}(-1) &= (\mu_{\underset{\sim}{\mathrm{B}}}(14) \wedge \mu_{\underset{\sim}{\mathrm{C}}}(15)) \vee (\mu_{\underset{\sim}{\mathrm{B}}}(15) \wedge \mu_{\underset{\sim}{\mathrm{C}}}(16)) \vee (\mu_{\underset{\sim}{\mathrm{B}}}(16) \wedge \mu_{\underset{\sim}{\mathrm{C}}}(17)) \\
&= (0 \wedge 0{,}8) \vee (0{,}7 \wedge 0{,}4) \vee (1 \wedge 0) = 0{,}4 \\
\mu_{\underset{\sim}{\mathrm{N}}}(0) &= (\mu_{\underset{\sim}{\mathrm{B}}}(14) \wedge \mu_{\underset{\sim}{\mathrm{C}}}(14)) \vee (\mu_{\underset{\sim}{\mathrm{B}}}(15) \wedge \mu_{\underset{\sim}{\mathrm{C}}}(15)) \vee (\mu_{\underset{\sim}{\mathrm{B}}}(16) \\
&\quad \wedge \mu_{\underset{\sim}{\mathrm{C}}}(16)) \vee (\mu_{\underset{\sim}{\mathrm{B}}}(17) \wedge \mu_{\underset{\sim}{\mathrm{C}}}(17)) \\
&= (0 \wedge 0{,}9) \vee (0{,}7 \wedge 0{,}8) \vee (1 \wedge 0{,}4) \vee (0{,}9 \wedge 0) = 0{,}7
\end{aligned}$$

Continuing with this operation we arrive at:

$$\begin{array}{lll}
\mu_{\underset{\sim}{\mathrm{N}}}(1) = 0{,}8 & \mu_{\underset{\sim}{\mathrm{N}}}(4) = 0{,}9 & \mu_{\underset{\sim}{\mathrm{N}}}(7) = 0{,}2 \\
\mu_{\underset{\sim}{\mathrm{N}}}(2) = 0{,}9 & \mu_{\underset{\sim}{\mathrm{N}}}(5) = 0{,}8 & \mu_{\underset{\sim}{\mathrm{N}}}(8) = 0{,}1 \\
\mu_{\underset{\sim}{\mathrm{N}}}(3) = 1 & \mu_{\underset{\sim}{\mathrm{N}}}(6) = 0{,}5 & \mu_{\underset{\sim}{\mathrm{N}}}(9) = 0
\end{array}$$

---

[3] As is sufficiently well known, a fuzzy number is a fuzzy sub-set of a referential of real numbers, with the characteristics of normality and convexity.

[4] See Kaufmann A and Gil Aluja J (1987) Técnicas operativas de gestión para el tratamiento de la incertidumbre. (Ed) Hispano Europea, Barcelona. p. 52.

Therefore, we can express the net profit $\underset{\sim}{\mathbf{N}}$ as follows:

	-2	-1	0	1	2	3	4	5	6	7	8	9
$\underset{\sim}{\mathbf{N}}$ =	0	0,4	0,7	0,8	0,9	1	0,9	0,8	0,5	0,2	0,1	0

All this tells us that there are no net losses greater than 2 nor net profits greater than 9. We can also see that the greatest probability is that the net profit will be 3 and that it is estimated, for example, that there little probability (with a degree of 0,2) that the profit will be 7 and, on the other hand, quite a lot of probability (with a degree of 0,7) that there will be no profit nor losses.

It can be seen that, if in place of operating with fuzzy numbers in Z, T.F.N. were to be considered, the result for the maximum probability would have been the same, and for the extremes it would have varied in one unit as a consequence of the passage from the sphere of caution to the continuous sphere. In fact, if $\underset{\sim}{\mathbf{B}} = (14; 16; 20)$ and $\underset{\sim}{\mathbf{C}} = (10; 13; 17)$, we arrive at:

$$\underset{\sim}{\mathbf{N}} = \underset{\sim}{\mathbf{B}}(-)\underset{\sim}{\mathbf{C}} = (-3; 3; 10)$$

Also in this case the general rule is complied with that states that as we do more and more operations the uncertainty of the result increases.

We now move on to estimate the possibility of insolvency. In a situation characterised by uncertainty, it can be said that insolvency exists in the future when gross profits are less than the cost of the debts. This conclusive "law of limitation" does not reflect, in our humble opinion, the impressions that take place in reality, but we feel that a law that reflects the fact that there is less possibility of insolvency as net losses get less or net profits grow, is far more adequate. What would be the form of this Law? Obviously it would be a "law of possibilities" in which, in this form, the following would be complied with:

$$\vee \mu_{\underset{\sim}{\mathbf{L}}}(x) = 1$$

Let us now establish a law of possibility of insolvency for our hypothetical business and for a future period. Let us assume that an expert estimates that when the cost of the debts exceed gross profits in 3 points or more, then it is certain that insolvency will occur; when they exceed then by 2 units the sensation that there will be insolvency is less (estimated at 0,8), ... and so on successively values will be estimated, for the characteristic function of membership, all the time getting smaller as the difference is reduced. This characteristic function of membership as it changes value the excess of gross profits over the cost of the debts increases. In this way we arrive at the following law of possibility expressed by means of the following fuzzy sub-set:

	≤ −3	-2	-1	0	1	2	≥ 3
$\underset{\sim}{\mathbf{L}}$ =	1	0,8	0,7	0,6	0,3	0,1	0

In order to find the possibility of the fuzzy number N relative to law L for each element $x \in Z$ we will arrive at the lowest value of the characteristic functions $\mu_{\underset{\sim}{\mathbf{N}}}(x)$ and $\mu_{\underset{\sim}{\mathbf{L}}}(x)$, choosing the highest from among them all. This can be expressed by means of the following formula:

$$\mathrm{Pos}_{\underset{\sim}{\mathbf{L}}} \underset{\sim}{\mathbf{N}} = \vee(\mu_{\underset{\sim}{\mathbf{L}}}(x) \wedge \mu_{\underset{\sim}{\mathbf{L}}}(x))$$

which in this case will be:

$$\begin{aligned}\mathrm{Pos}_{\underset{\sim}{\mathbf{L}}} \underset{\sim}{\mathbf{N}} = {} & (1 \wedge 0) \vee (0{,}8 \wedge 0) \vee (0{,}7 \wedge 0{,}4) \vee (0{,}6 \wedge 0{,}7) \vee (0{,}3 \wedge 0{,}8) \\ & \vee(0{,}1 \wedge 0{,}9) \vee (0 \wedge 1) \vee (0 \wedge 0{,}9) \vee (0 \wedge 0{,}8) \vee (0 \wedge 0{,}5) \\ & \vee(0 \wedge 0{,}2) \vee (0 \wedge 0{,}1) \vee (0 \wedge 0) = 0{,}6\end{aligned}$$

Therefore, the possibility of insolvency will be, in this case 0,6, which tends to indicate that a certain possibility exists, rather high, that insolvency will occur.

This scheme is also valid in the case that the formal hypothesis were to be considered that insolvency appears when gross profits is less than the cost of the debts, since that this would be a special case of the previous problem. In fact, the law would acquire the following expression:

	≤ −3	-2	-1	0	1	2	≥ 3
$\underset{\sim}{L}$ =	1	1	1	1	0	0	0

in which case the possibility of insolvency would obviously be 0,7 since:

$$\mathrm{Pos}_{\underset{\sim}{\mathbf{L}}} \underset{\sim}{\mathbf{N}} = (1\wedge 0)\vee(1\wedge 0)\vee(1\wedge 0{,}4)\vee(1\wedge 0{,}7)\vee(0\wedge 0{,}8)\vee(0\wedge 0{,}9)\vee(0\wedge 1) = 0{,}7$$

The estimate of insolvency is suitable for treatment but means of valuations that differ from "possibility" such as the "index of consent", among others. The use of one or the other will depend upon each situation and the context in which the business moves.

Finally, we should point out that in a modern society, characterised by profound changes, concepts acquire new dimensions far removed from the strictness that traditionally surrounded finance management studies.

9 Inference Chains in Financial Analysis

9.1 Valuation of Financial Phenomena

One of the concepts that most frequently crops up in the study of uncertainty is known by the name of "valuation". This is a numerical expression in an adequate scale of values that we affect to a phenomenon perceived by our senses or by our experience. It is a very primitive mechanism, that every living being posses, and that acquires very different characteristics from one person to another. Valuations can be expressed by means of absolutes, relatives, superlatives and even by means of values that are subjectively associated to words in a language.

Frequently for said valuations numbers between 0 and 1 are used. In this case "valuation" should not be confused with "probability". A valuation is "subjective" data provided by one person or several people. A probability is objective data and therefore theoretically accepted in a general nature. The notion of probability is linked to randomness. Valuation is linked to uncertainty, to subjectivity. It is fundamental not to confuse "probable" with "possible". What is probable is associated to the notion of measurement, while what is possible is associated to subjectivity in the absence of measurement.

As has been pointed out, the assignment of valuations can be done in every known scale of numerical values. We can take, therefore, as an interval of reference that of the numbers 0 and 1, including both 0 and 1. This is written as 0,1 and as an example we mention: $0{,}3; 1; 0; 0{,}68; 0{,}01$, etc. But also this can be expressed by two extremes in $[0; 1]$, for example, $[0{,}7; 0{,}9]$, $[0; 0{,}2]$, etc. when we are incapable of expressing our subjectivity by one and only one number. This then is a "confidence interval". If necessary we can go even further by placing between the two extremes a "maximum presumption" in order to form a triplet, always with numbers in $[0; 1]$. In this way we have triplets such as $(0{,}7; 0{,}9; 1)$ in which 0,9 is the maximum presumption; $(0{,}2; 0{,}5; 0{,}6)$, in which 0,5 is this maximum; $(0; 0; 0{,}4)$ in which 0 is this maximum. In this case the valuation is given by "confidence triplets". In the same way that a bricklayer or tailor promises to execute a house or a suit not before a certain date nor after another, and between these two is the most possible dates (note! We have not written the most probable), a subjective estimate can also be expressed by means of a valuation in which the maximum presumption can be an interval and, in this way, $(0{,}3; [0{,}4; 0{,}6]; 0{,}9)$

Anna Maria Gil-Lafuente: *Fuzzy Logic in Financial Analysis*, StudFuzz **175**, 135–157 (2005)
www.springerlink.com

is an extended triplet: equal or more than 0,3, equal or less than 0,9, and the maximum presumption between 0,4 and 0,6, including 0,4 and 0,6. This then is a "confidence quadruplet".

There are many ways of representing subjectivity by means of a number in $[0;1]$; nevertheless, the four ways we have shown are those that are most adequate for the treatment of financial problems.

9.2 Considerations on Multivalent Logic

It can be considered that a valuation expresses a level of truth expressed by means of a number, an interval, etc., between (0) false and (1) true. In a first approximation if a valuation is considered by means of a number, an infinity of semantic correspondences can be chosen which range from true to false. Among these we mention:

Binary:		
	0	false
	1	true
Ternary:		
	0	false
	0.5	neither false nor true (neutral)
	1	true
Hendecagonal:		
	0	false
	0.1	practically false
	0.2	nearly false
	0.3	quite false
	0.4	more false than true
	0.5	neither false nor true
	0.6	more true than false
	0.7	quite true
	0.8	nearly true
	0.9	practically true
	1	true

From this point on we are going to pay special attention to the hendecagonal system, since the corresponding numbers, as they have only one decimal point, make calculations very simple. When we write $[0{,}2; 0{,}5]$ in the hendecagonal system, we are expressing the fact that the level of truth is greater or equal to 0,2, and smaller or equal to 0,5. Its meaning is nearly false, neither false nor true. The use of confidence triplets does not change the nature of the problem. A general rule is established which admits no exceptions:

Every result must remain within $[0; 1]$.

The most used operators are:

- $\wedge$ that signifies "minimum" (take the lowest), and corresponds to "and" (the one and the other);
- $\vee$ that signifies "maximum" (take the greatest) and corresponds to "and/or" (the one, the other or both);
- $^{-}$ that above a number, interval, etc, means taking the complement to 1.

There also exist an infinity of other operators which usually appear in pairs, such as $\wedge$ and $\vee$. Operator $\wedge$ is included within the T-norms and $\vee$ within the T-co-norms.

With this we are introducing ourselves into logic called "multivalent" in which a property, although it may be true (1) or false (0), also allows for nuances by means of valuations between 0 and 1, as we have already stated. Another type of logical operator within [0; 1] is inference or implication. This the famous or well known "yes .... then". When considering binary algebra it will be seen that there only exists one inference operator. In fact, if we do $a = v(P)$, $b = v(Q)$, $C = v(P \to Q)$ where P, Q and $P \to Q$ are proposals or properties (also called predicates) and $P \to Q$ means that "if P is true then Q is true", we would have an inference:

$$\bar{a} \vee b = c$$

In this way, in binary logic, when we use the symbols of proposals, we have:.

$$(1) \quad \left.\begin{array}{l} v(P \to Q) = 1 \\ v(P) = 1 \end{array}\right\} \text{premises}$$
$$v(Q) = 1 \text{ conclusion}$$

$$(2) \quad \left.\begin{array}{l} v(P \to Q) = 1 \\ v(P) = 0 \end{array}\right\} \text{premises}$$
$$v(Q) = 0 \text{ conclusion}$$

These are two classical cases and at the same time unique of the logic of binary inferences. In problems of logic treated by computers, when starting out from c and from a to arrive at b what is done is "forward linking"; while establishing c and b in order to find a gives rise to "backward linking". In this way as many forward and backward logical reasoning chains are conceived as may be necessary in order to arrive at useable conclusions.

When we move over to the sphere of multivalent logic the existence of an unlimited number of inferences will be seen, and the problem that arises is the selection of the one or the ones that are better adapted to mental mechanisms and to the realities that intervene in the problems that are being attempted to resolve.

9.3 Multivalent Inferences

One of the most important aspects of multivalent inferences is that, contrary to binary logic, if one proposition implicates another, that is to say $P \to Q$, the valuation of $P \to Q$, that is $v(P \to Q)$ does not necessarily have to be equal to 1, but is estimated by means of a value in $[0; 1]$. Obviously, $v(P)$ and $v(Q)$ take their values in $[0; 1]$.

Now let us move on to describe some of the more used multivalent inferences. We will commence with the inference:

$$\bar{a} \vee b = c \text{ that is } \overline{v(P)} \vee v(Q) = v(P \to Q)$$

which we have already considered in the binary field. On this occasion, a, b and c take their values in $[0; 1]$. This inference is called the Lee[1] inference, in honour of this mathematician who studied it using the ternary system. Table 9.1 shows said inference when using the hendecagonal system.

Table 9.1.

		b										
	ā∨b	0	0,1	0,2	0,3	0,4	0,5	0,6	0,7	0,8	0,9	1
	0	1	1	1	1	1	1	1	1	1	1	1
	0,1	0,9	0,9	0,9	0,9	0,9	0,9	0,9	0,9	0,9	0,9	1
	0,2	0,8	0,8	0,8	0,8	0,8	0,8	0,8	0,8	0,8	0,9	1
	0,3	0,7	0,7	0,7	0,7	0,7	0,7	0,7	0,7	0,8	0,9	1
	0,4	0,6	0,6	0,6	0,6	0,6	0,6	0,6	0,7	0,8	0,9	1
a	0,5	0,5	0,5	0,5	0,5	0,5	0,5	0,6	0,7	0,8	0,9	1
	0,6	0,4	0,4	0,4	0,4	0,4	0,5	0,6	0,7	0,8	0,9	1
	0,7	0,3	0,3	0,3	0,3	0,4	0,5	0,6	0,7	0,8	0,9	1
	0,8	0,2	0,2	0,2	0,3	0,4	0,5	0,6	0,7	0,8	0,9	1
	0,9	0,1	0,1	0,2	0,3	0,4	0,5	0,6	0,7	0,8	0,9	1
	1	0	0,1	0,2	0,3	0,4	0,5	0,6	0,7	0,8	0,9	1

It will be seen, for example, that when doing $a = 0{,}4$ and $b = 0{,}3$, we arrive at 0,6, since $\overline{0{,}4} \vee 0{,}3 = 0{,}6$. The problem does not always arise in these terms but that, generally speaking, once having chosen a type of inference (such as $\bar{a} \vee b = c$, for example) a value $v(P \to Q) = c$ is affected and a forward linking is done selecting a in order to find b. In this way the premises are c and a, while the conclusion is b. For this we establish an inversion to the right from Table 9.1. Therefore if we take $a = 0{,}8$ and $c = \bar{a} \vee b = 0{,}6$, we arrive at the fact that $b = 0{,}6$. Now then, the inversion does not always come up with such simple results. In fact, if we consider $a = 0{,}3$ and $c = 0{,}6$, it so happens that in row 0,3 there is no box in which $c = 0{,}6$. This means that it is impossible for $a = 0{,}3$ and $c = 0{,}6$ to coexist. Let us now take a look at what happens if

[1] See Kaufmann A and Gil Aluja J La prevision subjetiva. This work is in the printing phase and we have been allowed to read several parts of its contents thanks to the authors.

we take $a = 0{,}4$ and $c = 0{,}6$. On analysing row 0,4 it will be seen that from 0 to 0,6, in all the boxes there is a 0,6, therefore $b = 0{,}6$, that is in this case b is a confidence interval.

In this way a table can be established the entries of which are a and $\bar{a} \vee b$ and the exit b, in which the solutions that are impossible have been underlined (see Table 9.2).

We would act in the same way in order to draw up a table in which we started out from b and $\bar{a} \vee b$ in order to find a.

In this process it should not be forgotten that, when operating with intervals, the result should provide the widest interval, with the object of it is containing all the solutions. Therefore if for example we start out from $a = [0{,}4; 0{,}8]$ and $c = [0{,}2; 0{,}8]$ this will be:

$$[0{,}2; 0{,}6](\vee)[x_1; x_2] = [0{,}2; 0{,}8]$$

Table 9.2.

$c = \bar{a} \vee \bar{b}$

a \ b	0	0,1	0,2	0,3	0,4	0,5	0,6	0,7	0,8	0,9	1
0											0, 1
0,1										0, 0,9	1
0,2									0, 0,8	0,9	1
0,3				Ø				0, 0,7	0,8	0,9	1
0,4							0, 0,6	0,7	0,8	0,9	1
0,5						0, 0,5	0,6	0,7	0,8	0,9	1
0,6					0, 0,4	0,5	0,6	0,7	0,8	0,9	1
0,7				0, 0,3	0,4	0,5	0,6	0,7	0,8	0,9	1
0,8			0, 0,2	0,3	0,4	0,5	0,6	0,7	0,8	0,9	1
0,9		0, 0,1	0,2	0,3	0,4	0,5	0,6	0,7	0,8	0,9	1
1	0	0,1	0,2	0,3	0,4	0,5	0,6	0,7	0,8	0,9	1

therefore:

$$[x_1; x_2] = [0; 0{,}8]$$

As can be seen, interval $[0; 0{,}8]$ is wider than $[0{,}2; 0{,}8]$ and therefore is the one that should be considered as the most suitable.

Some 60 years ago Lukaciewicz discovered an inference which we will show by means of the hendecagonal system. This is expressed as follows:

Table 9.3.

$1 \wedge (\bar{a} \vee b)$		b										
		0	0,1	0,2	0,3	0,4	0,5	0,6	0,7	0,8	0,9	1
	0	1										
	0,1	0,9	1									
	0,2	0,8	0,8	1								
	0,3	0,7	0,7	0,7	1				1			
	0,4	0,6	0,6	0,6	0,6	1						
a	0,5	0,5	0,5	0,5	0,5	0,5	1					
	0,6	0,4	0,4	0,4	0,4	0,4	0,5	1				
	0,7	0,3	0,3	0,3	0,3	0,4	0,5	0,6	1			
	0,8	0,2	0,2	0,2	0,3	0,4	0,5	0,6	0,7	1		
	0,9	0,1	0,1	0,2	0,3	0,4	0,5	0,6	0,7	0,8	1	
	1	0	0,1	0,2	0,3	0,4	0,5	0,6	0,7	0,8	0,9	1

$$1 \wedge (\bar{a} + b) = c$$

which indicates that if $\bar{a} + b$ is greater or equal to 1 we do $c = 1$; if $\bar{a} + b$ is less than 1 then we take $c = \bar{a} + b$.

In the same way as was done with the Lee inference an inference table can be constructed for the Lukaciewicz inference which shows c in function of a and b. In this way we arrive at Table 9.3.

When we need to make a "forward linking", that is given c and a find b, we look in the rows of a for the given values of c; the columns in which they are found provide b. If they exist b is arrived at, which can be a number or an interval as in the case in which $a = 0{,}4$ and $c = 1$, which results in $b = [0{,}4; 1]$; if they do not exist it means that the solution is impossible (this happens when $a = 0{,}6$ and $c = 0{,}3$, for example).

There are many other multivalent inferences, among them we could mention the Gödel inference which can be expressed as $c = 1 \wedge b$, and the Goguen inference, the expression for which is $c = 1 \wedge (b/a)$, which has the interesting property that a and b can be valuations or probabilities, since this is the formula for conditional probabilities.

9.4 Objectives in Inference Chains

In the financial sphere, the notion "inference" acquires special importance, taking into account the nexus existing between the values that make up the economic and financial structure of the balance sheet and the basic elements of management. In fact, it is perfectly well known that financial cost depends, among other things, on the volume reached by short "and" long term liabilities; as also by the fact that permanent capitals depend on the volume of long term liabilities "and" that for equity. Also it is elementary to bring to light the fact that the patrimonial masses of the financial structure of the balance sheet constitute the sources of financing, the application of which takes place within the patrimonial masses of the economic structure. Values such as invoicing, production and distribution costs, auto-financing, etc., and much used instruments of financial analysis such as working capital, cash-slow, financial risk, etc, are elements that determine in the end, the appreciation that the business warrants from the economic agents of its surroundings and which move the quotation of its shares in the market place and therefore, are the determining factors for the "wealth of the shareholders". It is known that one of the objectives which justify business activity is the maintenance or expansion of the "wealth" of those holding the capital of the same.

With the object of bringing to light an inference chain which, starting out from a determined financial structure, will lead us in the end to the share quotation, we have drawn up the diagram as appearing in Fig. 9.1.

It can be seen from this that there exist a number of knots or points, such as $D, F, J, \ldots, Q$, at which more than two or more arcs converge. In this case

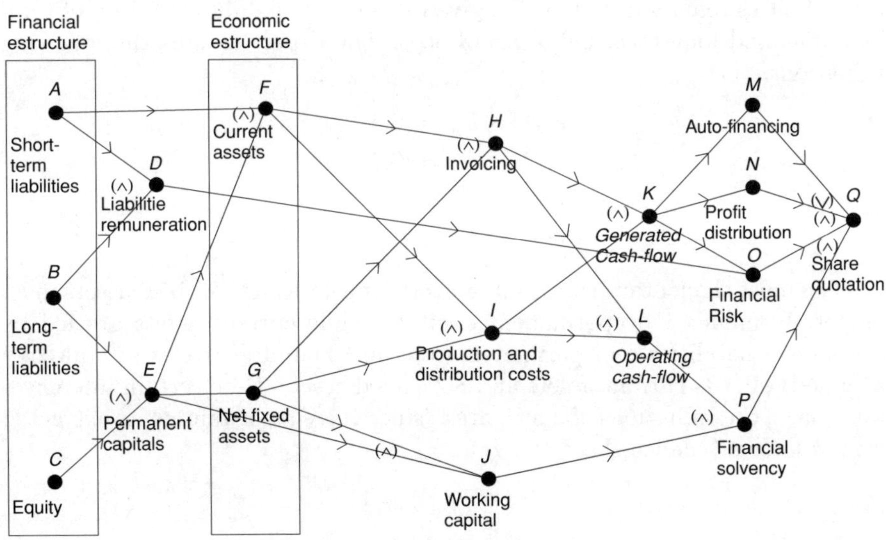

Fig. 9.1.

it is necessary to determine if for arriving at the desired result it is necessary to do one "and" the other of the events which give origin to the arc, in which case the operator ($\wedge$) will be used, or else it will be sufficient that either the one "and/or" the other take place, using then the operator ($\vee$). D, F and J, for example, belong to first case, Q to the second relative to the arc starting out from M and N.

We will now move on to explain the bases for the construction of the inference graph. We start out from a business that, at a certain point in time has a determined financial structure made up of the following patrimonial masses:

$A =$ Short-term liabilities	15%
$B =$ Long-term liabilities	30%
$C =$ Equity	55%
	100%

Taking into account the fact that this financial structure is existing in reality, and therefore, we are dealing with actual realisations, the valuations will be: $v(A) = 1$ and $v(B) = 1$ and $v(C) = 1$.

With these liabilities it is intended to arrive at:

$D =$ Liabilities remuneration	7%
$E =$ Permanent capitals	85%

With the existing economic structure the objective set for permanent capitals will be complied with automatically, since $v(B \rightarrow E) = 1$ and $v(C \rightarrow E) = 1$. On the other hand, an expert is asked to indicate, in accordance with his knowledge and experience, the valuations relative to attainment of the liabilities remuneration at 7%, given short-term liabilities of 15% of total liabilities and long-term liabilities of 30%. The expert assigns the following valuations:

$$v(A \rightarrow D) = [0{,}8; 0{,}9]$$
$$v(B \rightarrow D) = [0{,}7; 0{,}9]$$
$$v(C \rightarrow E) = 1$$
$$v(B \rightarrow E) = 1$$

The most elementary rules of management establish that fixed assets must be totally financed by permanent capitals, while current assets are fed by short-term liabilities and permanent capitals. The objective is set (already attained) of 20% current assets and 80% fixed assets. Here certain inferences take place, the valuations of which are 1, since these are attainments extracted from a known balance sheet. Therefore:

$$v(A \rightarrow F) = 1$$
$$v(E \rightarrow F) = 1$$
$$v(E \rightarrow G) = 1$$

With these bases it is hoped to attain higher invoicing or invoicing equal to 10% with increases in production and distribution costs not in excess of 7,5%, maintaining working capital in a volume equivalent to 5% of liabilities. Both the figure for invoicing and production and distribution costs basically depend on the industrial equipment and its physical and economic condition, labour, existing stocks, etc., that is to say that net fixed assets "and" current assets. As is well known, working capital is consequent upon the existence of certain permanent capitals "and" of net fixed assets. The expert consulted assigns the following valuations to the inferences:

$$\begin{aligned}
v(F \to H) &= [0{,}6; 0{,}6] \\
v(F \to I) &= [0{,}5; 0{,}6] \\
v(G \to H) &= 0{,}9 \\
v(G \to I) &= [0{,}7; 0{,}9] \\
v(G \to J) &= [0{,}6; 0{,}7] \\
v(E \to J) &= [0{,}9; 1]
\end{aligned}$$

Financial analysis studies place special attention to the concepts of "cash-flow generated" and "cash-flow operating". The first brings to light the excess that the business has attained by means of its activity and which will permit among others a contribution to depreciation and profit distribution to the shareholders. The second brings to light the cash flows or runs that are a consequence on the inflows and outflows of payment means in the cash and bank accounts. An objective is established of arriving at a cash-flow generated of 22% of equity and cash-flow operating to exceed the last year level in 5%. Either one or the other concept stem from the attainment of invoicing "and" production and distribution costs. The expert gives an opinion that the corresponding inferences can be assigned the following valuations:

$$\begin{aligned}
v(H \to K) &= [0{,}6; 0{,}8] \\
v(I \to K) &= [0{,}4; 0{,}7] \\
v(H \to L) &= [0{,}9; 1] \\
v(I \to L) &= [0{,}8; 0{,}9]
\end{aligned}$$

Arriving at a determined volume of cash-flow generated allows us to contribute to the accounts that form auto-financing and proceed to distribute profits. The intention for the accounting period under study is to increase equity in an amount equal to or in excess of 8% as well as to proceed to distribute profits in an amount no less than 6%.

On the other hand, the consideration of the cash-flow generated "and" the cost of remuneration of liabilities give rise to the existence of financial risk. Given the context of uncertainty in which we are acting we will assume a grading of the risk in the hendecagonal scale of segment $[0; 1]$. A low risk $R_F = 0{,}3$ is hoped for. Another important factor, financial solvency, can also be expressed by means of an estimate in the hendecagonal scale of segment [0; 1]. This is consequent upon the existence of a cash-flow operating (short term solvency) and working capital (long term solvency). The objective aimed for is a high level of solvency $S_F = 0{,}8$. The expert consulted assigns the following valuations to the inferences we have listed:

$$v(K \to M) = [0{,}8; 1]$$
$$v(K \to N) = [0{,}8; 1]$$
$$v(K \to O) = [0{,}8; 1]$$
$$v(D \to O) = [0{,}7; 0{,}9]$$
$$v(L \to P) = [0{,}5; 0{,}7]$$
$$v(J \to P) = [0{,}4; 0{,}7]$$

Let us recall that one of the fundamental objectives of the activity of businesses is to maintain or increase the wealth of the shareholders, and an expression of this is the share quotation on the stock market. In this case the aim is to arrive at quotations that are no less than 300%. It can be accepted (we will not discuss the existence of other elements which could even become determining) that the price in the market for a share is conditioned on its fundamental value (which varies with the contributions for auto-financing) "and/or" the remuneration of equity (profit distribution). The assignment of an amount to one and the other concept can have an alternative nature, since one and the same cash-flow generated can give rise to greater auto-financing and less distribution of profits, or the reverse. Others among the many elements that can be considered as having an influence on the share quotations, are financial risk and financial solvency. The expert provides the valuations that are listed below:

$$v(M \to Q) = [0{,}4; 0{,}5]$$
$$v(N \to Q) = [0{,}5; 0{,}7]$$
$$v(O \to Q) = [0{,}7; 0{,}9]$$
$$v(P \to Q) = [0{,}8; 1]$$

From this information we propose following a process that, at the same time that it permits estimating the possibilities of attaining the final objective, it could constitute an instrument for management and control of partial objectives. In general lines the scheme to be followed can be summarised as in the organisation chart shown in Fig. 9.2.

9.5 The Result by Means of the Lukaciewicz Inference

With the object of arriving at the valuation of the possibilities of attaining the objective proposed at vertex Q of the graph, that is a quotation for the

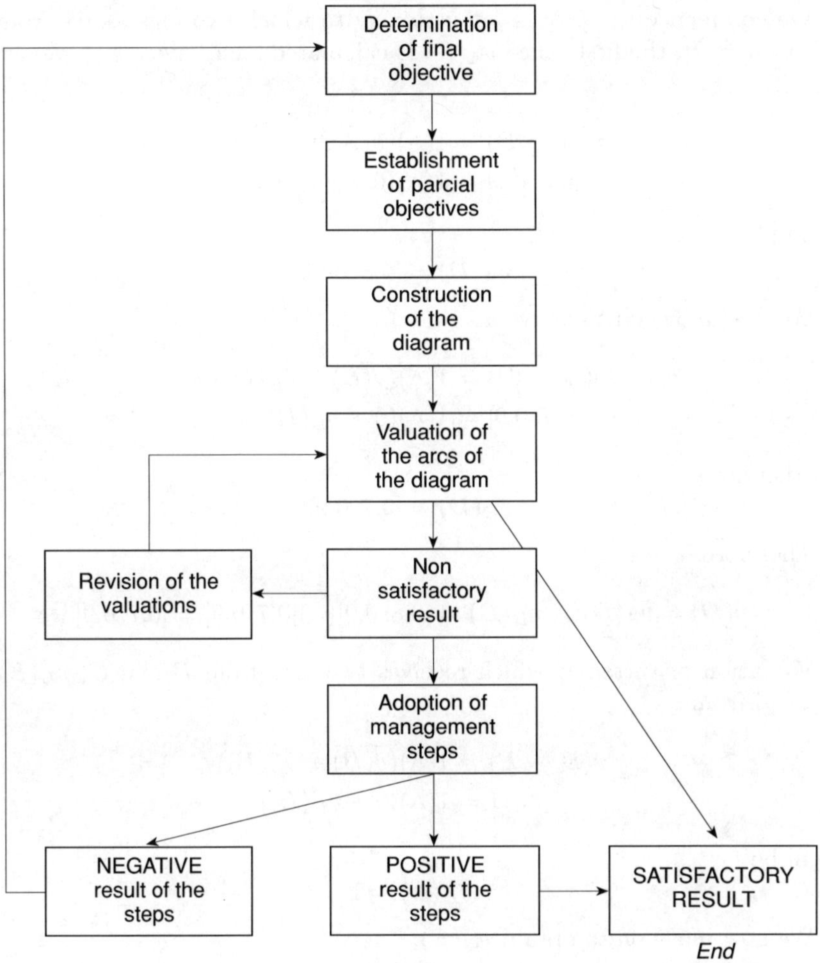

Fig. 9.2.

shares equal to or higher than 300% at a future moment, we are going to use the Lukaciewicz inference, the expression of which is:

$$\boxed{c = 1 \wedge (\bar{a} + b)}$$

where a, b and c are always intervals[2] in $[0; 1]$ and the operator $+$ has the meaning of an ordinary addition.

[2] We should remember that a crisp number can also be considered as a confidence interval in which the extremes are so close that they become confused in a single value.

We commence the process at vertex D to which two arcs arrive from A and from B. In the first place $v_A(D)$ is calculated[3] and given that we arrive at:

$$v(A \to D) = 1(\wedge)(\overline{v(A)} + v_A(D))$$
$$[0{,}8; 0{,}9] = 1(\wedge)(0 + v_A(D))$$

it will be

$$v_A(D) = [0{,}8; 0{,}9]$$

We now move on to arrive at $v_B(D)$:

$$v(B \to D) = 1(\wedge)(\overline{v(B)} + v_B(D))$$
$$[0{,}7; 0{,}9] = 1(\wedge)(0 + v_B(D))$$

and therefore

$$v_B(D) = [0{,}7; 0{,}9]$$

Therefore:

$$v(D) = v_A(D)(\wedge)v_B(D) = [0{,}8; 0{,}9](\wedge)[0{,}7; 0{,}9] = [0{,}7; 0{,}9]$$

Valuation of vertex E which receives two arcs from B and C, $v_B(E)$ is obtained from:

$$v(B \to E) = 1(\wedge)(\overline{v(B)} + v_B(E))$$
$$1 = 1(\wedge)(0 + v_B(E))$$

it will be

$$v_B(E) = 1$$

We now move on to obtain $v_C(E)$:

$$v(C \to E) = 1(\wedge)(\overline{v(C)} + v_C(E))$$
$$1 = 1(\wedge)(0 + v_C(E))$$

where:

$$v_C(E) = 1$$

Therefore:

$$v(E) = v_B(E)(\wedge)v_C(E) = 1(\wedge)1 = 1$$

Valuation of vertex F. Arriving at this vertex are two arcs proceeding from A and E. After doing the corresponding operations we arrive at:

$$v(F) = 1$$

[3] We designate the valuation in D that comes from vertex A by $v_A(D)$.

Valuation of vertex G. The existence of a single arc arriving at this vertex gives rise to:

$$v(G) = 1$$

Valuation of vertex H. Two arcs arrive at this vertex from F and G. $v_F(H)$ is found as follows:

$$v(F \to H) = 1(\wedge)(\overline{v(F)} + v_F(H))$$
$$[0{,}6; 0{,}8] = 1(\wedge)(0 + v_F(H))$$

which results in:

$$v_F(H) = [0{,}6; 0{,}8]$$

In order to arrive at $v_G(H)$ we do:

$$v(G \to H) = 1(\wedge)(\overline{v(G)} + v_G(H))$$
$$0{,}9 = 1(\wedge)(0 + v_G(H))$$

Thus:

$$v_G(H) = 0{,}9$$

Will be:

$$v(H) = v_F(H)(\wedge)v_G(H) = [0{,}6; 0{,}8](\wedge)0{,}9 = [0{,}6; 0{,}8]$$

Valuation of vertex I. Two arcs arrive at this vertex from F and G. $v_F(I)$ is arrived at:

$$v(F \to I) = 1(\wedge)(\overline{v(F)} + v_F(I))$$
$$[0{,}5; 0{,}6] = 1(\wedge)(0 + v_F(I))$$

therefore:

$$v_F(I) = [0{,}5; 0{,}6]$$

Now we move on to find $v_G(I)$:

$$v(G \to I) = 1(\wedge)(\overline{v(G)} + v_G(I))$$
$$[0{,}7; 0{,}9] = 1(\wedge)(0 + v_G(I))$$

therefore:

$$v_G(I) = [0{,}7; 0{,}9]$$

The result will be that:

$$v(I) = v_F(I)(\wedge)v_G(I) = [0{,}5; 0{,}6](\wedge)[0{,}7; 0{,}9] = [0{,}5; 0{,}6]$$

Valuation of vertex J. Two arcs arrive here that commence in G and E. $v_G(J)$ is calculated:

$$v(G \to J) = 1(\wedge)(\overline{v(G)} + v_G(J))$$
$$[0{,}6; 0{,}7] = 1(\wedge)(0 + v_G(J))$$

the result being:

$$v_G(J) = [0{,}5; 0{,}7]$$

We now obtain $v_E(J)$:

$$v(E \to J) = 1(\wedge)(\overline{v(E)} + v_E(J))$$
$$[0{,}9; 1] = 1(\wedge)(0 + v_E(J))$$

This will be:

$$v_E(J) = [0{,}9; 1]$$

Therefore we arrive at:

$$v(J) = v_G(J)(\wedge)v_E(J) = [0{,}6; 0{,}7](\wedge)[0{,}9; 1] = [0{,}6; 0{,}7]$$

Valuation of vertex K. Two arcs arrive here from H and I. We find $v_H(K)$:

$$v(H \to K) = 1(\wedge)(\overline{v(H)} + v_H(K))$$
$$[0{,}6; 0{,}8] = 1(\wedge)([0{,}2; 0{,}4] + v_H(K))$$

therefore:

$$v_H(K) = [0{,}4; 0{,}4] = 0{,}4$$

In order to find $v_I(K)$ we do:

$$v(I \to K) = 1(\wedge)(\overline{v(I)} + v_I(K))$$
$$[0{,}4; 0{,}7] = 1(\wedge)([0{,}4; 0{,}5] + v_I(K))$$

Thus:

$$v_I(K) = [0; 0{,}2]$$

The result being:

$$v(K) = v_H(K)(\wedge)v_I(K) = 0{,}4(\wedge)[0; 0{,}2] = [0; 0{,}2]$$

Valuation of vertex L. Here two arcs arrive originating at H and I. In order to arrive at $v_H(L)$ we do;

$$v(H \to L) = 1(\wedge)(\overline{v(H)} + v_H(L))$$
$$[0{,}9; 1] = 1(\wedge)([0{,}2; 0{,}4] + v_H(L))$$

with which we obtain:

$$v_H(L) = [0{,}7; 1]$$

For the upper extreme of $v_H(L)$ we choose 1, since $0{,}7; 0{,}8; 0{,}9$ comply with the equality, the "widest" interval will occur when considering $[0{,}7; 1]$. We must not forget that:

$$1(\wedge)[0{,}2;0{,}4]+[0{,}7;1]=[1;1](\wedge)[0{,}9;1]=[0{,}9;1]$$

We now calculate $v_I(L)$:

$$v(I \to L)=1(\wedge)(\overline{v(I)}+v_I(L))$$
$$[0{,}8;0{,}9]=1(\wedge)([0{,}4;0{,}5]+v_I(L))$$

the result being:

$$v_I(L)=0{,}4$$

we arrive at:

$$v(L)=v_H(L)(\wedge)v_I(L)=[0{,}6;1](\wedge)0{,}4=0{,}4$$

Valuation of vertex M. Arriving at this point is a single arc originating in K. We arrive at $v(M)$ as follows:

$$v(K \to M)=1(\wedge)(\overline{v(K)}(+)v(M))$$
$$[0{,}8;1]=1(\wedge)([0{,}8;1](+)v(M))$$

therefore:

$$v(M)=[0;1]$$

With regard to this valuation once again take into account that it is necessary to select a wider interval. This result can be seen in the table included in the second section.

Valuation of vertex N. The only arc arriving at this point commences at K. In order to arrive at $v(N)$ we do:

$$v(K \to N)=1(\wedge)(\overline{v(K)}(+)v(N))$$
$$[0{,}8;1]=1(\wedge)([0{,}8;1](+)v(N))$$

Thus:

$$v(N)=[0;1]$$

The same reasoning is valid as for vertex M.

Valuation of vertex O. Two arcs converge at this point from K and D. In the first place we find $v_K(O)$:

$$v(K \to O)=1(\wedge)(\overline{v(K)}+v_K(O))$$
$$[0{,}8;1]=1(\wedge)([0{,}8;1]+v_K(O))$$

The result is:

$$v_K(O)=[0;1]$$

We now find the valuation for $v_D(O)$:

$$v(D \to O)=1(\wedge)(\overline{v(D)}+v_D(O))$$
$$[0{,}7;0{,}9]=1(\wedge)([0{,}1;0{,}3]+v_D(O))$$

The result being:

$$v_D(O) = [0{,}6; 0{,}6] = 0{,}6$$

and in short:

$$v(O) = v_K(O)(\wedge)v_D(O) = [0; 1](\wedge)0{,}6 = [0; 0{,}6]$$

Valuation of vertex P. Two arcs arrive at this point initiating from L and J. In order to obtain $v_L(P)$ we do:

$$\begin{aligned} v(L \to P) &= 1(\wedge)(\overline{v(L)}(+)v_L(P)) \\ [0{,}5; 0{,}7] &= 1(\wedge)(0{,}6(+)v_L(P)) \end{aligned}$$

It will be seen that $v_L(P) =$ impossible, since there cannot be an lower extreme of $v_L(P)$ in $[0; 1]$ that shows, by adding to 0,6, a total of 0,5. Now let us imagine that the expert considers the valuation $v(L \to P)$ and places it in $[0{,}6; 0{,}8]$ for example: In this case we would have:

$$[0{,}6; 0{,}8] = 1(\wedge)(0{,}6(+)v_L(P))$$

and therefore:

$$v_L(P) = [0; 0{,}2]$$

We move on to obtain the valuation of $v_J(P)$:

$$\begin{aligned} v(J \to P) &= 1(\wedge)(\overline{v(J)}(+)v_J(P)) \\ [0{,}4; 0{,}7] &= 1(\wedge)([0{,}3; 0{,}4](+)v_J(P)) \end{aligned}$$

where:

$$v_J(P) = [0{,}1; 0{,}3]$$

We arrive at:

$$v(P) = v_L(P)(\wedge)v_J(P) = [0; 0{,}2](\wedge)[0{,}1; 0{,}3] = [0; 0{,}2]$$

Given the valuations of vertices M, N, O and P we are going to value the final vertex Q. For this we do:

$$\begin{aligned} v(M \to Q) &= 1(\wedge)(\overline{v(M)}(+)v_M(Q)) \\ [0{,}4; 0{,}5] &= 1(\wedge)([0; 1](+)v_M(Q)) \\ v_M(Q) &= [0{,}4; \text{impossible}] \end{aligned}$$

$$\begin{aligned} v(N \to Q) &= 1(\wedge)(\overline{v(N)}(+)v_N(Q)) \\ [0{,}5; 0{,}7] &= 1(\wedge)([0; 1](+)v_N(Q)) \\ v_N(Q) &= [0{,}5; \text{impossible}] \end{aligned}$$

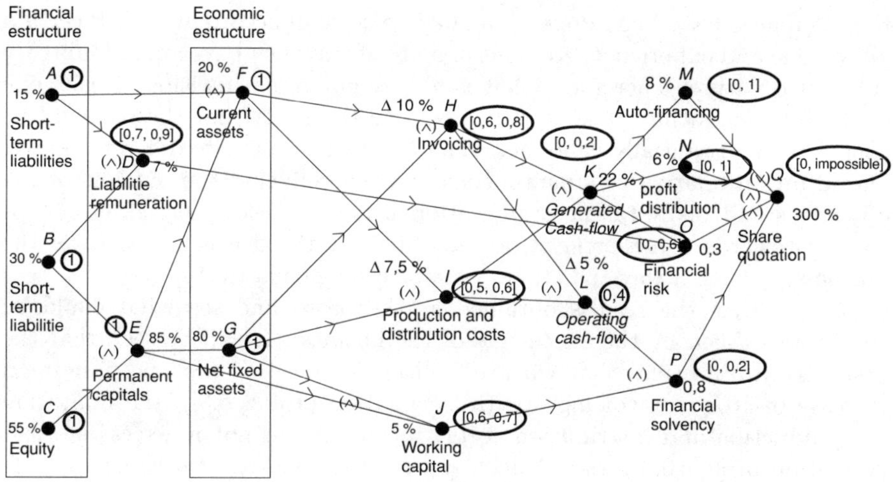

Fig. 9.3.

$$
\begin{aligned}
v(O \to Q) &= 1(\wedge)(\overline{v(O)}(+)v_O(Q)) \\
[0{,}7; 0{,}9] &= 1(\wedge)([0{,}4; 1](+)v_O(Q)) \\
v_O(Q) &= [0{,}3; \text{impossible}] \\
v(P \to Q) &= 1(\wedge)(\overline{v(P)}(+)v_P(Q)) \\
[0{,}8; 1] &= 1(\wedge)([0{,}8; 1](+)v_P(Q)) \\
v_P(Q) &= [0; 1]
\end{aligned}
$$

And finally:

$$((v_M(Q)(\vee)v_N(Q))(\wedge)v_O(Q)(\wedge)v_P(Q) = [0; \text{impossible}]$$

The drawing in Fig. 9.3 brings to light the primary objectives as well as the valuations corresponding to each vertex. The final result, by means of this example, allows us to make determined comments.

9.6 Inference Chains as an Element for Decision

The valuations that are obtained as one advances in inference chains bring to light the difficulties existing for arriving at a final objective, that is to say, that the shares be quoted at a rate equal to or higher than 300%. The requirement of attaining a cash-flow generated of 22% and an increase in the cash-flow operating of 5% leave this objective with no possibility whatsoever or in the best of cases impossible.

Faced with a situation such as this, very frequent in reality, it is necessary to totally reconsider the objectives and means for attaining the same. Nevertheless, it should not be forgotten that the valuations relative to each one of the inferences have been done by a single expert in accordance with certain knowledge and experience, etc., but in spite of this they have an undoubtedly subjective nature. Therefore, what is also required is a revision of each one of the inferences that make up the chain. In the event that these should be considered substantially valid, action should be taken in two different senses, one in an alternative or simultaneous way, by reducing the amount of the objectives and/or by taking organisational, financial, commercial, etc. steps leading towards a higher level of operability in the decision centres of the business, in such a way that they permit better management.

Let us take the results obtained as the basis, and see what could be the actions taken by the executives of the business. It will be seen that the first partial objectives that with difficulties for meeting are the "minimum increase of 10% in invoicing" (with a valuation of [0,6; 0,8]), the limitation in "production and distribution costs" to an amount not in excess of 7,5% (valuation of [0,5; 0,6]) and to achieve the objective that "working capital" be 5% of the financial structure (valuation [0,5; 0,7]).

With regard to invoicing and costs it is possible to improve by means of a modification and improvement of commercial and production policies (strengthen the distribution network, more suitable advertising, better stock management, changes in personnel stimulations, management and use of fixed assets, and a lengthy etcetera). With these prospects it may be possible to make changes in the valuation of the inferences: $v(F \rightarrow H)$, $v(F \rightarrow I)$, $v(G \rightarrow H)$ and $v(G \rightarrow I)$. Let us assume that the expert, after the acceptance of new steps, estimates that:

$$\begin{aligned} v(F \rightarrow H)^{(1)} &= [0{,}8; 0{,}9] \\ v(F \rightarrow I)^{(1)} &= [0{,}7; 0{,}8] \\ v(G \rightarrow H)^{(1)} &= 0{,}9 \\ v(G \rightarrow I)^{(1)} &= [0{,}8; 0{,}9] \end{aligned}$$

Then the new valuations for vertices H and I would be:
Vertex H:

$$\begin{aligned} [0{,}8; 0{,}9] &= 1(\wedge)(0 + v_F(H)) \\ v_F(H) &= [0{,}8; 0{,}9] \\ 0{,}9 &= 1(\wedge)(0 + v_G(H)) \\ v_G(H) &= 0{,}9 \\ v(H) &= [0{,}8; 0{,}9](\wedge)0{,}9 = [0{,}8; 0{,}9] \end{aligned}$$

Vertex I:

$$\begin{aligned} [0{,}7; 0{,}8] &= 1(\wedge)(0 + v_F(I)) \\ v_F(I) &= [0{,}7; 0{,}8] \\ [0{,}8; 0{,}9] &= 1(\wedge)(0 + v_G(I)) \end{aligned}$$

$$v_G(I) = [0{,}8; 0{,}9]$$
$$v(I) = [0{,}7; 0{,}8](\wedge)[0{,}8; 0{,}9] = [0{,}7; 0{,}8]$$

Now we take a look at working capital. The question arises as to why the expert has thought that with permanent capitals standing at 85% and fixed assets at 80% in the current balance sheet, it should be so difficult that in the future working capital of 5% should be difficult to attain. It is obvious that included in the estimate are the changes which over time will occur above all in fixed assets, since an inference of $G \to J$ has been assigned a valuation of $[0{,}5; 0{,}7]$. If it can be ensured that the percentages will be maintained throughout the economic horizon, inferences $G \to J$ and $E \to J$ will have a valuation equal to 1. For this the executives of the business must undertake to adopt the necessary measures in order to maintain these patrimonial masses at the set levels. Only in this way will the valuation of vertex J be:

$$v(J) = 1$$

Continuing with the inference chain we arrive at the objective of reaching a "cash-flow generated" of 22% and an increase in the "cash-flow operating" of 5%. The expert has considered that it is quite a complex matter reaching this first objective with the set invoicing and costs. It would appear that in this case a reduction must be made in the aimed for objective, without excluding the complementary management measures. But this in all certainty will mean the modification of the final objective. The management of the business sets as the new intermediary objective arriving at a cash-flow generated of 20%. In this way the expert assigns the following valuations to inferences $H \to K$ and $I \to K$:

$$v(H \to K)^{(1)} = [0{,}8; 0{,}9]$$
$$v(I \to K)^{(1)} = [0{,}7; 0{,}9]$$
$$v(H \to L) = [0{,}8; 1]$$
$$v(I \to L) = [0{,}8; 0{,}9]$$

where it can be seen that inferences $H \to L$ and $I \to L$ have the same initial valuations. The new valuations for vertices K and L will be:

Vertex K:

$$[0{,}8; 0{,}9] = 1(\wedge)([0{,}1; 0{,}2](+)v_H(K))$$
$$v_H(K) = [0{,}7; 0{,}7] = 0{,}7$$
$$[0{,}7; 0{,}9] = 1(\wedge)([0{,}2; 0{,}3](+)v_I(K))$$
$$v_I(K) = [0{,}5; 0{,}6]$$
$$v(K) = 0{,}7(\wedge)[0{,}5; 0{,}6] = [0{,}5; 0{,}6]$$

Vertex L:

$$
\begin{aligned}
[0{,}8;1] &= 1(\wedge)([0{,}1;0{,}2](+)v_H(L))\\
v_H(L) &= [0{,}7;1]\\
[0{,}8;0{,}9] &= 1(\wedge)([0{,}2;0{,}3](+)v_I(L))\\
v_I(K) &= [0{,}5;0{,}6]\\
v(K) &= 0{,}7(\wedge)[0{,}5;0{,}6] = [0{,}5;0{,}6]
\end{aligned}
$$

It is obvious that with the new steps to be taken and adjustment in the objectives, the possibilities of reaching the intermediary objectives has increased substantially. It would seem, nevertheless, that a new revision is in order. Therefore a revision should be made of objectives K and L downward, establishing a cash-flow generated of 18% and an increase in the cah-flow operating of 3%. The expert then supplies the following new valuations:

$$
\begin{aligned}
v(H \rightarrow K)^{(2)} &= [0{,}9;1]\\
v(I \rightarrow K)^{(2)} &= [0{,}9;1]\\
v(H \rightarrow L)^{(1)} &= [0{,}9;1]\\
v(I \rightarrow L)^{(1)} &= [0{,}9;1]
\end{aligned}
$$

where for vertex K we have:

$$
\begin{aligned}
[0{,}9;1] &= 1(\wedge)([0{,}1;0{,}2](+)v_H(K))\\
v_H(K) &= [0{,}8;1]\\
[0{,}9;1] &= 1(\wedge)([0{,}2;0{,}3](+)v_H(K))\\
v_I(K) &= [0{,}7;1]\\
v(K) &= [0{,}8;1](\wedge)[0{,}7;1] = [0{,}7;1]
\end{aligned}
$$

and for vertex L:

$$
\begin{aligned}
[0{,}9;1] &= 1(\wedge)([0{,}1;0{,}2](+)v_H(L))\\
v_H(L) &= [0{,}8;1]\\
[0{,}9;1] &= 1(\wedge)([0{,}2;0{,}3](+)v_H(L))\\
v_I(L) &= [0{,}7;1]\\
v(L) &= [0{,}8;1](\wedge)[0{,}7;1] = [0{,}7;1]
\end{aligned}
$$

Now we move on to revise the objectives of vertices M, N, O and P which will now be:

- Auto-financing $\geq$5%
- Distribution profits $\geq$6% (constant)
- Financial risk = 0.4
- Financial solvency = 0.7

It would seem that the executives entrusted with establishing the objectives have wanted to ensure that they are complied with to a maximum, moderating the corresponding figures, which has permitted the following valuations:

$$\begin{aligned} v(K \to M)^{(1)} &= 1 \\ v(K \to N)^{(1)} &= [0{,}6; 0{,}9] \\ v(K \to O)^{(1)} &= [0{,}6; 1] \\ v(D \to O)^{(1)} &= [0{,}8; 1] \\ v(L \to P)^{(1)} &= [0{,}6; 0{,}9] \\ v(J \to P)^{(1)} &= [0{,}8; 0{,}9] \end{aligned}$$

With these estimates we find, for vertex M:

$$\begin{aligned} 1 &= 1(\wedge)([0; 0{,}3](+)v(M)) \\ v(M) &= 1 \end{aligned}$$

For vertex N:

$$\begin{aligned} [0{,}6; 0{,}9] &= 1(\wedge)([0; 0{,}3](+)v(N)) \\ v(N) &= [0{,}6; 0{,}6] = 0{,}6 \end{aligned}$$

For vertex O:

$$\begin{aligned} [0{,}6; 1] &= 1(\wedge)([0; 0{,}3](+)v_K(O)) \\ v_K(O) &= [0{,}6; 1] \\ [0{,}8; 1] &= 1(\wedge)([1; 0{,}3](+)v_D(O)) \\ v_D(O) &= [0{,}7; 1] \\ v(O) &= [0{,}6; 1](\wedge)[0{,}7; 1] = [0{,}6; 1] \end{aligned}$$

For vertex P:

$$\begin{aligned} [0{,}6; 0{,}9] &= 1(\wedge)([0; 0{,}3](+)v_L(P)) \\ v_L(P) &= [0{,}6; 0{,}6] = 0{,}6 \\ [0{,}8; 0{,}9] &= 1(\wedge)(0(+)v_J(P)) \\ v_J(P) &= [0{,}8; 0{,}9] \\ v(P) &= [0{,}6; 0{,}6](\wedge)[0{,}8; 0{,}9] = [0{,}6; 0{,}6] = 0{,}6 \end{aligned}$$

Finally we are going to proceed to a revision of the final objective Q. Given the partial objectives that have been set, a reduction in the expectations for the quotation of the shares in order to situate these in a limit of no less than 260%, which would seem, however taken, to be an easily attainable objective if the preceding objectives are met. The inferences than are estimated as follows:

$$v(M \to Q) = [0{,}8; 0{,}9]$$
$$v(N \to Q) = [0{,}7; 0{,}9]$$
$$v(O \to Q) = [0{,}9; 1]$$
$$v(P \to Q) = [0{,}9; 1]$$

In this way:

$$\begin{aligned}
[0{,}8; 0{,}9] &= 1(\wedge)(0(+)v_M(Q)) \\
v_M(Q) &= [0{,}8; 0{,}9] \\
[0{,}7; 0{,}9] &= 1(\wedge)(0{,}4(+)v_N(Q)) \\
v_N(Q) &= [0{,}3; 0{,}5] \\
[0{,}9; 1] &= 1(\wedge)([0; 0{,}4](+)v_O(Q)) \\
v_O(Q) &= [0{,}9; 1] \\
[0{,}9; 1] &= 1(\wedge)(0{,}4(+)v_P(Q)) \\
v_P(Q) &= [0{,}6; 1] \\
v(Q) &= ([0{,}8; 0{,}9](\vee)[0{,}3; 0{,}5](\wedge)[0{,}8; 0{,}9](\wedge)[0{,}6; 1]) \\
&= [0{,}6; 0{,}9](\wedge)[0{,}8; 0{,}9](\wedge)[0{,}6; 1] \\
&= [0{,}6; 0{,}9]
\end{aligned}$$

Thus, we reach the conclusion that the objective that was set, after all the corresponding revisions, means a quotation for the shares of the business at a rate of 260% at a valuation of no less than 0,6 and no more than 0,9.

In the graph shown in Fig. 9.4 we can see, within a circle, the valuations corresponding to each vertex, after the variations that were made. Also

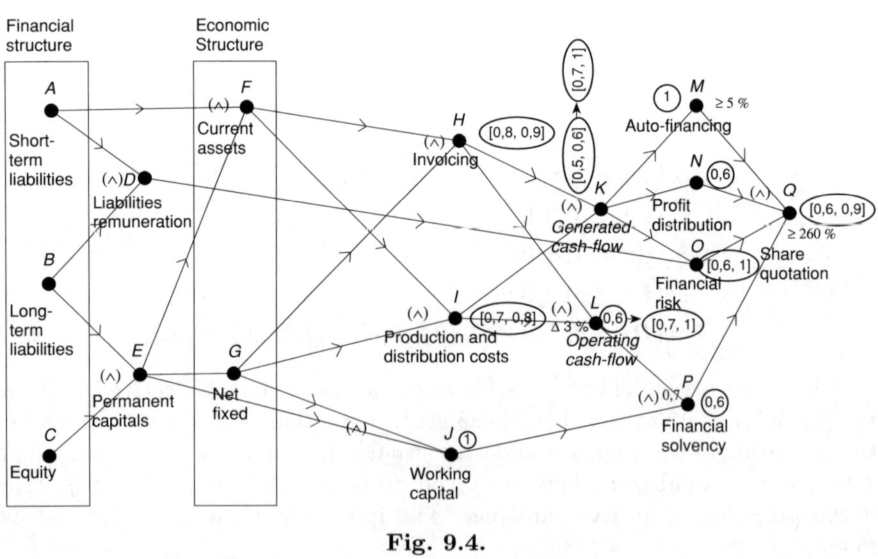

Fig. 9.4.

explicitly shown are the new objectives, only where the initial objectives were modified.

For resolving the problem that arose we opted to the use of the Lukaciewicz inference. Obviously there are no conclusive reasons that recommend, with a general nature, the use of this inference, but each case should mean the study of the circumstances relative to the specific problem in order to establish the most adequate logic for resolving the problem. The process to be followed would be parallel to the one we have described, both if we used the Lee inference or any other, such as the Gödel or Goguen inferences, for example. For this reason we conclude this section here with the hope that the interested reader will find no difficulties for applying this technique to the varied cases which may arise due to the reality of management of institutions.

10 The Effect of Fixed Costs on Profit

10.1 The Breakeven Point as a Threshold for Profitability

From the very moment that a business commences its productive activity, it incurs certain fixed costs as a consequence of acquiring production equipment (considering this concept in the broadest possible sense). Its amount coincides with the loss that would be accounted for if the production process were not to commence and, in consequence, there were no variable costs incurred.

Once the production process has commenced and as the amounts of assets and/or services increase at the same time as sales, provided the sales price were to be higher than the average variable costs, the corresponding income absorbs an ever increasing part of the fixed costs, slowly reducing losses until such time, for a determined production volume and sale, income covers variable and fixed costs. This point, when there is neither profit nor loss is called the "break-even point".

The break-even point appears, therefore, as a situation of economic balance. Successive increases in production will permit the difference between income and variable costs to give rise to a profit, since fixed costs have been totally covered.

The hypothesis established that the sales price be higher than the average variable cost, requires the income curve $I(x)$ to be always over that of variable costs $C_V(x)$, which implies that this is only complied with up to a determined production volume (in the graph shown in Fig. 10.1, $0M$).

From this point on, increases in production not only cease to absorb fixed costs, but losses will be increased by variable costs.

Nevertheless there should be no confusion between production volume and sales M and the break-even point, since in this scheme variable costs are used and the analysis of the break-even point requires the consideration of total costs.

The difference in ordinates at each point of the segment $[0; M]$ represents the excess of income over variable costs and the businessman hopes that at whatever level of production this difference covers fixed costs plus a possible profit.

Anna Maria Gil-Lafuente: *Fuzzy Logic in Financial Analysis*, StudFuzz **175**, 159–178 (2005)
www.springerlink.com

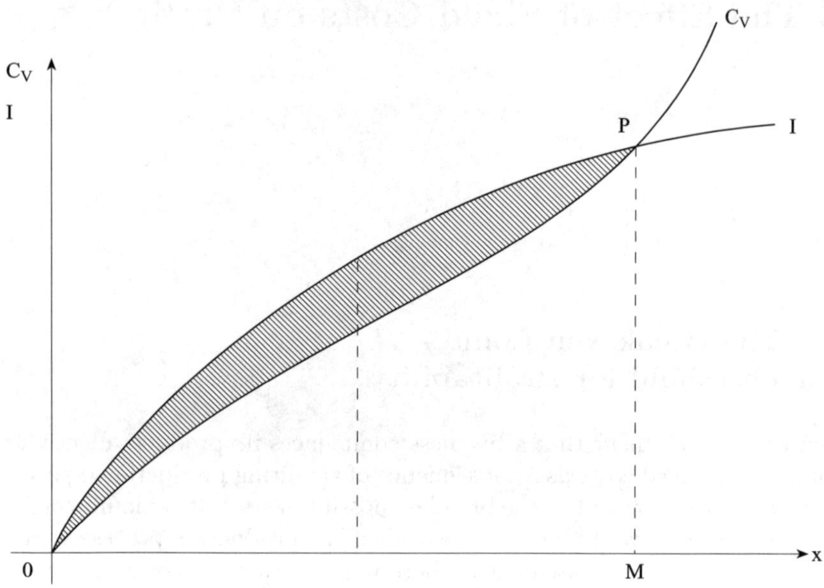

Fig. 10.1.

If we now consider fixed costs and we obtain the total costs $C_T(x)$, it will be seen that there is an area, given by the production segment $[0; P]$ in which the total costs are higher than income, while on the other hand, the area given by segment $[P; N]$ not only are the variable cost covered but also fixed costs, but that as from P profits commence. P is then the threshold of profitability (see Fig. 10.2).

With the object of simplifying wherever possible the analysis of the break-even point certain restrictive hypotheses are adopted. Relative to the function of income, it is assumed that the sales price remains constant, whatever the amount that is launched in the market, and relative to costs it is assumed that average variable costs are also independent of the amount produced. In this case both functions will take on the linear form as:

$$I(x) = x \cdot p_X$$
$$C_V(x) = C_V^* \cdot x$$
$$C_T(x) = C_V^* \cdot x + C_F$$

where:

x = amount produced and sold
p_X = sales price
C_F = fixed costs
C_V = variable costs
C_V^* = average variable cost

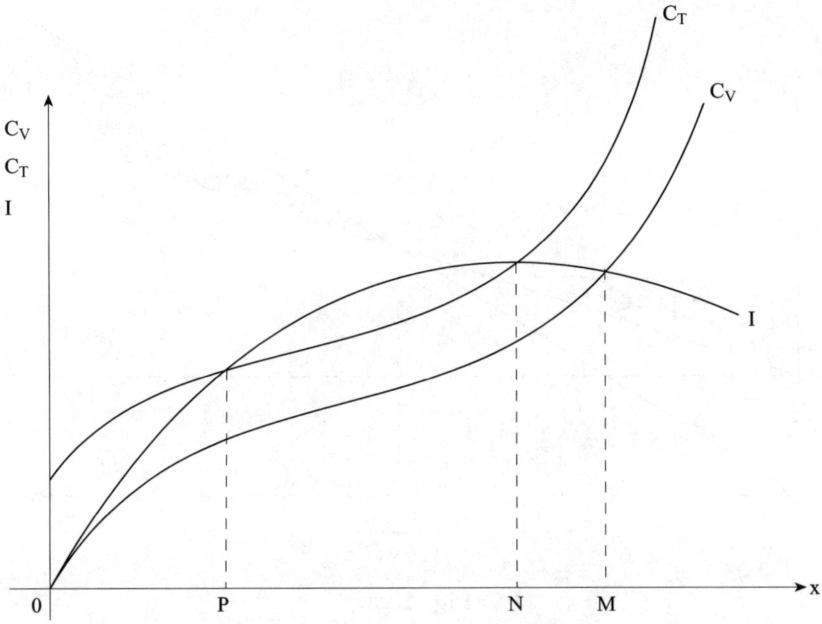

Fig. 10.2.

Given the fact that at the break-even point $I_x = C_T(x)$ will be:

$$P_x \cdot x = C_T(x) = C_V^* \cdot x + C_F \Rightarrow x = \frac{C_F}{P_x - C_V^*}$$

In Fig. 10.3 these functions and the breakeven point are shown.

Analysis of the break-even point is highly significant in the short term, since any long-term analysis would require re-stating the separation between fixed costs and variable costs. If the possibility is considered that fixed costs vary due to an increase in the level of activity, the analysis of the break-even point will give rise to new interpretations. In fact, let us assume[1] that in order to increase production to an amount in excess of 20.000 units it is necessary to increase fixed costs:

- Of 1.500.000 pesetas in order to pass over from 20.000 units to 22.500 units.
- Of 18.500.000 pesetas in order to produce more than 22.500 units. This increase could be justified by the need to construct a second production work-shop in order to reach said production levels.

---

[1] Depallens G. (1970) Gestion financière de l'entreprise (in French). (Ed) Sirey, Paris, 5th Edition., Vol. 1, p. 278.

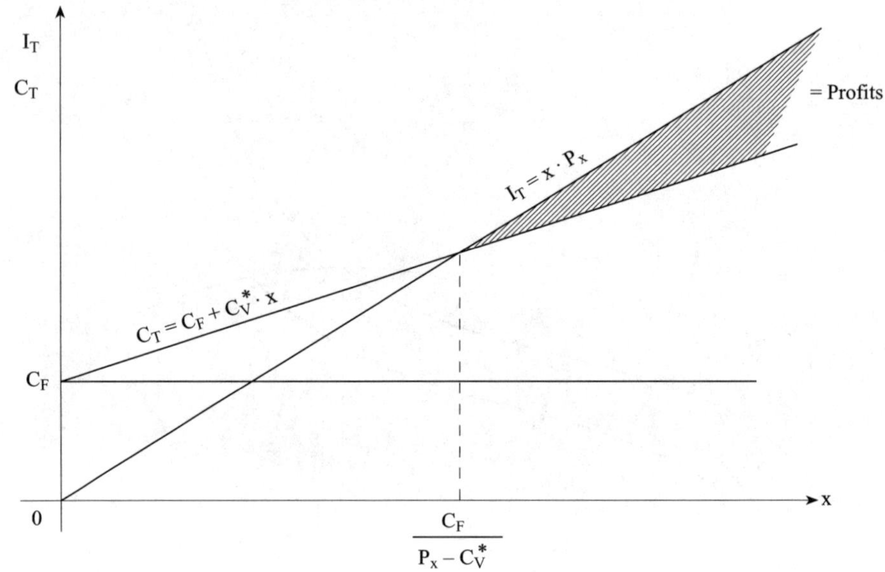

Fig. 10.3.

In the graph in Fig. 10.4 is shown a scheme in which two breakeven points appear, as a consequence of modifying fixed costs (long-term hypothesis).

10.2 The Threshold of Profitability. Point or Interval?

Determination of the break-even point only has sense if reference is made to a future period. In fact, if a financial group intends to commence a productive activity, one of the elements it should take into account is the size of the business, and, in accordance with this, attempt to get to know what production level is required to be reached to cover total costs (fixed and variable). In other words, find the estimated break-even point.

Now, within a context of uncertainty, is the term "break-even point" correct? From this point on we are going to develop a scheme that is suitable for arriving at the threshold of profitability within the sphere of uncertainty. For this we will resort to a very simple example, but particularly useful for didactic effects.

Let us assume that in a project for the creation or expansion of a business certain variables have been estimated in the future by means of confidence intervals. Thus it is hoped that:

- Fixed costs are included in the interval [700.000; 800.000] monetary units

$$\underset{\sim}{\mathbf{C}}_F = [700.000; 800.000]$$

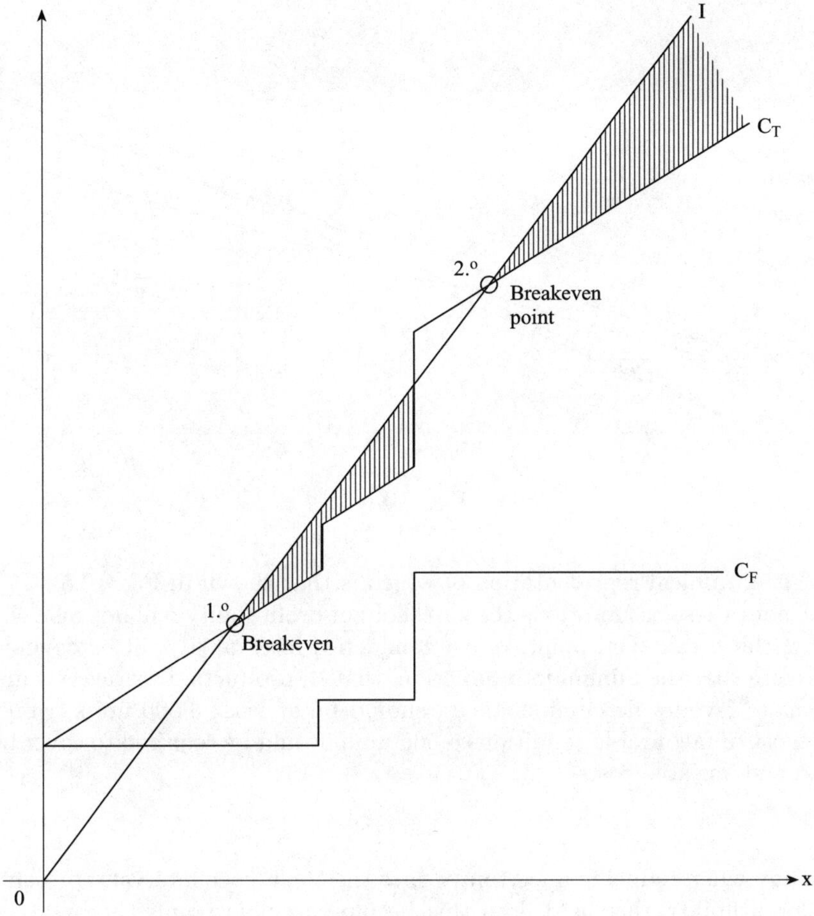

Fig. 10.4.

– Average variable costs will be found in the interval [300; 330] monetary units.

$$\underset{\sim}{\mathbf{C}}^{*}_{\underset{\sim}{\mathbf{V}}} = [300; 330]$$

Sales price will not be less than 450 nor higher than 500.

$$\underset{\sim}{\mathbf{P}}_{x} = [450; 500]$$

These valuations permit arriving at the threshold of profitability. Which is:

$$\text{Threesold of profitability} = \frac{[700.000; 800.000]}{[450; 500] - [300; 330]}$$

$$= \frac{[700.000; 800.000]}{[120; 200]} = [3.500; 6.666]$$

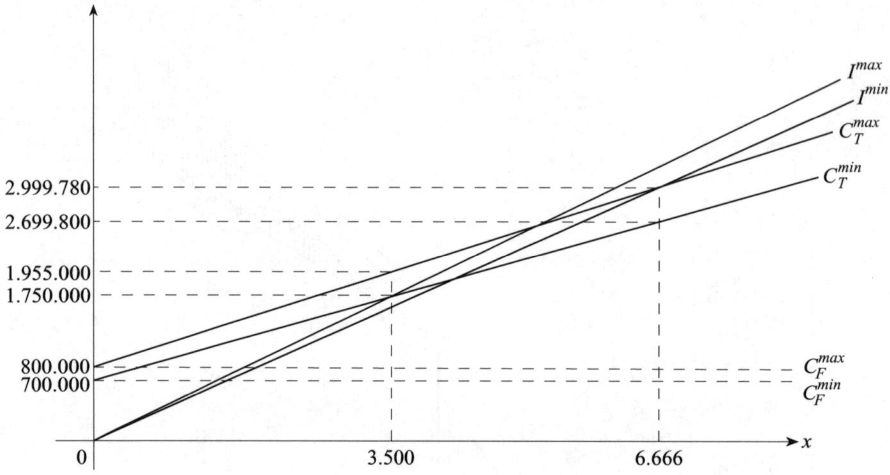

Fig. 10.5.

The graphical representation of which is that shown in Fig. 10.5.

Under these assumptions the threshold of profitability will not be a single point, the break-even point, but a confidence interval. It will be concluded therefore that the minimum production, that is, production carried out under the most favourable circumstances, should be at least 3.500 units, while in the most unfavourable conditions 6.666 units would be required to cover both fixed and variable costs.

Only when reducing uncertainty does the confidence interval representing the profitability threshold does this become a single point, the break-even point. In this example it would be:

$$U_R = \frac{3.500 + 6.666}{2} = 5.083$$

The process that we have followed is the result of the application of the principle that governs confidence intervals, in the sense that any possible situation must be found between the lower and upper limits. Verifying this is simple:

- Income under the worst conditions will continue to be an evolution represented by:

$$I_T^{(\min)} = 450 \cdot x$$

- Costs in the worst of cases will evolve according to function:

$$C_T^{(\max)} = 800.000 + 330 \cdot x$$

Therefore, the profitability threshold in the most unfavourable circumstances will be found in the position that complies with:

$$I_T^{(\min)} = C_T^{(\max)}$$

Therefore:

$$450 \cdot x = 800.000 + 330 \cdot x \sqcap 120.\ x = 800.000$$
$$\square\ x = \frac{800.000}{120} = 6.666$$

– Income in the best of circumstances will follow the evolution expressed by:

$$I_T^{(\max)} = 500 \cdot x$$

– The best cost will be shown by:

$$C_T^{(\min)} = 700.000 + 300 \cdot x$$

Therefore, the profitability threshold in the most optimistic case will be arrived at by resolving the equation:

$$I_T^{(\max)} = C_T^{(\min)}$$

therefore:

$$500 \cdot x = 700.000 + 300 \cdot x \sqcap 200 \cdot x = 700.000$$
$$\square\ x = \frac{700.000}{200} = 3.500$$

Finally, if the lowest extreme is taken as the most optimistic position and the upper as the most pessimistic, we arrive at the following confidence interval:

$$[3.500; 6.666]$$

which was arrived at before.

10.3 The Reduction of the Interval Representing the Threshold of Profitability

The confidence interval which is arrived at as the threshold of profitability can be so wide that it even becomes not particularly representative, as a consequence of its uncertainty. The example we have just shown is an illustration of this fact. It may seem of little use to say that the threshold of profitability will be located between 3.500 and 6.666 units. Obviously the uncertainty has been controlled with the two extremes of the interval, but it is just possible that this is not sufficient.

With the object of making an attempt at providing a solution to this problem we propose the path that we have described below.

Once we have arrived at the confidence interval representing the threshold of profitability, an expert is asked to express an opinion, by means of the hendecagonal system, pointing out if it gets closer to or further removed from the extremes in the following manner:

0	for	3.500
0.1	practically	3.500
0.2	nearly	3.500
0.3	close to	3.500
0.4	closer to	3.500 than to 6.666
0.5	as close	3.500 than to 6.666
0.6	closer to	6.666 than to 3.500
0.7	close to	6.666
0.8	nearly	6.666
0.9	practically	6.666
1	for	6.666

We now admit the following correspondence. If the expert provides an opinion of the imprecise number $[3.500; 6.666]$ by means of $\alpha \in [0; 1]$, this means that the estimate of this imprecise number is:

$$3.500 + (6.666 - 3.500 \cdot \alpha) = 3.500 + 3.166 \cdot \alpha$$

In this way, if the number is estimated as "close to 3.500" we arrive at:

$$3.500 + 3.166 \times 0{,}3 = 4.449{,}8$$

We are now going to extend the process by making a determined number (obviously limited) of experts intervene, who will give their opinions relative to the interval we obtained as the threshold of profitability. In order to simplify let us assume we resort to 10 experts (the case can be generalised to n experts). Once we have obtained the corresponding information we can write a table such as the one shown in Table 10.1

We now immediately transform these estimates by means of an accumulation, noting the number of times that one and the same opinion is repeated (see Table 10.2).

We now immediately obtain the accumulated probabilities, if we can talk about probability with just 10 experts. Nevertheless it should be remembered that the scheme is valid for a large number of experts (this here is only a didactic example). For this we divide each value of a by 10 (see Table 10.3).

Finally for interval $[3.500; 6.666]$ we arrive at Table 10.4

In order to make the entropy fall we look for the mathematically expected value by adding the accumulated probabilities for all the α of the hendecagonal system except for $\alpha = 0$, dividing it all by 10. We then arrive at:

Table 10.1.

	[3.500; 6.666]
1	0,3
2	0,5
3	0,4
4	0,6
5	0,3
6	0,4
7	0,2
8	0,4
9	0,3
10	0,5

Table 10.2.

	[3.500; 6.666]	Number of times
0	10	0
0,1	10	0
0,2	10	1
0,3	9	3
0,4	6	3
0,5	3	2
0,6	1	1
0,7	0	0
0,8	0	0
0,9	0	0
1	0	0

$$\varepsilon = \frac{1 + 1 + 0{,}9 + 0{,}6 + 0{,}3 + 0{,}1}{10} = 0{,}39$$

In this way we will arrive at the fact that the estimated breakeven point will be as follows:

$$U_R = 3.500 + 3.166 \times 0{,}39 = 4.734{,}74$$

We will obviously arrive at the same result if we were to calculate the arithmetical median of the estimated values for each level α (of course, excepting $\alpha = 0$ as we did before). In this way we arrive at:

Table 10.3.

	[3.500; 6.666]
0	1
0,1	1
0,2	1
0,3	0,9
0,4	0,6
0,5	0,3
0,6	0,1
0,7	0
0,8	0
0,9	0
1	0

Table 10.4.

	0	1		0	6.666,0
	0,1	1		0,1	6.666,0
	0,2	1		0,2	6.666,0
	0,3	0,9		0,3	6.349,4
	0,4	0,6		0,4	5.399,6
3.500+3.166 x	0,5	0,3	=	0,5	4.449,8
	0,6	0,1		0,6	3.816,6
	0,7	0		0,7	3.500,0
	0,8	0		0,8	3.500,0
	0,9	0		0,9	3.500,0
	1	0		1	3.500,0

$$U_R = \frac{1}{10}\begin{pmatrix} 6.666 + 6.666 + 6.349{,}4 + 5.399{,}6 + 4.449{,}8 \\ +3.816{,}6 + 3.500 + 3.500 + 3.500 + 3.500 \end{pmatrix} = 4.734{,}74$$

For better comprehension of the process we have followed, resort can be made to de-accumulation in order to arrive at the mathematically expected value.

In the first place in order to proceed with de-accumulation, the difference must be sought between the value taken the function for a determined α and

Table 10.5.

	F(α)-F(α+0,1)
0	1 - 1 = 0
0,1	1 - 1 = 0
0,2	1 - 0,9 = 0 ,1
0,3	0,9 - 0,6 = 0,3
0,4	0,6 - 0,3 = 0,3
0,5	0,3 - 0,1 = 0,2
0,6	0,1 - 0 = 0,1
0,7	0 - 0 = 0
0,8	0 - 0 = 0
0,9	0 - 0 = 0
1	0 - 0 = 0

the value taken by the function for $\alpha + 0{,}1$, respectively, doing this same operation for each one of the α-cuts. In this way we arrive at Table 10.5.

Finally, in order to arrive at the mathematically expected value we proceed to do the addition of each one of the values of α multiplied by its corresponding value found in the de-accumulation. That is, in order to arrive at the mathematically expected value the following will be used:

$$\varepsilon(F(\alpha)) = \sum_{\alpha=0}^{1} \alpha \cdot [F(\alpha) - F(\alpha + 0{,}1)]$$

which in this case will be:

$$\varepsilon(F(\alpha)) = 0{,}2 \times 0{,}1 + 0{,}3 \times 0{,}3 + 0{,}4 \times 0{,}3 + 0{,}5 \times 0{,}2 + 0{,}6 \times 0{,}1 = 0{,}39$$

result which we had arrived at before, and which with the use of the corresponding equation would give rise to the aggregate opinion of the experts of α "breakeven point" corresponding to a production and sale of 4.734.74 units.

The scheme we have proposed is susceptible to generalization. Now then in order to present a different way we are going to take advantage of this generalising process in order to use a new element: the R-expertons[2]. For this we establish, in order not to force the opinion of the experts too much, that they give their opinion on the situation of the threshold of profitability

[2] Kaufmann A (July 1988) R-expertons (expertons dans R) (in French). Note de Travail, La Tronche France, p. 186

Table 10.6.

Expert	[3.500; 6.666]
1	[0,3; 0,5]
2	0,5
3	0,4
4	[0,5; 0,7]
5	[0,3; 0,4]
6	[0,4; 0,6]
7	0,2
8	0,4
9	[0,2; 0,5]
10	0,5

within the interval $[\alpha_*; \alpha^*]$ (in our example $[3.500; 6.666]$ by means of another interval $[\alpha_1; \alpha_2]$ included in the segment $[0; 1]$, that is:

$$[\alpha_1; \alpha_2] \subset [0; 1]$$

Each one of the n experts will provide, therefore, a confidence interval $[\alpha_{i1}; \alpha_{i2}]$, $i = 1; 2; \ldots; n$, which in determined cases can result in being a crisp number if $\alpha_{i1} = \alpha_{i2}$. Starting from these n confidence intervals we arrive at an experton $\underset{\sim}{\mathbf{a}}$ and knowing this experton, we find the R-experton $\underset{\sim}{\mathbf{A}}$ as follows:

$$\underset{\sim}{\mathbf{A}} = a_* + (a^* - a_*) \cdot \underset{\sim}{\mathbf{a}}$$

If we now recuperate the case we were developing and we ask the ten experts to give their opinion relative to the threshold of profitability $[3.500; 6.666]$ their answers, for example, could be those listed in Table 10.6.

We are now going to transform these estimates into expertons, for which we will make the same calculations as we did before, but duplicating the operations, for the lower extreme and for the upper extreme. In this way we arrive at Table 10.7

By means of this experton we will arrive at the R-experton (see Table 10.8).

The mathematically expected value of the experton is obtained by adding the lower extremes for each level (except for level 0) and dividing the result by 10, repeating this operation for the upper levels. In this case the result is:

$$\begin{aligned}\varepsilon(\text{Exp}) &= \frac{[1 + 1 + 0{,}8 + 0{,}6 + 0{,}3; 1 + 1 + 0{,}9 + 0{,}9 + 0{,}6 + 0{,}2 + 0{,}1]}{10} \\ &= [0{,}37; 0{,}47]\end{aligned}$$

Table 10.7.

Number of times	α	[3.500; 6.666]		Number of times
	0	1	1	
	0,1	1	1	
2	0,2	1	1	1
2	0,3	0,8	0,9	
3	0,4	0,6	0,9	3
3	0,5	0,3	0,6	4
	0,6	0	0,2	1
	0,7	0	0,1	1
	0,8	0	0	
	0,9	0	0	
	1	0	0	

Table 10.8.

	α	Experton			α	R-Experton	
	0	1	1		0	6.666,0	6.666,0
	0,1	1	1		0,1	6.666,0	6.666,0
	0,2	1	1		0,2	6.666,0	6.666,0
	0,3	0,8	0,9		0,3	6.032,8	6.032,8
	0,4	0,6	0,9		0,4	5.399,6	5.399,6
3.500+3.166 x	0,5	0,3	0,6	=	0,5	4.449,8	4.449,8
	0,6	0	0,2		0,6	3.500,0	3.500,0
	0,7	0	0,1		0,7	3.500,0	3.500,0
	0,8	0	0		0,8	3.500,0	3.500,0
	0,9	0	0		0,9	3.500,0	3.500,0
	1	0	0		1	3.500,0	3.500,0

which represents saying [closer to 3.500 than to 6.666; as close to 3.500 as to 6.666].

By using the corresponding equation it can be accepted that:

$$3.500 + 3.166 \cdot [0{,}37; 0{,}47] = [4.671{,}42; 4.988{,}02]$$

If what is desired is a single estimate corresponding to the aggregate of the 10 experts we could do:

$$\overline{U_R} = \frac{4.671{,}42; 4.988{,}02}{2} = 4.829{,}72$$

In short, it can be stated that only by reducing the uncertainty can the threshold of profitability become converted, expressed by means of a confidence interval, into the "breakeven point".

It has also been seen that the R-expertons acquire special interest when requiring to convert the expertons into data in R. Obviously the operations we have used for aggregation also allow for multiple variations, each of which may result as useful and suitable in the wide spectrum of situations in which they can be applied.

Also we should remember that as a consequence of the mathematically expected value in a linear operator, the following can be written:

$$\varepsilon(\underset{\sim}{\mathbf{A}}) = A_* + (A^* - A_*) \cdot \varepsilon(\underset{\sim}{\mathbf{a}})$$

Now then, when requiring to arrive at $\varepsilon(\underset{\sim}{\mathbf{a}})$ starting out from the previous equation it is necessary to resort to the Minkowski difference, as from the expression:

$$\varepsilon(\underset{\sim}{\mathbf{a}}) = \frac{\varepsilon(\underset{\sim}{\mathbf{A}})(\bar{m})A_*}{A^* - A_*}$$

These schemes that we have developed for the treatment of the "breakeven point" in the field of uncertainty can be extended with the use of confidence triplets in R. We will leave the application of this concept for the resolution of other problems.

10.4 The "Lever Effect" or Leverage

The "lever effect" or leverage permits the study of the effect of certain fixed costs (capital equipment or financial costs) on certain aspects of the profit (operating profit or profit for the shareholders). Its denomination stems from the fact that fixed costs exercise an effect similar to a lever on profits.

When speaking of "fixed costs" the scope of the study is automatically established, since these only exist in the short term. This is the reason why the analysis of leverage is a short-term study.

If we consider the "fixed nature" of the costs in a wide sense, it can be said that two types of costs exist: those relative to the economic activity of production and those referring to the financial activity. The first give rise to a process of fixed asset depreciation, the second to the repayment of debts. In this way the difference between operating leverage and financial leverage arises.

Operating leverage is defined by means of the elasticity of gross profit (operating profit) relative to the quantities produced and sold[3].

[3] It is common knowledge that the derivative elastic or elasticity is a measurement of the relative variation of a value with regard to the variation of another, in infinitesimal terms.

If the operating profit (gross profit) is designated B, and x is the variable representing the amount produced and sold, we can write:

$$\varepsilon_0 = \lim_{x \to 0} \frac{\frac{\Delta B}{B}}{\frac{\Delta x}{x}} = \frac{dB}{dx} \cdot \frac{x}{B}$$

But as:

$$B = x \cdot (P_x - C_V^*) - C_F$$

the result will be, in short, that the elasticity of the profit relative to production and sales, will be:

$$\varepsilon_0 = (P_x - C_V^*) \frac{x}{x \cdot (P_x - C_V^*) - C_F} = \frac{x \cdot (P_x - C_V^*)}{x \cdot (P_x - C_V^*) - C_F}$$

When it is considered that the only independent variable is x it will be seen that operating leverage ε_0 takes the form of a hyperbola the asymptotes of which are $\varepsilon_0 = 1$ and $x = C_F/(P_x - C_V^*)$, that is to say when the elasticity is equal to the unit and when the amount that constitutes the breakeven point is produced and sold. The graphical representation of this can be seen in Fig. 10.6.

Therefore if a relation is established between the amount produced and sold and the operating leverage, the remaining values remaining invariable. It will be seen that, in the area of profits, the operating leverage becomes so much higher the more production is reduced, becoming $\varepsilon_0 = \infty$ when $x = C_F/(P_x - C_V^*)$, that is to say at the breakeven point. In the opposite sense, successive increases in production will decrease the leverage effect, as a consequence of the increasing dilution of fixed costs in the amounts produced and sold, therefore, when $x = \infty$, $\varepsilon_0 \to 1$. In this way it can be seen how important fixed costs are in reduced production volumes and the decreasing incidence as production increases.

Having reached this point, the possibility can be approached for the introduction of the concept of leverage in a context of uncertainty. For this we are going to continue with the previous didactic example.

We should recall that certain fixed costs $\underset{\sim}{\mathbf{C}}_F = [700.000; 800.000]$, sales price $\underset{\sim}{\mathbf{P}}_* = [450; 500]$ and an average variable cost of $\underset{\sim}{\mathbf{C}}_V^* = [300; 330]$ had been considered and that the threshold of profitability that was arrived at was $\underset{\sim}{\mathbf{U}}_R = [3.500; 6.666]$.

These estimates permit arriving at the difference between the sales price and the variable cost per unit:

$$\underset{\sim}{\mathbf{B}}_V^* = [450; 500] - [300; 330] = [120; 200]$$

as well as the estimated profit for the lower extreme and the upper extreme of the profitability threshold. Thus:

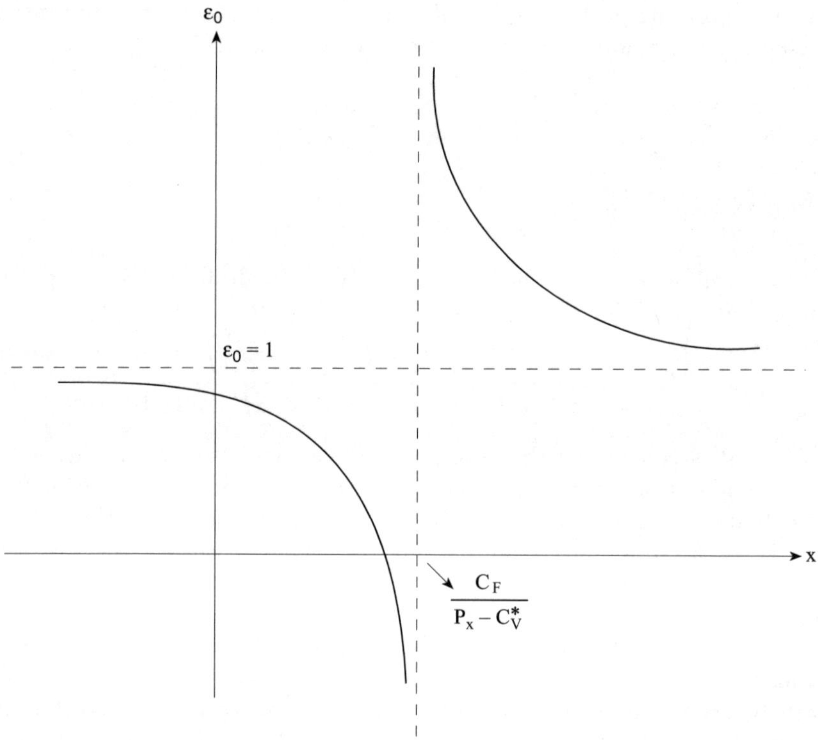

Fig. 10.6.

– For $x = 3{,}500$ we will have:

$$\begin{aligned}(P_x - C_V^*) \cdot x &= [120; 200](\cdot)3.500 + [420.000; 700.000] \\ (P_x - C_V^*) \cdot x - C_F &= [420.000; 700.000](-)[700.000; 800.000] \\ &= [-380.000; 0]\end{aligned}$$

where it will be seen that only under the very best of conditions will there be no losses (nor, obviously, profits).

– For $x = 6{,}666$ we will have:

$$\begin{aligned}(P_x - C_V^*) \cdot x &= [120; 200](\cdot)6.666 \cong [800.000; 1.333.200] \\ (P_x - C_V^*) \cdot x - C_F &= [800.000; 1.333.200](-)[700.000; 800.000] \\ &= [0; 633.200]\end{aligned}$$

Likewise it will be seen that only under the very worst of conditions will no profits be made.

Now let us consider the reason for operating leverage for each one of these situations. For a production of $x = 3.500$ there will be two positions:

1. In the most pessimistic, when there will be losses (less profits) equal to:

$$(P_x - C_V^*) \cdot x = 120 \times 3.500 - 800.000 = -380.000$$

leverage will be:

$$\varepsilon_0^{(1)} = \frac{(450 - 330) \cdot x}{x \cdot (450 - 330) - 800.000} = \frac{420.000}{-380.000} = -1{,}105$$

which corresponds to point A in Fig. 10.7.

2. In the most optimistic, when no losses (nor profits) are incurred:

$$(P_x - C_V^*) \cdot x = 200 \times 3.500 - 700.000 = 0$$

With a leverage of:

$$\varepsilon_0^{(2)} = \frac{(500 - 300) \cdot x}{x \cdot (500 - 300) - 700.000} = \frac{700.000}{0} = \infty$$

which corresponds to the asymptote of point 3.500 corresponding to Fig. 10.7.

If the leverage is studied for a production of $x = 6{,}666$ there will also be two positions:

1. The most pessimistic, when there are neither losses nor profits, given the fact that:

$$(P_x - C_V^*) \cdot x = 120 \times 6.666 - 800.000 = 0$$

leverage will be:

$$\varepsilon_0^{(3)} = \frac{(450 - 330) \cdot x}{x \cdot (450 - 330) - 800.000} = \frac{800.000}{0} = \infty$$

which corresponds to the asymptote of point 6,666 of Fig. 10.7.

2. The most optimistic, when there are profits equal to:

$$(P_x - C_V^*) \cdot x = 200 \times 6.666 - 800.000 = 633.200$$

leverage will be:

$$\varepsilon_0^{(4)} = \frac{(500 - 300) \cdot x}{x \cdot (500 - 300) - 700.000} = \frac{1.333.200}{633.200} = 2{,}105$$

corresponding to point B of Fig. 10.7.

The study of leverage has been done with a confidence interval corresponding to the threshold of profitability in uncertainty, the extremes of which correspond to the intersection of the straight line of higher income

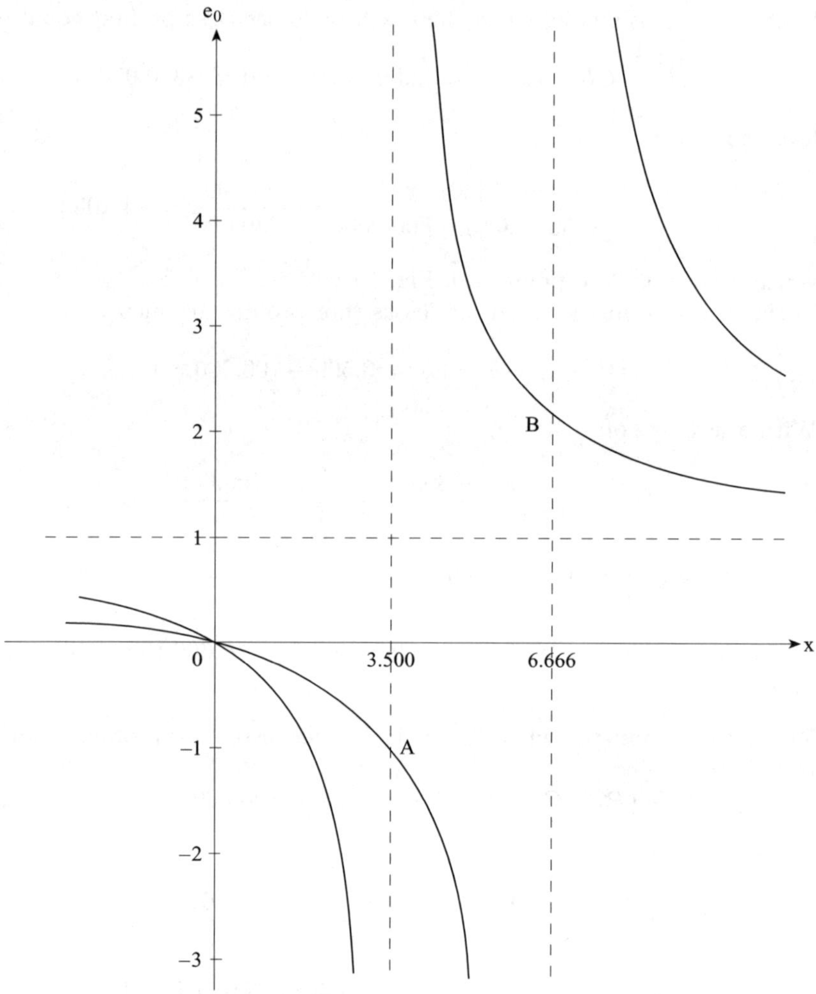

Fig. 10.7.

with the straight line of lower cost, on the one hand, and the straight line of less income and higher cost, on the other as can be seen from Fig. 10.8.

As we pointed out at the beginning of this section, the notion of leverage can be extended to the financial aspect. Financial leverage can be defined as an instrument that permits bringing to light the relation existing between net profit and operating profit, as a consequence of the existence of financial "fixed costs".

In other words, financial leverage brings to light the repercussion of fixed financial costs, that is, the cost of the medium and long-term debts and the profit or profitability available for the shareholders.

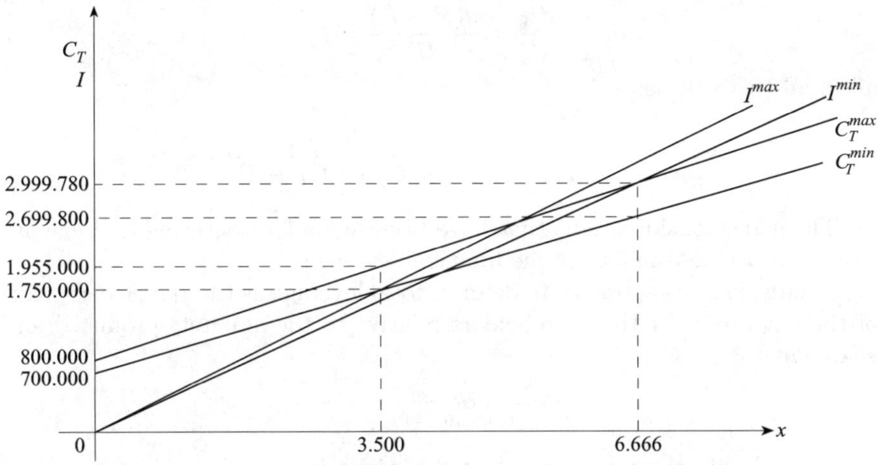

Fig. 10.8.

Therefore, financial leverage has as its objective the highlighting of how profit that is available for the shareholders is modified when waiving the financing of projects with equity by using finance based on medium and long-term indebtedness.

Standing out, in this analysis is the fact that outside resources have a cost, which is represented by the interest the business has to pay in order to obtain these financial resources.

When production yield is higher than the costs of the debt, the business can (with that difference between yield and cost) not only pay the interest, but also obtain additional profits with which the profits available for the shareholders will be increased. It is this increase which in short is measured by the financial leverage.

Financial leverage is also expressed by means of an elastic derivative. In this case the relative variation of profit available for the shareholders is due to a relative variation in the profit prior to deducting financial charges:

$$\varepsilon_f = \lim_{\Delta B \to 0} \frac{\frac{\Delta R}{R}}{\frac{\Delta B}{B}}$$

where R is the profit available for the shareholders and F the interest on the debts. This then will be:

$$R = B - F$$

From here that:

$$\varepsilon_f = \lim_{\Delta B \to 0} \frac{\frac{\Delta R}{R}}{\frac{\Delta B}{B}} = \frac{dR}{dB} \cdot \frac{B}{R}$$

But as:

$$\frac{dR}{dB} = \frac{d(B-F)}{dB} = 1$$

all in all this will be:

$$\varepsilon_f = \frac{B}{R} = \frac{x \cdot (P_x - C_V^*) - C_F}{x \cdot (P_x - C_V^*) - C_F - F}$$

The same considerations that have been made for operating leverage are valid, mutatis mutandis, for the financial leverage.

Finally, it is also possible to define a total leverage as the elastic derivative of the net profit for the shareholders relative to the amounts produced and sold. Thus:

$$\varepsilon_t = \lim_{\Delta x \to 0} \frac{\frac{\Delta R}{R}}{\frac{\Delta x}{x}} = \frac{\frac{dR}{R}}{\frac{dx}{x}}$$

If we divide the numerator and the denominator of this expression by dB/B we will arrive at:

$$\varepsilon_t = \frac{\frac{dR/R}{dB/B}}{\frac{dx/x}{dB/B}} = \frac{\varepsilon_f}{1/\varepsilon_0} = \varepsilon_f \cdot \varepsilon_0$$

So that:

$$\varepsilon_t = \frac{x \cdot (P_x - C_V^*) - C_F}{x \cdot (P_x - C_V^*) - C_F - F} \cdot \frac{x \cdot (P_x - C_V^*)}{x \cdot (P_x - C_V^*) - C_F} = \frac{x \cdot (P_x - C_V^*)}{x \cdot (P_x - C_V^*) - C_F - F}$$

In this way we have arrived at a new management instrument the use of which in the sphere of uncertainty does not entail any additional problems from those shown for operating leverage.

11 Capital Cost

11.1 Concept and Characteristics of Capital Cost

It has been repeatedly pointed out that the financial structure of a business is the basis on which the resources for covering payments arising out of exploitation or investment in fixed assets stands. The degree of demand and the cost of the sources of financing constitute the basic elements for good financial management.

These financing sources may have different origins:

- Internal financing or auto-financing. This includes the financial resources that the business generates itself. It is customary to include under this heading: reserves, depreciation, provisions and estimates, although by their very nature and destination it is not possible to give them the same meaning. Thus, while reserves mean an increase in patrimony, the object of depreciation is to maintain production potential unalterable.
- External financing. This includes those resources that the business incorporates from outside as a consequence of the fact that auto-financing is not always sufficient to cover requirements. For this debentures are issued, or resort is made to credit at the short, medium or long-term, in many different guises and with numerous financial products.

The existence of sources of financing also implies a different cost of usage for each one of these. Capital Cost is the rate that measures the charge represented by the use of financial resources by a business.

According to Mao, capital cost can be defined as "the price paid by a business for funds obtained from its capital suppliers".[1]

In a first approximation, a result will be obtained by considering the cost of each one of the different sources, in order later to continue by calculating the weighted average cost of all the financial operations to which the business may resort.

Whichever of the credit instruments were to be considered (debenture, policy, bill discounting, etc.) these have specific conditions attached that define one or several entries of financial means and certain outlays that include

[1] Mao, James G T (1974) Financial Analysis. (Ed) El Ateneo, Buenos Aires, p. 325.

Anna Maria Gil-Lafuente: *Fuzzy Logic in Financial Analysis*, StudFuzz **175**, 179–201 (2005)
www.springerlink.com

payment of interest and return of principal. The expenses derived from the payment of interest, brokerage, etc., are fiscally deductible, so that the tax effect plays an occasionally important role in the financial decisions for seeking funds.

The fact that a "financial instrument" means inflows and outflows of payment means makes it obligatory to bring to light the difference between a finance operation and an investment operation, also characterised by payments and collections. For this the following criterion could be established: when payments "overall" are prior to collections, also overall, we are faced with an investment operation. In this case, the "central moment of payments" will be less than the "central moment of collections".[2] On the other hand, when collections (considered overall) are prior to payments (also considered overall) we are faced with a finance operation. Jean proposes "the convention that a project is an investment if the first cash run is a payment and financing if the first cash run is an inflow into the cash position of a business".[3]

Being aware of the effective cost of each one of the options has a double sense for the business:

- It allows the selection of resources of the least cost, in this way contributing to improve profits, which is compatible with reaching its general objectives (maximization of the value of the shares, growth, increase market share, etc.).
- Constitutes an essential element for the decisions to invest, since it must always be complied with provided the average cost of capital is not higher than the profitability of the investments for which these do not generate losses. This is then a minimum limit of profitability, for which it is also known as "required rate of return".

Philippatos[4] established a distinction between explicit cost of capital, as a discount rate that equals the current values of inflows and outflows associated with the opportunity for financing, and implicit cost or of opportunity, as the profitability rate associated to the best alternative of that investment which is rejected if a determined project is accepted.

The studies that are normally carried out relative to the concept of capital cost introduce, implicitly or explicitly, three restrictive hypotheses that are pointed out by Suárez[5]:

The financial structure of a business remains constant, that is to say, the participation relative to each source in the overall financing does not vary.

---

[2] In the following pages the "central moment of payments and collections" will be defined.

[3] Jean, William H (1976) Teoría analítica de la financiación, (in Spanish). (Ed) Anaya, Barcelona. p. 161.

[4] Philippatos G C (1979) Fundamentos de administración financiera. Texto y casos. (Ed) McGraw-Hill, México. pp. 182–183.

[5] Suárez Suárez Andrés S (1983) Decisiones óptimas de inversión y financiación en la empresa (in Spanish). (Ed) Pirámide, Madrid. 5th Edition, p. 498.

- The economic risk remains unaltered, that is, the project or combination of investment projects that are carried out does not modify the risk of the business.
- The dividend policy also remains constant, since a change in the dividend policy would affect the financial structure as it would have an incidence on reserves and, as a consequence of this, on the financial risk of the business.

The elimination of one or several of these hypotheses permits the expansion of the sphere of use of these schemes as this will permit the establishment of relations between the variation of the financial structure on capital cost, for example.

In a first approximation, it can be stated that the cost of capital will be the result of the weighting carried out between the different costs relative to each source of financing.

The rate that represents the cost of a source of financing will be the interest rate that equals the net current of financial means received (after deducting expenses) with the payment means delivered as repayment of the principal, interest or dividends, and expenses.

We shall now move on to study the cost of debts.

11.2 The Cost of Outside Means of Financing

The cost of a debt, i_d, if "corporation tax" is not taken into consideration, will be shown by the following expression:

$$I_0 = G_1(1+i_d)^{-1} + G_2(1+i_d)^{-2} + \cdots + G_n(1+i_d)^{-n} = \sum_{j=1}^{n} G_j(1+i_d)^{-j}$$

where:

$I_0 =$ monetary means received by the business at the present time (cause of the debt);

$G_j =$ monetary means delivered at moment j for covering payments arising from the debt

$n =$ duration of the finance operation (in years).

The calculation of the cost rate of a financial means, i_d, brings up certain difficulties arising from the resolution of the previous equation and the need to clear i_d. For this resort is made to an approximating calculation starting out from three possibilities:

1. A serial development of the expression $(1+i_d)^{-j}$ as follows:

$$(1+i_d)^{-j} = 1 - j \cdot i_d + \frac{j \cdot (j-1)}{2!} \cdot i_d^2 - \frac{j \cdot (j-1) \cdot (j-2)}{3!} \cdot i_d^3 + \cdots$$

Normally the third term and following are not considered. On substituting this equation with the previous one we arrive at:

$$I_0 = G_1(1 - i_d) + G_2(1 - 2 \cdot i_d) + \cdots + G_n(1 - n \cdot i_d) = \sum_{j=1}^{n} G_j - \sum_{j=1}^{n} j \cdot G_j$$

Finally arriving at:

$$i_d = \frac{\sum_{j=1}^{n} G_j - I_0}{\sum_{j=1}^{n} j \cdot G_j}$$

This expression only provides an approximate value of the cost of the debt. This approximation will be all the more correct when the value of i_d is lower, since in this way the value of the depreciated terms will also be lower. It is simple to verify that the method used means the substitution of a compound discount for a simple discount, so that any error made can be quite significant.

2. Introduction of the "central moment" for repayments (t_G). To be understood as such the deferment of a single financial capital to the overall hoped for payments. This can be represented graphically as seen in Fig. 11.1.

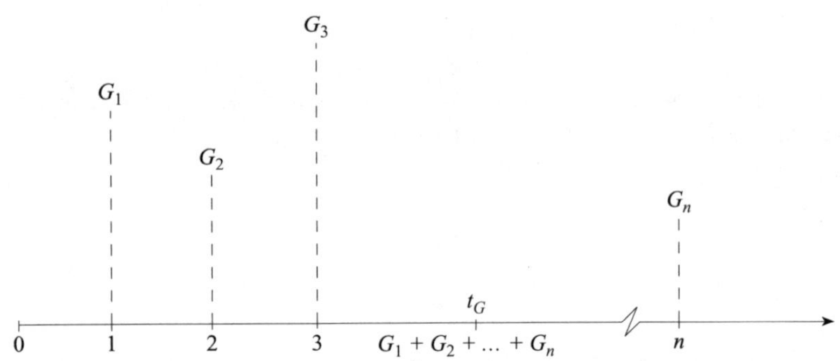

Fig. 11.1.

It can be accepted that equivalence exists between payments of each of the moments $1, 2, \ldots, n$ and a single payment at moment t_G so that in such an event central moment t_G will comply with the following condition:

$$\sum_{j=1}^{n} G_j \cdot (1 + i_d)^{-j} = \frac{\sum_{j=1}^{n} G_j}{(1 + i_d)^{t_G}} = I_0$$

that is, it would be quite indifferent to pay the sum of all expected payments at a single moment t_G as to pay each G_j at its moment j.

The criterion of equivalency that we use consists then, in equalling the current value of the expected current of payments with the current value

of a single payment at moment t_G and both discounted at a rate of i_d which will correspond to the price of the monetary means received from the source of financing.
Knowing t_G, we can proceed to clear i_d from the previous equation:

$$(1+i_d)^{t_G} = \frac{\sum_{j=1}^{n} G_j}{I_0} \; \square \; i_d = \sqrt[t_G]{\frac{\sum j = 1^n G_j}{I_0}} - 1$$

For the calculation of t_G resort can also be made to the known approximation:

$$(1+i_d)^{-j} \cong 1 - j \cdot i_d$$

which by substituting it, we arrive at:

$$(1 - t_g \cdot i_d) \sum_{j=1}^{n} G_j = G_1(1-i_d) + G_2(1 - 2 \cdot i_d) + \cdots + G_n(1 - n \cdot i_d)$$

$$\sum_{j=1}^{n} G_j - t_G \cdot i_d \cdot \sum_{j=1}^{n} G_j = \sum_{j=1}^{n} G_j - i_d \cdot \sum_{j=1}^{n} j \cdot G_j$$

Finally arriving at the expression:

$$t_G = \frac{\sum_{j=1}^{n} j \cdot G_j}{\sum_{j=1}^{n} G_j}$$

In the majority of cases that arise it will be seen that this second method allows for a greater approximation than with the method used previously.

3. "Trial and error" procedure. This consists in observing how the different values of i_d perform by substituting them in the equation and observing the error that is committed, until the value of i_d is reached that satisfies the corresponding equation. Let us take a look at a didactic example:

$$3.500 = \frac{1.000}{(1+i_d)} + \frac{2.000}{(1+i_d)^2} + \frac{3.000}{(1+i_d)^3}$$

If we do $i_d = 0{,}26$ in the second member of the equation, we arrive at 3.552,78 and when $i_d = 0{,}27$, we find 3.491,97. Consequently, it must be assumed that the true value of Id appears somewhere between 26% and 27%. We could continue to try values for Id in order to arrive at a shorter distance, or simply to do a linear interpolation 6 of the form shown in Fig. 11.2.

By similarity between the triangles the following is complied with:

$$\left|\frac{3.552{,}78 - 3.500}{x}\right| = \left|\frac{3.500 - 3.491{,}97}{1-x}\right| \; \square \; \frac{52{,}78}{x} = \frac{8{,}03}{1-x}$$

$$\square \; 52{,}78 = 60{,}81 \cdot x$$

$$\square \; x = 0{,}86\% = 0{,}0086$$

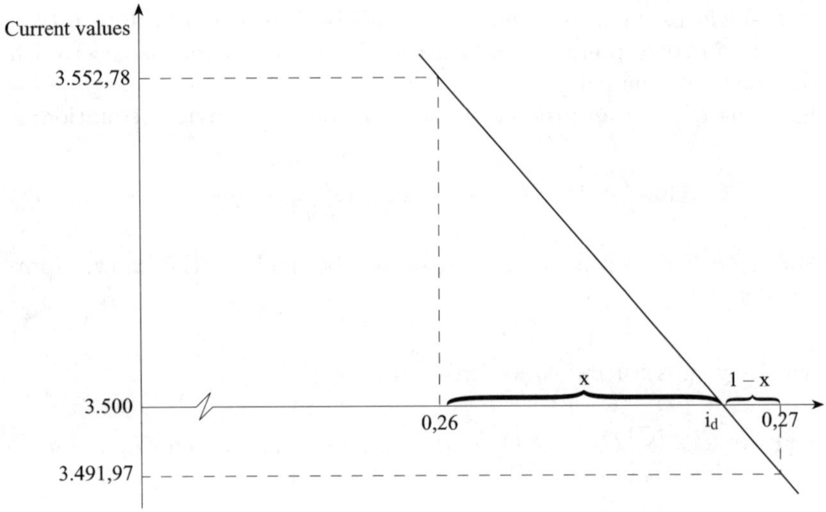

Fig. 11.2.

Finally obtaining:

$$i_d = 0{,}26 + x = 0{,}26 + 0{,}0086 = 0{,}2686 = 26{,}86\%$$

Scholars on the subject are perfectly aware of the possibility that several i_d may exist for one and the same financial operation, as well as that i_d may not exist. We leave it up to the reader interested in this subject to consult the very good texts that exist on the subject[6].

Up to this point we have started out from the case that the business is not submitted to any fiscal pressure. Nevertheless, those studies that intend to draw closer to the actual financial systems must incorporate taxes as a normal phenomenon in current business. To these effects it is normal to consider "Corporate Income Tax", given that the cost of outside capitals are admitted as tax relief for the effect of arriving at the fiscal profit.

If it assumed that the unitary tax rate is f, capital cost will, in reality, be:

$$i_d^{(f)} = i_d \cdot (1 - f)$$

Obviously the capital cost $i_d^{(f)}$ is only valid in the event that the business makes a profit, on the contrary i_d should be used.

We now move on to consider that business is subject to a system governed by uncertainty.

[6] Among others we mention: Mao, James C.T., Análisis financiero, Buenos Aires, El Ateneo, 1977, Chap. 7; Suárez Suárez, Andrés S., Decisiones òptimas de la inversión y financiación de la empresa, 5th Edition, Madrid, Pirámide, 1983, Chaps. 5 and 7; Gil Aluja, J., Apuntes de inversiones, Barcelona, Facultad de Ciencias Económicas y Empresariales, 1989, Chap. 5.

11.3 Outside Capital Cost in Uncertainty

A business goes to a financing institution in order to obtain funds, by means of a loan agreement with a variable rate of interest, for example. This case is fully valid at a time in which the evolution of the price of money in the future is unknown. This type of agreement means the incorporation of uncertainty with regard to the amount of outflow of funds for making the payments arising from the debt. It will be necessary then to use operating techniques for the treatment of uncertainty.

With the object of illustrating a situation just as this, we are going to resort to an example. A business contracts a loan at this time for 12.000 monetary units with a pay back period of 4 years. The amount to be returned at the end of each of these years will depend on the interest rate, commissions and other expenses current at each moment, therefore it is not possible to foresee them with certainty. A financial expert estimates that:

- For year 1, on the assumption that the interest rate will be very low, the amount to be returned will be no less than 3.500 m.u., and if the interest is very high, then the amount for return will be no higher than 3.800 m.u.
- For year 2, the amount to be returned will be no less than 4.400 and no more than 4.800
- For year 3, the amount to be returned is estimated at no less than 4.500 m.u. and under the very worst conditions, 4.800 m.u.
- For year 4, the least that will be paid is estimated at 5.000 m.u. and as a maximum 5.400.

It is assumed that the business obtains sufficient profit and that corporate tax has a single rate of 36%.

For the calculation of the cost of the debt it will be necessary to start out from the following expression, given in thousands of units in which the consequence of taxation has no yet been considered:

$$12 = \begin{pmatrix} [3{,}5; 3{,}8] \cdot \dfrac{1}{1+[i_{d_*}; i_{d^*}]} + [4{,}4; 4{,}8] \cdot \dfrac{1}{(1+[i_{d_*}; i_{d^*}])^2} \\ +[4{,}5; 4{,}8] \cdot \dfrac{1}{(1+[i_{d_*}; i_{d^*}])^3} + [5{,}0; 5{,}4] \cdot \dfrac{1}{(1+[i_{d_*}; i_{d^*}])^4} \end{pmatrix}$$

It is quite obvious that interdependence exists between the lower extreme of the interval corresponding to the cost of the debt i_{d_*} and the lower extreme of the interval for payments for each year, 3.500,4.400, 4.500 and 5.000: as well as between the upper extreme of the interval relative to the cost of the debt i_{d^*} and the upper extreme of the interval for payments, 3.800, 4.800, 4.800 and 5.400. Therefore in order to obtain the lower and upper payments between which the cost of the debts will be found it will be necessary to resolve the following equations.

For the lower extreme:

$$12 = \frac{3{,}5}{1+i_{d_*}} + \frac{4{,}4}{(1+i_{d_*})^2} + \frac{4{,}5}{(1+i_{d_*})^3} + \frac{5{,}0}{(1+i_{d_*})^4}$$

For the upper extreme:

$$12 = \frac{3{,}8}{1+i_{d^*}} + \frac{4{,}8}{(1+i_{d^*})^2} + \frac{4{,}8}{(1+i_{d^*})^3} + \frac{5{,}4}{(1+i_{d^*})^4}$$

Now let us see what results are arrived at when using each of the procedures for arriving at $[i_{d_*}; i_{d^*}]$:

1. With the example of $i_d = \frac{\sum_{j=1}^{n} G_j - I_0}{\sum_{j=1}^{n} j \cdot G_j}$

$$i_{d_*} = \frac{17{,}4 - 12}{45{,}8} = 0{,}1179 \quad \text{that is } 11.79\%$$

$$i_{d^*} = \frac{18{,}8 - 12}{49{,}4} = 0{,}1376 \quad \text{that is } 13.76\%$$

We arrive at the fact that $[i_{d_*}; i_{d^*}] = [0{,}1179; 0{,}1376]$

With the object of checking this we are going to substitute the results arrived at in the second members of the equations relative to the lower and upper extremes. For the lower extreme we arrive at 13,0743 and for the upper 14,0440, very far removed from the value 12 which should be arrived at, so they become difficult to admit.

2. With the use of the central moment:

$$t_G = \frac{\sum_{j=1}^{n} j \cdot G_j}{\sum_{j=1}^{n} G_j}$$

and the expression:

$$i_d = \sqrt[t_G]{\frac{\sum_{j=1}^{n} G_j}{I_0}} - 1$$

we arrive at:

$$t_{G_*} = \frac{45{,}8}{17{,}4} = 2{,}6321 \quad i_{d_*} = \sqrt[2{,}6321]{\frac{17{,}4}{12}} - 1 = 0{,}1516$$

$$t_{G^*} = \frac{49{,}4}{18{,}8} = 2{,}6276 \quad i_{d^*} = \sqrt[2{,}6276]{\frac{18{,}8}{12}} - 1 = 0{,}1863$$

With this procedure then, we arrive at the $[i_{d_*}; i_{d^*}] = [0{,}1516; 0{,}1863]$. On carrying out the corresponding verification we arrive at 12,146 for the lower extreme and at 12,2157 for the upper. Either one of these figures is sufficiently close to 12 for us to admit as valid the result of $[i_{d_*}; i_{d^*}] = [0{,}1516; 0{,}1863]$. But if greater precision is required not quite so necessary in the sphere of uncertainty as is required for the sphere of certainty, the trial and error procedure can be used.

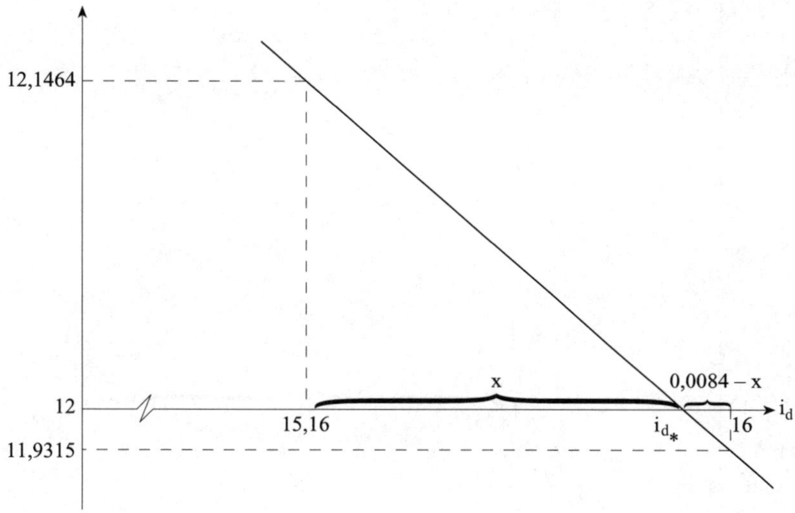

Fig. 11.3.

3. We use the figures that have been arrived at 0,1516 and 0, 1863 and as the results have been higher than the required Fig. 11.12, a reasonable amount is added, for example, 0,0084 and 0,0097. Thus for the lower extreme:
   – On doing $i_{d_*} = 0{,}1516$ an updated value is obtained of 12,1464
   – On doing $i_{d_*} = 0{,}1600$ an updated value is obtained of 11,9315
   The solution by this procedure, by means of a linear interpolation, can be represented graphically as in Fig. 11.3.

   By means of the similarity of the triangles we arrive at:

$$\frac{x}{0{,}1464} = \frac{0{,}0084 - x}{0{,}0685}$$

by which $x = 0{,}00572$ and, therefore:

$$i_{d_*} = 0{,}1573$$

Which gives rise to an updated value of 11,99.
For the upper extreme:

– On doing $i_{d^*} = 0{,}1863$ an updated value is obtained of 12,2157
– On doing $i_{d^*} = 0{,}1960$ an updated value is obtained of 11,9778

The solution is graphically represented as in Fig. 11.4.
i_{d^*} is arrived at as follows:

$$\frac{x}{0{,}2157} = \frac{0{,}0097 - x}{0{,}0222}$$

by which $x = 0{,}0088$ and, therefore:

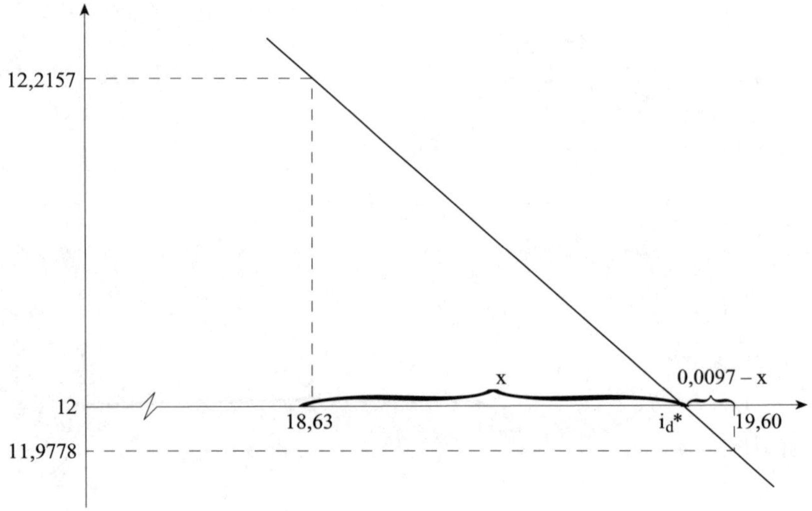

Fig. 11.4.

$$i_{d^*} = 0{,}1911$$

which gives rise to an updated value of 12,097.

In short, the cost of the debt will be estimated by means of the interval $[i_{d_*}; i_{d^*}] = [15{,}79; 19{,}48]$ expressed by means of percentages.

If as is logical it is hoped that the business will obtain sufficient profits, it is necessary to consider that it will not have to pay corporate tax relative to the interest paid for the debt, since these are deductible for tax effects. Therefore, if the tax rate is, for example, 36% we finally arrive at:

$$\left[i_{d_*}^{(f)}; i_{d^*}^{(f)}\right] = (1-f)(\cdot)[15{,}73; 19{,}51] = 0{,}64(\cdot)[15{,}73; 19{,}51] = [10{,}06; 12,48]$$

which brings to light that under the very worst conditions the cost of this outside capital would be 12,46% and under the most favourable conditions 10,10% would be obtained.

We have arrived then at the opinion of a financial expert by means of a confidence interval which, therefore, contains a certain amount of uncertainty. Also, the opinion of an expert, however well founded it may be, is highly charged with subjectivity. These two reasons may make it suitable to resort to other experts so that, by means of what we could call counter-expertise, we could reduce the subjectivity and the uncertainty.

The path to be followed could be as follows. A determined number of experts (in this case, in order to simplify six of them have been consulted) are requested to answer a hendecagonal[7] questionnaire:

[7] It is the same proposal as made in the preceding chapter, although for this subject we are going to modify the form of the replies.

Table 11.1.

Experts	Triplets
1	(0,3; 0,5; 0,6)
2	(0,7; 0,9; 0,9)
3	(0,4; 0,5; 0,7)
4	1
5	(0; 0,4; 0,5)
6	(0,6; 0,6; 0,7)

0	the best hypothesis is 10,10%
0.1	practically 10,10%
0.2	nearly 10,10%
0.3	close to 10,10%
0.4	closer to 10,10% that to 12,46%
0.5	as close to 10,10% as to 12,46%
0.6	closer to 12,46% than to 10,10%
0.7	close to 12,46%
0.8	nearly 12,46%
0.9	practically 12,46%
1	the best hypothesis is 12,46%

In order to give then greater freedom of expression and at the same time gather more information the experts are going to be asked to express their opinions by means of confidence triplets in [0; 1], that is to say, that between the two extremes they should place a value representing the maximum presumption. Once the information is received Table 11.1 is drawn up.

It may occur that one or several experts felt that their opinion falls outside the interval [10,10; 12,46]. In this event it could be widened after having made the necessary observations by offering the six experts a new widened interval with which to express their opinions.

With the data included in Table 11.1 we proceed to draw up the experton in confidence triplets $\underset{\sim}{a}$ gathering for every level $\alpha \in [0; 1]$ the number of times that each estimate is repeated, both for the lower extreme as for the upper extreme, as well as for the maximum presumption. In this way we find Tables 11.2 and 11.3.

In order to obtain the R-experton in triplets, $\underset{\sim}{A}$, it is sufficient to use the known formula[8]:

[8] Remember that A_* is the lower extreme of the interval representing in this case tha cost of the debts, and A^* its upper extreme. This formula is the immediate

Table 11.2.

α			
0	6/6	6/6	6/6
0,1	5/6	6/6	6/6
0,2	5/6	6/6	6/6
0,3	5/6	6/6	6/6
0,4	4/6	6/6	6/6
0,5	3/6	5/6	6/6
0,6	3/6	3/6	5/6
0,7	2/6	2/6	4/6
0,8	1/6	2/6	2/6
0,9	1/6	2/6	2/6
1	1/6	1/6	1/6

Table 11.3.

0	1	1	1
0,1	0,833	1	1
0,2	0,833	1	1
0,3	0,833	1	1
0,4	0,666	1	1
$\underset{\sim}{a}$ = 0,5	0,500	0,833	1
0,6	0,500	0,500	0,833
0,7	0,333	0,333	0,666
0,8	0,166	0,333	0,333
0,9	0,166	0,333	0,333
1	0,166	0,166	0,166

$$\left[i_{d_*}^{(f)}; i_{d^*}^{(f)}\right] = (1 - f)(\cdot)[15{,}79; 19{,}48]$$

which in this case acquires the form shown in Table 11.4 from where the mathematically expected value is obtained:

In this way triplet (11,27%; 11,63%; 11,82%) could be accepted as the aggregate estimate of the experts for the cost of the debt.

---

consequence of admitting that if the imprecise number $[A_*; A^*]$ is qualified by $\alpha \in [0; 1]$ the imprecise number is estimated by means of: $A_* + (A^* - A_*) \cdot \alpha$

Table 11.4.

$\underset{\sim}{A} = 10{,}10\ (+)\ (12{,}46 - 10{,}10)\ (\cdot) =$ (left table) $=$ (right table)

0		1		0		12,46	
0,1	0,833	1	1	0,1	12,06	12,46	12,46
0,2	0,833	1	1	0,2	12,06	12,46	12,46
0,3	0,833	1	1	0,3	12,06	12,46	12,46
0,4	0,666	1	1	0,4	11,67	12,46	12,46
0,5	0,500	0,833	1	0,5	11,28	12,06	12,46
0,6	0,500	0,500	0,833	0,6	11,28	11,28	12,06
0,7	0,333	0,333	0,666	0,7	10,88	10,88	11,67
0,8	0,166	0,333	0,333	0,8	10,49	10,88	10,88
0,9	0,166	0,333	0,333	0,9	10,49	10,88	10,88
1	0,166	0,166	0,166	1		10,49	

The use of the R-expertons in confidence triplets also allows for the construction of a triangular fuzzy number (T.F.N.) that expresses the opinion of n experts with all the advantages that this entails for later calculations.

Obviously, this procedure is not unique, but many others can be imagined such as the one which would mean taking the mathematically expected value of $\underset{\sim}{a}$ and later moving on to that of $\underset{\sim}{A}$ by means of the expression:

$$\varepsilon(\underset{\sim}{A}) = A_* + (A^* - A_*) \cdot \varepsilon(\underset{\sim}{a})$$

This same procedure can be used for confidence quadruples if what is required is a trapezoidal fuzzy number.

11.4 The Cost of Own (in House) Financial Means

For carrying out their activities businesses normally use their own or outside financial means. Up to this point we have studied the cost of certain financial means that have to be returned to the financial institutions in conjunction with interest (price for their use) and other expenses. Now we are going to refer to financial means that through various channels (foundation capital, capital increases, non-distributed profits) have been incorporated into the business as equity.

The fact that the use of these resources does not require any fixed and direct payment (as in the case of outside capital) could lead us to think that their cost was nothing. We do not feel it is necessary to insist on the commonly accepted fact, that the use of these means that gave rise to a lower yield at a certain level would mean a certain breakdown in the ultimate objectives of the business.

It is for this reason that in order to determine the cost of these own financial means it will be necessary first to define the objective of the financial policy of the business.

It would be trite on our part to state that different objectives can be established for a financial policy, nevertheless it would not appear to be incorrect to consider that it is good for the business that its market value be as high as possible, that is, that the shareholders maintain or increase the quotation for their shares to the maximum. Having accepted this objective it can be said that the cost of equity will be given by the minimum rate that can be obtained in the placing of financial means in order for the wealth of the shareholders to be maintained, that is to say, the quotation for the shares.

Therefore to find the cost of own capital (equity), it will be necessary to establish a relation that relates this variable with the price of the shares. This relation should express the behaviour of purchasers of securities and, therefore, bring to light what it is that moves them to decide to acquire the same.

Briefly, it is not difficult to admit that a person resorting to a market to purchase shares wishes to obtain the maximum of two components: dividends every year (symbolised by D_j) and the sales price for the shares (symbolised by A_n), on ending their economic horizon (symbolised by n). Under these circumstances, the cost of equity i_a will be that which equals the sales value of a share at the current time A_0 and the updated value, at rate i_a of the dividends for each year plus the updating of the sales value of the share at moment n, that is:

$$A_0 = \frac{D_1}{1+i_a} + \frac{D_1}{(1+i_a)^2} + \cdots + \frac{D_n}{(1+i_a)^n} + \frac{A_n}{(1+i_a)^n}$$

The incorporation into this formula of D_j and A_n, explains the fact that shares that do not allow for obtaining high dividends or even no dividend, sometimes have high quotations, because their holders hope to get a high sales price sometime in the future.

It should be pointed out that both the estimate of future dividends and the sales price, in n, does not depend on the projects and data in the hands of the executives of the business, but on the "sensations" of the eventual purchasers on the future evolution of the same, even though these sensations may be influenced by those projects and data.

In a sequence of logical reasoning, it can be admitted that the quotation for the share at moment n, that is to say, A_n, will also depend on the dividends as from n and on the quotation for the share at a future time $n+m$ economic horizon of the new purchaser. A_n so on successively, if it is assumed that the business will have an unlimited life, we arrive at:

$$A_0 = \sum_{j=1}^{n} D_j \cdot (1+i_a)^{-j} + \sum_{j=n+1}^{n+m} D_j \cdot (1+i_a)^{-j} + \cdots + \sum_{j=q+1}^{\infty} D_j \cdot (1+i_a)^{-j}$$

If it is assumed that the business has a limited life span T, even though for a long period, we arrive at:

$$A_0 = \sum_{j=1}^{T} D_j \cdot (1 + i_a)^{-j}$$

For arriving at the cost of equity i_a the same schemes could be followed as those described in previous paragraphs of this chapter, all considerations made in the same being valid.

The eventual purchasers of the shares, on making estimates for the future of the businesses, often think in average terms, that is, they make their estimates of dividends assuming they will receive an amount D, constantly equal throughout the length of the different future periods. This is equal to assuming that the business will have a policy of constant dividends. In this case we arrive at:

– For a unlimited horizon:

$$A_0 = D \cdot \sum_{j=1}^{\infty} (1 + i_a)^{-j} = \frac{D}{i_a}$$

where:

$$i_a = \frac{D}{A_0}$$

– For a horizon limited to T periods:

$$A_0 = \sum_{j=1}^{T} D_j \cdot (1 + i_a)^{-j} = D \cdot \frac{(1 + i_a)^T - 1}{(1 + i_a)^T \cdot i_a}$$

where i_a would be arrived at by approximation, in accordance with previously described schemes.

Now, as we have repeatedly pointed out, estimating variables relative to the future cannot be normally done by means of crisp numbers, and in this case even more so, given the fact that the dividends to be considered do not even have the basis of studies carried out in the business itself, but in the sensations of eventual purchasers, without, obviously, excluding information received from the business itself.

11.5 The Determination of the Cost of the Capital

Every business has, at any given time, a financial structure that is the result of a composition between equity and outside capital, and in each of these categories different masses coming from various sources. The study of the cost of this capital, as we have developed it, starts out from the supposition that the business estimates for the future, a certain financial structure that will not undergo substantial changes. The elimination of this restriction gives rise to the analysis of the optimum financial structure for the business. We are going to continue here with the hypothesis of the constant financial structure.

It should be said, nevertheless, that this constancy does not automatically mean that the masses the make up each one of the elements of this financial structure are know with certainty, but that they can be expressed by means of any uncertain number.

Therefore, the financial means available to a business have a varied origin and, therefore, a different cost. For this reason it is necessary to arrive at a figure that is the representation and compendium of the relative costs of each source of financing.

For this a criterion of aggregation of the different costs must be established. The criterion most used consists in arriving at a weighted average that implies assigning to each partial cost a weight equal to the percentage of each source in the total financing. If, in order to simplify matters, it is assumed that only one source of financing exists for outside capital and that equity has the same cost whatever its source[9], it is customary to accept the following formula for calculating the cost of capital:

$$i_c = i_a \cdot \frac{A}{A+D} + i_d \cdot (1-f) \cdot \frac{D}{A+D}$$

where, as we established in Chap. 8:

D = outside capital
A = equity.

But as:

$$\frac{A}{A+D} = 1 - \frac{D}{A+D}$$

by calling the coefficient of indebtedness g, we arrive at the fact that the average weighted cost of the capital will be:

$$i_c = i_a \cdot (1-\gamma) + i_d \cdot (1-f) \cdot \gamma$$

This expression corresponds perfectly to the idea of the cost of capital as the weighted average of the different costs relative to each source. However, there are other forms of arriving at the cost of capital[10], which we are not going to develop so as not to make this subject too long and not to deviate the attention from the principal objective of this book, centred on financial analysis in uncertainty.

---

[9] The generalization to more sources with varied cost is immediate.

[10] See, among others, Suárez Suárez Andrés S (1985) Decisiones óptimas de inversión y financiación en la empresa (in Spanish). (Ed) Pirámide, Madrid.

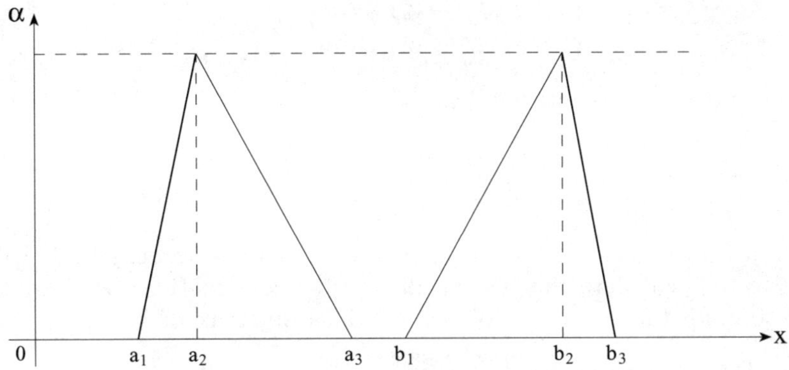

Fig. 11.5.

We move on then to study the cost of capital when the variables located in the future are estimated in an imprecise manner. For this we are going to assume that a business estimates for the future a financial structure that is formed by certain percentages of outside and own capital shown by means of triangular fuzzy numbers:

$$\underset{\sim}{\mathbf{A}} = (a_1, a_2, a_3) \quad \text{and } \underset{\sim}{\mathbf{B}} = (b_1, b_2, b_3) \text{ respectively}$$

It is obvious that the possible combinations between the rate of own financing and outside financing should add up to the unit, since we are dealing with so many by one.

If these T.F.N. are expressed in the form of α-cuts we arrive at:

$$A_\alpha = [a_1 + (a_2 - a_1) \cdot \alpha; a_3 - (a_3 - a_2) \cdot \alpha], \quad \text{outside capital}$$
$$B_\alpha = [b_1 + (b_2 - b_1) \cdot \alpha; b_3 - (b_3 - b_2) \cdot \alpha], \quad \text{equity}$$

the graphical representation of which could be as shown in Fig. 11.5.

These triangles are characterised, among others, by the fact that if added, for the same value of a, the value of the abscissa of the right hand side of one and the left hand side of the other will always give the unit (thick line of a triangle with the thick line of the other triangle, and thin line of one with the thin line of the other).

Therefore, for each α (that is, for each level of supposition) two combinations will be arrived at between equity and outside capital, which will give rise to two costs for the capital, which are normally different. What appears then, with total clarity is that for each level of supposition α there exist two combinations between own and outside capitals and it is possible to consider as many combinations as are necessary, on giving to α as many values as were to be required.

Now let us suppose that certain estimates have been arrived at for the costs of own and outside capitals, by means of T.F.N.:

$$\underset{\sim}{\mathbf{i}}_a = (p_1; p_2; p_3)$$
$$\underset{\sim}{\mathbf{i}}_d = (d_1; d_2; d_3)$$

which expressed as α-cuts will be:

$$\underset{\sim}{\mathbf{i}}_a^{(\alpha)} = [p_1 + (p_2 - p_1) \cdot \alpha; p_3 - (p_3 - p_2) \cdot \alpha]$$
$$\underset{\sim}{\mathbf{i}}_d^{(\alpha)} = [d_1 + (d_2 - d_1) \cdot \alpha; d_3 - (d_3 - d_2) \cdot \alpha]$$

In order to arrive at the weighted average cost of the capital it will be necessary to calculate the cost of all possible combinations, level by level. Therefore, for the first group of combinations we arrive at:

$$\begin{pmatrix} (a_1 + (a_2 - a_1) \cdot \alpha)(\cdot)[d_1 + (d_2 - d_1) \cdot \alpha; d_3 - (d_3 - d_2) \cdot \alpha] \\ (+)(b_3 - (b_3 - b_2) \cdot \alpha)(\cdot)[p_1 + (p_2 - p_1) \cdot \alpha; p_3 - (p_3 - p_2) \cdot \alpha] \end{pmatrix}$$
$$= \begin{pmatrix} [(a_1 + (a_2 - a_1) \cdot \alpha) \cdot (d_1 + (d_2 - d_1) \cdot \alpha); (a_1 + (a_2 - a_1) \cdot \alpha) \\ \cdot (d_3 - (d_3 - d_2) \cdot \alpha] \\ (+)[(b_3 - (b_3 - b_2) \cdot \alpha) \cdot (p_1 + (p_2 - p_1) \cdot \alpha); (b_3 - (b_3 - b_2) \cdot \alpha) \\ \cdot (p_3 - (p_3 - p_2) \cdot \alpha)] \end{pmatrix}$$

For the second group of combinations this will be:

$$\begin{pmatrix} (a_3 + (a_3 - a_2) \cdot \alpha)(\cdot)[d_1 + (d_2 - d_1) \cdot \alpha; d_3 - (d_3 - d_2) \cdot \alpha] \\ (+)(b_1 - (b_2 - b_1) \cdot \alpha)(\cdot)[p_1 + (p_2 - p_1) \cdot \alpha; p_3 - (p_3 - p_2) \cdot \alpha] \end{pmatrix}$$
$$= \begin{pmatrix} [(a_3 + (a_3 - a_2) \cdot \alpha) \cdot (d_1 + (d_2 - d_1) \cdot \alpha; (a_3 + (a_3 - a_2) \cdot \alpha) \cdot (d_3 - (d_3 - d_2) \cdot \alpha] \\ (+)[(b_1 - (b_2 - b_1) \cdot (p_1 + (p_2 - p_1) \cdot \alpha); (b_1 - (b_2 - b_1) \cdot \alpha) \cdot (p_3 - (p_3 - p_2) \cdot \alpha)] \end{pmatrix}$$

Finally, in order to show a result that has some significance of the cost of capital, a fuzzy number (not triangular) can be accepted, which expressed in the form of α-cuts (by level of probability-confidence interval) would have:

1. As its lower extreme the smallest of the lower extremes of each group:

$$\begin{aligned} &(a_1 + (a_2 - a_1) \cdot \alpha) \cdot (d_1 + (d_2 - d_1) \cdot \alpha) + (b_3 - (b_3 - b_2) \cdot \alpha) \cdot \\ &\quad (p_1 + (p_2 - p_1) \cdot \alpha) \\ &\wedge (b_1 + (b_2 - b_1) \cdot \alpha) \cdot (p_1 + (p_2 - p_1) \cdot \alpha) + (a_3 - (a_3 - a_2) \cdot \alpha) \cdot \\ &\quad (d_1 + (d_2 - d_1) \cdot \alpha) \end{aligned}$$

2. As the upper extreme, the largest of the upper extremes of each group:

$$\begin{aligned} &(a_1 + (a_2 - a_1) \cdot \alpha) \cdot (d_3 - (d_3 - d_2) \cdot \alpha) + (b_3 - (b_3 - b_2) \cdot \alpha) \cdot (p_1 \\ &\quad + (p_2 - p_1) \cdot \alpha) \\ &\vee (b_1 + (b_2 - b_1) \cdot \alpha) \cdot (p_3 - (p_3 - p_2) \cdot \alpha) + (a_3 - (a_3 - a_2) \cdot \alpha) \cdot (d_3 \\ &\quad - (d_3 - d_2) \cdot \alpha) \end{aligned}$$

If the triangular approximation of the cost of the capital could be accepted as representative, the resulting T.F.N. would be made up as follows:

1. The result of using the formula expressed in point 1. above giving to α the value of 0 would be considered as the lower extreme.
2. The result of making $\alpha = 0$ in the formula expressed in point 2 would be considered as the upper extreme.
3. The result of using either of the formula 1 and 2 above making $\alpha = 1$ would be considered as the maximum probability.

For better understanding of this algorithm we are now going to develop a didactic example in the fullest detail.

Let us assume that the financial structure of a business is comprised of:

$$\% \text{ of outside capital} = \underset{\sim}{\mathbf{A}} = (0{,}3\widehat{3}; 0,40; 0,50)$$

$$\% \text{ equity} = \underset{\sim}{\mathbf{B}} = (0,50; 0,60; 0,6\widehat{6})$$

which expressed in α-cuts would be:

$$A_\alpha = [0{,}3\widehat{3} + 0{,}0\widehat{6}\alpha; 0,50 - 0{,}10\alpha]$$

$$B_\alpha = [0{,}50 + 0{,}10\alpha; 0{,}6\widehat{6} - 0{,}0\widehat{6}\alpha]$$

and the graphical expression of which can be seen in Fig. 11.6.

In this figure it can be seen that, for each level of probability α, the sum of the left side of the percentage of outside capital and of the right side of equity give the unit as a result. The same occurs with the left and right hand sides of equity and outside capital, respectively.

In order to arrive at all the possible combinations the lower extreme of one T.F.N and the upper of the other must be taken. Then we arrive at:

Thick lines:

$$0{,}3\widehat{3} + 0{,}0\widehat{6}\alpha(\cdot) \text{ cost of outside capital } (+) 0{,}6\widehat{6} - 0{,}0\widehat{6}\alpha(\cdot) \text{ cost of equity}$$

Thin lines:

$$0{,}50 - 0{,}10\alpha(\cdot) \text{ cost of outside capital } (+) 0{,}50 + 0{,}10\alpha(\cdot) \text{ cost of equity}$$

Let us also assume that the following costs have been estimated.

Costs of equity

$$\underset{\sim}{\mathbf{i}}_a = (15; 18; 20) = [15 + 3\alpha; 20 - 2\alpha]$$

Costs outside capital:

$$\underset{\sim}{\mathbf{i}}_d = (10; 11; 14) = [10 + \alpha; 14 - 3\alpha]$$

On substituting these costs in the previous expression we arrive at:

For the first group:

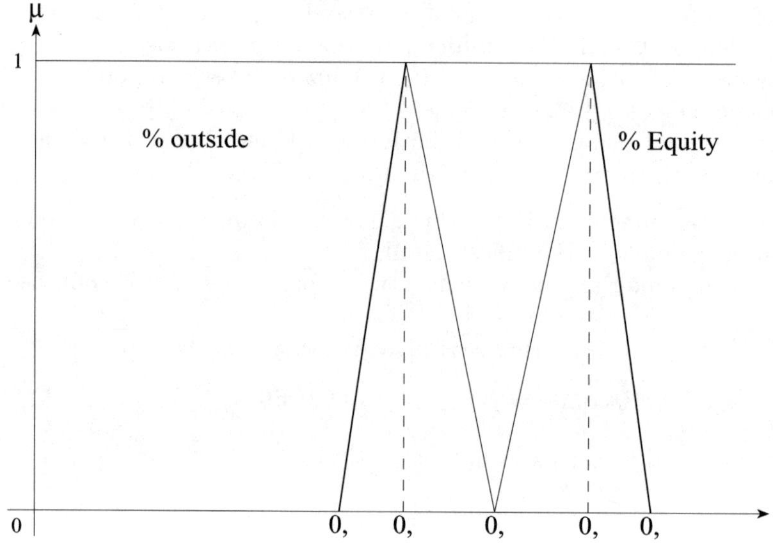

Fig. 11.6.

$$\begin{aligned}
&(0{,}3\widehat{3} + 0{,}0\widehat{6}\alpha)(\cdot)[10+\alpha; 14-3\alpha](+)(0{,}6\widehat{6} - 0{,}0\widehat{6}\alpha)(\cdot)[15+3\alpha; 20-2\alpha] \\
&\quad = [(0{,}3\widehat{3} + 0{,}0\widehat{6}\alpha)(\cdot)(10+\alpha); (0{,}3\widehat{3} + 0{,}0\widehat{6}\alpha)(\cdot)(14-3\alpha)] \\
&\quad\quad (+)[(0{,}6\widehat{6} - 0{,}0\widehat{6}\alpha)(\cdot)(15+3\alpha); (0{,}6\widehat{6} - 0{,}0\widehat{6}\alpha)(\cdot)(20-2\alpha)] \\
&\quad = [(0{,}3\widehat{3} + 0{,}0\widehat{6}\alpha)(\cdot)(10+\alpha) + (0{,}6\widehat{6} - 0{,}0\widehat{6}\alpha)(\cdot)(15+3\alpha); \\
&\quad\quad (0{,}3\widehat{3} + 0{,}0\widehat{6}\alpha)(\cdot)(14-3\alpha) + (0{,}6\widehat{6} - 0{,}0\widehat{6}\alpha)(\cdot)(20-2\alpha)]
\end{aligned}$$

For the second group:

$$\begin{aligned}
&(0{,}50 + 0{,}10\alpha)(\cdot)[15+3\alpha; 20-2\alpha](+)(0{,}50 - 0{,}10\alpha)(\cdot)[10+\alpha; 14-3\alpha] \\
&\quad = [(0{,}50 + 0{,}10\alpha)(\cdot)(15+3\alpha); (0{,}50 + 0{,}10\alpha)(\cdot)(20-2\alpha)] \\
&\quad\quad (+)[(0{,}50 - 0{,}10\alpha)(\cdot)(10+\alpha); (0{,}50 - 0{,}10\alpha)(\cdot)(14-3\alpha)] \\
&\quad = [(0{,}50 + 0{,}10\alpha)(\cdot)(15+3\alpha) + (0{,}50 - 0{,}10\alpha)(\cdot)(10+\alpha); \\
&\quad\quad (0{,}50 + 0{,}10\alpha)(\cdot)(20-2\alpha) + (0{,}50 - 0{,}10\alpha)(\cdot)(14-3\alpha)]
\end{aligned}$$

In order to represent the fuzzy number that we accept as the cost of the capital in the form of α-cuts, we take as the lower extreme:

$$\begin{aligned}
&(0{,}3\widehat{3} + 0{,}0\widehat{6}\alpha)(10+\alpha) + (0{,}6\widehat{6} - 0{,}0\widehat{6}\alpha)(15+3\alpha) \\
&\wedge(0{,}50 + 0{,}10\alpha)(15+3\alpha) + (0{,}50 - 0{,}10\alpha)(10+\alpha)
\end{aligned}$$

and as the upper extreme:

$$(0{,}3\widehat{3} + 0{,}0\widehat{6}\alpha)(14 - 3\alpha) + (0{,}6\widehat{6} - 0{,}0\widehat{6}\alpha)(20 - 2\alpha)$$
$$\vee(0{,}50 + 0{,}10\alpha)(20 - 2\alpha) + (0{,}50 - 0{,}10\alpha)(14 - 3\alpha)$$

If the triangular approximation is admitted, on doing $\alpha = 0$ in the lower extreme we arrive at 12,5; $\alpha = 0$ for the upper extreme then 17,98; and finally $\alpha = 1$ for either of the two extremes of the interval we arrive at 15,2. Therefore the triangular approximation will be:

$$\underset{\sim c}{\mathbf{i}} = (12{,}50; 15{,}20; 17{,}98)$$

Now let us see how to combine equity with outside capital, on a level by level basis, using the hendecagonal system. For the first group we will arrive at Table 11.5 and for the second Table 11.6.

Table 11.5.

			A	B			A+B
α	$0,3\widehat{3}+0,0\widehat{6}\alpha$	$10+\alpha$	$(0,3\widehat{3}+0,0\widehat{6}\alpha)\ (10+\alpha)$	$0,66-0,0\widehat{6}\alpha$	$15+3\alpha$	$(0,66-0,0\widehat{6}\alpha)\ (15+3\alpha)$	TOTAL
0	0,333	10	3,33	0,666	15	9,99	13,32
0,1	0,340	10,1	3,43	0,659	15,3	10,08	13,51
0,2	0,346	10,2	3,52	0,653	15,6	10,18	13,7
0,3	0,353	10,3	3,63	0,646	15,9	10,27	13,9
0,4	0,359	10,4	3,73	0,640	16,2	10,36	14,09
0,5	0,366	10,5	3,84	0,633	16,5	10,44	14,28
0,6	0,373	10,6	3,95	0,626	16,8	10,51	14,46
0,7	0,379	10,7	4,05	0,620	17,1	10,60	14,65
0,8	0,386	10,8	4,16	0,613	17,4	10,66	14,82
0,9	0,393	10,9	4,28	0,607	17,7	10,72	15
1	0,399	11	4,38	0,600	18	10,80	15,18

Lower extreme

			A	B			A+B
α	$0,3\widehat{3}+0,0\widehat{6}\alpha$	$14-3\alpha$	$(0,3\widehat{3}+0,0\widehat{6}\alpha)\ (14-3\alpha)$	$0,66-0,0\widehat{6}\alpha$	$20-2\alpha$	$(0,66-0,0\widehat{6}\alpha)\ (20-2\alpha)$	TOTAL
0	0,333	14	4,66	0,666	20	13,32	17,98
0,1	0,340	13,7	4,65	0,659	19,8	13,04	17,69
0,2	0,346	13,4	4,63	0,653	19,6	12,79	17,42
0,3	0,353	13,1	4,62	0,646	19,4	12,53	17,15
0,4	0,359	12,8	4,59	0,640	19,2	12,28	16,87
0,5	0,366	12,5	4,57	0,633	19	12,02	16,59
0,6	0,373	12,2	4,55	0,626	18,8	11,76	16,31
0,7	0,379	11,9	4,51	0,620	18,6	11,53	16,04
0,8	0,386	11,6	4,47	0,613	18,4	11,27	15,74
0,9	0,393	11,3	4,44	0,606	18,2	11,02	15,46
1	0,399	11	4,38	0,600	18	10,80	15,18

Upper extreme

Table 11.6.

		A				B	A+B
α	0,50+0,10α	15+3α	(0,50+0,10α) (15+3α)	0,50 – 0,10α	10+α	(0,50 – 0,10α) (10+α)	TOTAL
0	0,500	15	7,50	0,500	10	5,00	12,5
0,1	0,510	15,3	7,80	0,490	10,1	4,94	12,74
0,2	0,520	15,6	8,11	0,480	10,2	4,89	13
0,3	0,530	15,9	8,42	0,470	10,3	4,84	13,26
0,4	0,540	16,2	8,74	0,460	10,4	4,78	13,52
0,5	0,550	16,5	9,07	0,450	10,5	4,72	13,79
0,6	0,560	16,8	9,40	0,440	10,6	4,66	14,06
0,7	0,570	17,1	9,74	0,430	10,7	4,60	14,34
0,8	0,580	17,4	10,09	0,420	10,8	4,53	14,62
0,9	0,590	17,7	10,44	0,410	10,9	4,46	14,9
1	0,600	18	10,80	0,400	11	4,40	15,2

Lower extreme

		A				B	A+B
	0,50+0,10α	20 – 2α	(0,50+0,10α) (20 – 2α)	0,50 – 0,10α	14 – 3α	(0,50 – 0,10α) (14 – 3α)	TOTAL
0	0,500	20	10,00	0,500	14	7,00	17
0,1	0,510	19,8	10,09	0,490	13,7	6,71	16,8
0,2	0,520	19,6	10,19	0,480	13,4	6,43	16,62
0,3	0,530	19,4	10,28	0,470	13,1	6,15	16,43
0,4	0,540	19,2	10,36	0,460	12,8	5,88	16,24
0,5	0,550	19	10,45	0,450	12,5	5,62	16,07
0,6	0,560	18,8	10,52	0,440	12,2	5,36	15,88
0,7	0,570	18,6	10,60	0,430	11,9	5,11	15,71
0,8	0,580	18,4	10,67	0,420	11,6	4,87	15,54
0,9	0,590	18,2	10,73	0,410	11,3	4,63	15,36
1	0,600	18	10,80	0,400	11	4,40	15,2

Upper extreme

As can be seen, for each group we arrive at a fuzzy number (non triangular) the graphic representation of which is shown in the Fig. 11.7, differentiating one group from the other with a thick and thin line. Our solution has been to accept as the result the fuzzy number formed by the side closest to the origin as the lower extreme and the furthest away as the upper extreme since, in this way, the result, although more uncertain, includes the other possibilities.

If the evolution of the results is observed on variation of the level of probability α, which in this case coincides with the variations in the combination of equity and outside capitals, a sufficiently well know fact will become quite obvious: when, in this case, the cost of debts is lower than the cost of equity, that is $i_d \prec i_a$ the lower cost of capital is arrived at with higher indebtedness. In this case, the lower cost of the capital is shown by the interval [12,50; 17,98] when the combination is 50% of outside capital and 50% equity.

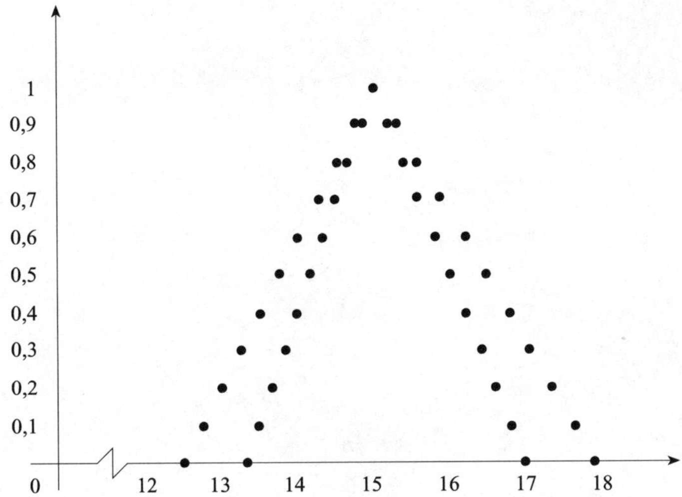

Fig. 11.7.

It is quite obvious that the process that has been followed in not the only one, neither do we claim it to be the best. Neither is it the fastest to calculate but, on the other hand, we feel that it allows for the establishment of certain conclusions that can be quite interesting.

Just one small observation before we finish this chapter, and that is that the use of T.F.N. in order to express the estimate of the financial structure of a business is rather like a study in sensitivity of the variation in the cost of capital on varying the financial structure, with the advantage of assigning levels of probability a to the different combinations.

Many other considerations could be made relative to the fascinating subject of capital cost, above all in what it means for taking decisions on investments and other deposits. The requirements of this work oblige us to put a full stop to this chapter, with the hope that on other occasions we can continue to delve deeper into the study.

12 Relation between Financial Structure and Capital Cost

12.1 An Overview of the Different Doctrinal Positions

When a ration of indebtedness is accepted as a good representation of the financial structure and, once estimated, the change that will occur by the mere fact of passing from one moment to another, one of the basic problems of financial analysis can be tackled: the existence and determination, if it exists, of the optimum financial structure.

The importance of this subject justifies the fact that practically all the works that have been published on financial management make a more or less extensive reference to the optimum financial structure of a business. In this sense, the controversy on the very existence of an optimum structure has filled thousands of pages and divided the specialists into more or less radical positions, which range from the categorical affirmation that it is possible to find a relation between the different patrimonial masses of liabilities that maximises the objectives of the business, right up to the out and out belief that these objectives can be attained independently of the fact that there is one or other financial structure.

Obviously opinions that are so different can only be justified if different objectives are being sought and/or the starting out points are different hypotheses. For this reason, in a first approximation the problem of the financial structure can be divided up according to the different objectives of the business, and for each of these, the varied range of cases on which the different authors base their works chosen. In this sense it can be assumed, for example, that the objective that it is intended to reach is the "maximisation of the wealth of the shareholders" and what is required is to see what influence (if any) the changes in the financial structure exercise on the capital cost and on the value the business merits for the shareholders.

Faced with this approach, which we feel can be accepted without too many reservations, a long list and later description could be drawn up of the position adopted by the specialists on the subject over time.

We are not going to fall into this temptation, which we feel would draw out this work unnecessarily and in itself would require us to dedicate hundreds of pages to a monographic treatise. We are going to limit ourselves to a brief description of some of the better known positions, trusting that the reader will understand that this is not an exhaustive list and that neither is it the

Anna Maria Gil-Lafuente: *Fuzzy Logic in Financial Analysis*, StudFuzz **175**, 203–216 (2005)
www.springerlink.com

intention to include the most important contributions All we want to do is to present a sample that allows us to tackle the considerations we require to make in a management study within the sphere of uncertainty. Even with the risk of falling into a certain amount of arbitrariness (classification always contain an arbitrary element), we will make a separation into three sections:

1. The hypotheses that are established, either implicitly or explicitly, are basically the following: the cost of the debts i_d is lower than the cost of equity i_a and both are independent of the ratio of indebtedness, that is to say, however much the debts increase proportionally, their cost does not vary. The market value of the shares A is obtained by the capitalisation of the net results (result of the exploitation less financial expenses). The conclusion is reached then, that as the ratio of indebtedness increases, the cost of capital increases and the market value of the business increases (shares and debts). This method has been designated with the initial letters NI (net income[1]) and also by RN (net result[2]). This concludes with the existence of an optimum financial structure.
2. The starting out point is the constancy of capital cost i_c relative to the ratio of indebtedness, as also the cost of debts i_d, that is, that also in this case the price of money does not increase however much the debts increase relative to equity. The value of the business for the shareholders depends on the capacity of assets to generate profits and is independent, therefore, from the ratio of indebtedness. With this hypothesis it is established a priori that no optimum financial structure exists. The justification appears because the advantages that are gained on substituting equity for outside capital (cheaper) are compensated by the increase of equity, since the share buying market requires, in order to maintain the quotation, higher profitability in order to compensate the increased financial risk. This method is known under the initial letter NOI (Net Operating Income) according to Durand or else RE (Exploitation Result) according to Suárez.
3. Special mention should be made of the thesis by Modigliani and Miller[3], whose hypotheses are as follows: capital markets are perfect. Investors behave in a rational manner in the sense that it is indifferent to them whether there is an increase in the dividends or an equivalent increase in the value of the shares. Constant future profits throughout time are assumed. Businesses can be grouped into "classes of equivalent yield" in the sense that the price of each unit of expected yield is the same for each

[1] Durand David (1952) Cost of Debt and Equity Funds for Business: Trend and Problems of Measurements in Conference and Research on Business Finance, N.B.E.R., New York, pp. 215–247.

[2] Suárez Suárez Andrés S (1985) Decisiones óptimas de inversión y financiación en la empresa, (in spanish). (Ed) Pirámide Madrid, pp. 517–523.

[3] Modigliani F and Miller M (1958) The Cost of Capital. Corporation finance and the Theory of Investments, The American Economic Review, Vol. 48, No. 3, June, pp. 261–297.

type of share within one and the same "class". The capital cost i_c and the value of the business for the shareholders V are independent from the ratio of indebtedness r_d. An optimum capital structure then does not exist: all are equally good.

This sample then should be sufficient for bringing to light the wide range of opposing positions that surround the problem of the existence and obtaining an optimum financial structure for the business, when it is assumed that the values that intervene in the problem are known with certainty. Faced with this diversity the question that could be asked is if it is possible, within the sphere of uncertainty, to draw up a procedure that would allow us to explain the incidence of the financial structure on capital cost and the value of the business for its shareholders, with the object of finding the form that the optimum structure takes, on the assumption that it exists.

With all that has been stated above, it is not difficult to admit that a single answer to this matter does not exist other than in the case that a revision could be established for the position for the generalisation of all the different approaches. It would appear that this is not possible at this time, given the current status of knowledge in the formal field. But what we do feel can be done is to re draw the models[4] we have seen by means of a transformation of the values that make them up, passing over from precision to fuzziness.

12.2 Uncertain Capital Cost and Financial Structure

With this background we now move on to make certain considerations relative to the ratio of the financial structure and capital cost. For this we are going to expound on one of the possible representations of capital cost by means of an expression that permits quantifying the variations that occur in it as a consequence of increases and decreases in the ratio of indebtedness.

In this sense, one of the forms under which capital cost is normally expressed in the sphere of certainty is the following:

$$i_c = i_a \cdot (1 - \gamma) + i_d \cdot (1 - f) \cdot \gamma$$

where, as we already stated:

$$i_c = \text{capital cost} ;$$
$$i_a = \text{cost of equity} ;$$
$$i_d = \text{cost of debts} ;$$

[4] We could add, among others, the model established by Schwartz E. (March 1959) Theory of the Capital Structure of the Firm, The Journal of Finance, Vol. 14 No. 1, pp. 18–39.

$$\gamma = \frac{D}{A+D} = \frac{\text{Total debts}}{\text{Total debts + Equity}}$$
$$f = \text{rate of corporate tax}$$

We saw in previous chapters that these values can be expressed by means of imprecise numbers and that the T.F.N. is a convenient and very adequate form for the manner of thinking of the businessman. For greater simplicity in showing this we are going to use the numerical examples with the same data as in previous chapters.

In a first approximation we are going to start out from the constancy of the cost of equity i_a and also of outside capital cost i_d. These will be as follows:

$$\underset{\sim}{\mathbf{i}}_a = (p_1; p_2; p_3) = (15; 18; 20)$$
$$\underset{\sim}{\mathbf{i}}_d = (d_1; d_2; d_3) = (10; 11; 14)$$

the expression of which in α-cuts is as follows:

$$i_a^{(\alpha)} = [15 + 3\alpha; 20 - 2\alpha]$$
$$i_d^{(\alpha)} = [10 + \alpha; 14 - 3\alpha]$$

On the other hand, the ratio of indebtedness that may be considered can also be estimated by means of T.F.N. Let us assume that at moment zero the business is in a position for which the ratio of indebtedness, expressed in α-cuts, will be:

$$r_d^{(\alpha)} = \left[\frac{0{,}30 + 0{,}03\alpha}{0{,}70 - 0{,}03\alpha}; \frac{0{,}35 - 0{,}016\alpha}{0{,}65 + 0{,}016\alpha}\right]$$

and its triangular approximation will be:

$$\underset{\sim}{\mathbf{r}}_d = (0{,}428; 0{,}500; 0{,}538)$$

which at the same time is equivalent to:

$$r_d^{(\alpha)} = [0{,}428 + 0{,}72\alpha; 0{,}538 - 0{,}038\alpha]$$

We have made this example coincide with the example of the previous chapter.

From a general point of view, it can be seen that the estimate of the ratio of indebtedness by means of a fuzzy number, be this or not a triangular approximation, raises the consideration of a whole range of ratios of indebtedness, expressed in confidence intervals, as a consequence of giving α values in $[0; 1]$, that is to say, on accepting as valid each of the different levels of presumption. Thus, in the previous example, corresponding to level $\alpha = 0{,}3$ is a ratio of indebtedness of $[0{,}449; 0{,}526]$, at level $a = 0.6$ the ratio is [0,470; 0,515] and at level $\alpha = 1$, the ratio is [0,500; 0,500] which represents the estimate with the maximum presumption.

This simple example allows us to make the following statement: when uncertainty is structured and is expressed by means of a fuzzy number, one single ratio of indebtedness brings to light as many combinations of equity and outside capital as there are levels of presumption under consideration.

Now then the different ratios (expressed in intervals) that are arrived at on accepting different levels of one and the same fuzzy number are not the result of varying the uncertain estimate, but that they form part of the debt that presides one and the same estimate. Therefore when an attempt is being made to investigate the incidence of the ratio of indebtedness on other values, such as capital cost, or the value of the business for the shareholders, this should be taken into account in a unitary manner (that is to say, at all levels).

The consideration of different levels of presumption has sense when analysing the repercussion of a value (in this case the ratio of indebtedness) on another (capital cost, for example) for the different "sensitivities" of businessmen faced with one and the same estimate, who take more or less of a risk when deciding to accept a higher or lower level of presumption.

Finally, and before moving on to the hub of this chapter, let us recall that, in order to study capital cost in a sphere of uncertainty as a result of weighting the volume of equity and outside capitals by the respective percentages with which they intervene in the financial structure, different procedures can be used. In previous chapters we adopted a scheme that consisted of accepting as the lower extreme of the resulting number the lowest of the two possible lower extremes, and as the upper extreme the greatest[5]. For greater simplicity we will do:

$$\begin{aligned}
\gamma_\alpha &= [a_1(\alpha); a_2(\alpha)] \\
1 - \gamma_\alpha &= [b_1(\alpha); b_2(\alpha)] \\
r_d^{(\alpha)} &= \left[\frac{a_1(\alpha)}{b_2(\alpha)}; \frac{a_2(\alpha)}{b_1(\alpha)}\right] \\
i_d^{(\alpha)} &= [x_1(\alpha); x_2(\alpha)] \\
i_a^{(\alpha)} &= [y_1(\alpha); y_2(\alpha)]
\end{aligned}$$

The lower extreme will be considered as:

$$a_1(\alpha) \cdot x_1(\alpha) + b_2(\alpha) \cdot y_1(\alpha) \wedge a_2(\alpha) \cdot x_1(\alpha) + b_1(\alpha) \cdot y_1(\alpha)$$

and the upper as:

$$a_1(\alpha) \cdot x_2(\alpha) + b_2(\alpha) \cdot y_2(\alpha) \vee a_2(\alpha) \cdot x_2(\alpha) + b_1(\alpha) \cdot y_2(\alpha)$$

Substituting the data of the example we will arrive at:

[5] We should remember that this aspect of the problem has already been developed in the chapter on capital cost.

Lower extreme:

$$(0{,}30 + 0{,}0\widehat{3}\alpha)(10-\alpha) + (0{,}70 - 0{,}0\widehat{3}\alpha)(15 - 3\alpha)$$
$$\wedge(0{,}35 + 0{,}01\widehat{6}\alpha)(10+\alpha) + (0{,}65 + 0{,}01\widehat{6}\alpha)(15 + 3\alpha)$$

Upper extreme:

$$(0{,}30 + 0{,}0\widehat{3}\alpha)(14 - 3\alpha) + (0{,}70 - 0{,}0\widehat{3}\alpha)(20 - 2\alpha)$$
$$\vee(0{,}35 - 0{,}01\widehat{6}\alpha)(14 - 3\alpha) + (0{,}65 + 0{,}01\widehat{6}\alpha)(20 - 2\alpha)$$

In uncertainty it is customary to accept as sufficient the triangular approximation which, in this case, provides us with the following capital cost:

$$\underset{\sim}{\mathbf{i}}_c = (13{,}25; 15{,}66; 18, 20)$$

Again in this case, and for greater simplicity, the tax effect has not been considered.

12.3 The Constancy of the Cost of Equity and Outside Capital

We now move on to carry out an initial study of the influence of the variation of the ratio of indebtedness on the cost of capital, as a consequence of the transition from one period to the next. For this we are going to establish the following hypotheses:

1. The cost of equity $\underset{\sim}{\mathbf{i}}_a$ and the cost of the debts $\underset{\sim}{\mathbf{i}}_d$ are constants, whatever the ratio of indebtedness. Therefore, the transformations that have occurred over the length of a period in the financial structure of the balance sheet are not sufficient for creditors and shareholders to require higher remuneration for their capitals, or on the other hand allow this to be reduced.
2. The cost of equity is, in any event, higher than the cost of the debts. Therefore the following will be complied with: $\underset{\sim}{\mathbf{i}}_u > \underset{\sim}{\mathbf{i}}_d$. In a sphere of uncertainty, when the cost of equity and the cost of outside capitals is given by fuzzy numbers (in our example by T.F.N.) the problem arises of the comparison between both with the object of establishing an order from highest to lowest, taking into account the fact that not always does a complete order exist.

With the object of providing a solution to this problem we propose a first priority criterion, which is as follows:

$$(\underset{\sim}{\mathbf{i}}_a = (p_1; p_2; p_3)\ \underset{\sim}{\mathbf{i}}_d = (d_1; d_2; d_3)) \Rightarrow \left(\frac{p_1 + 2p_2 + p_3}{4} > \frac{d_1 + 2d_2 + d_3}{4}\right)$$

In the event that:

$$\frac{p_1 + 2p_2 + p_3}{4} = \frac{d_1 + 2d_2 + d_3}{4}$$

the following can be taken as a complementary criterion:

$$((p_1; p_2; p_3) \succ (d_1; d_2; d_3)) \Rightarrow (p_2 > d_2)$$

Obviously other criteria can be adopted according to the nature and circumstances of the problem under study. In the example we are developing it is obvious that: $\underset{\sim}{\mathbf{i}}_a \succ \underset{\sim}{\mathbf{i}}_d$ as:

$$(15; 18; 20) \succ (10; 11; 14)$$

therefore this second hypothesis is complied with.

We now move and take a look at what occurs when throughout the period a variation takes place in the ratio of indebtedness on passing over from position a to position b, that is, on doing the transfer from:

$$\underset{\sim}{\mathbf{r}}_d^{(a)} = \frac{(0{,}30; 0{,}3\widehat{3}; 0{,}35)}{(0{,}65; 0{,}6\widehat{6}; 0{,}70)}$$

to

$$\underset{\sim}{\mathbf{r}}_d^{(b)} = \frac{(0{,}36; 0{,}40; 0{,}42)}{(0{,}58; 0{,}60; 0{,}64)}$$

increasing, thereby, the percentage of outside capitals and consequently reducing the percentage of equity.

This change has occurred as a result of passing over to:

$$\underset{\sim}{\gamma}^{(b)} = (0{,}30 + 0{,}066; 0{,}3\widehat{3} + 0{,}0\widehat{6}; 0{,}35 + 0{,}07) = (0{,}36; 0{,}40; 0{,}42)$$

and, therefore:

$$1 - \underset{\sim}{\gamma}^{(b)} = (0{,}65 - 0{,}07; 0{,}6\widehat{6} - 0{,}0\widehat{6}; 0{,}70 - 0{,}06) = (0{,}58; 0{,}60; 0{,}64)$$

When the ratio of indebtedness is expressed in position b in the form of α-cuts we get:

$$r_d^{(b)}(\alpha) = \frac{[0{,}36 + 0{,}04\alpha; 0{,}42 - 0{,}02\alpha]}{[0{,}58 + 0{,}02\alpha; 0{,}64 - 0{,}04\alpha]} = \left[\frac{0{,}36 + 0{,}04\alpha}{0{,}64 - 0{,}04\alpha}; \frac{0{,}42 - 0{,}02\alpha}{0{,}58 + 0{,}02\alpha}\right]$$

This ratio of indebtedness represents a relative increase of the debts with a relative decrease in equity. With this new financial structure we arrive at a capital cost the formal representation of which will be:

Lower extreme:

$$(0{,}36+0{,}04\alpha)(10+\alpha)+(0{,}64-0{,}04\alpha)(15+3\alpha)$$
$$\wedge(0{,}42-0{,}02\alpha)(14-3\alpha)+(0{,}58+0{,}02\alpha)(15+3\alpha)$$

Upper extreme:

$$(0{,}36+0{,}04\alpha)(14-3\alpha)+(0{,}64-0{,}04\alpha)(20-2\alpha)$$
$$\vee(0{,}42-0{,}02\alpha)(14-3\alpha)+(0{,}58+0{,}02\alpha)(20-2\alpha)$$

If the triangular approximation is accepted as sufficient, we will have that:

$$\underset{\sim}{\mathbf{i}}_c^{(b)}=(13{,}20;15{,}20;17{,}84)$$

And, given that:

$$(13{,}20;15{,}20;17{,}84)\prec(13{,}25;15{,}66;18,20)$$

the following will be complied with:

$$\underset{\sim}{\mathbf{i}}_c^{(b)}\prec\underset{\sim}{\mathbf{i}}_c^{(a)}$$

Therefore, it can be stated, that under the admitted hypotheses, an increase in the ratio of indebtedness gives rise to a reduction in capital cost. This conclusion, arrived at within the sphere of uncertainty, coincides with the conclusion reached traditionally when starting out from a determinist system.

Now, it becomes necessary to point out that, even when the results coincide in both spheres, it is not possible in uncertainty to continue along the same path which would lead to the direct use of the determinist formula transformed in fuzzy terms. Indeed, under the simplified case that the fiscal position is not considered, the following is accepted as the expression of capital cost:

$$i_c^{(a)}=\gamma-i_d+(1-\gamma)+i_a=\frac{D}{A+D}\cdot i_d+\frac{A}{A+D}\cdot i_a$$

A decrease in A and the corresponding increase in D will give rise to the new capital cost $i_c^{(b)}$:

$$\begin{aligned}i_c^{(b)}&=\frac{D+\Delta D}{A-\Delta A+D+\Delta D}\cdot i_d\\&\quad+\frac{A-\Delta A}{A-\Delta A+D+\Delta D}\cdot i_a=\gamma^{(b)}-i_d+(1-\gamma^{(b)})+i_a\\&=\frac{D}{A-\Delta A+D+\Delta D}\cdot i_d+\frac{A}{A-\Delta A+D+\Delta D}\cdot i_a\\&\quad+\frac{\Delta D\cdot i_d-\Delta A\cdot i_a}{A-\Delta A+D+\Delta D}\end{aligned}$$

With the object of reaching a conclusion in a simplified manner, it is customary to assume that the increase in outside capitals is equal, in absolute value, to the decrease of equity, in this way leaving the overall volume of the financial structure unaltered. That is it is assumed that $|\Delta D| = |-\Delta A|$. In which case:

$$i_c^{(b)} = \frac{D}{A+D} \cdot i_d + \frac{A}{A+D} \cdot i_a + \frac{\Delta D \cdot (i_d - i_a)}{A+D}$$

with which we arrive at the fact that when $i_a > i_d$, $i_c^{(b)} < i_c^{(a)}$ will be.

If we now pass over to an uncertain environment and continue with the same example, we arrive at the fact that the transformed formula of capital cost will be:

$$\underset{\sim}{\mathbf{i}}_c = \gamma(\cdot)\underset{\sim}{\mathbf{i}}_d(+)(1-\gamma)(\cdot)\underset{\sim}{\mathbf{i}}_a$$

which would give rise to:

$$\underset{\sim}{i}_c = [0{,}36 + 0{,}04 \cdot \alpha;\ 0{,}42 + 0{,}02 \cdot \alpha]\ (\cdot)\ [0{,}36 + 0{,}04 \cdot \alpha;\ 0{,}42 + 0{,}02 \cdot \alpha]$$

$$(+)\ [0{,}58 + 0{,}02 \cdot \alpha;\ 0{,}64 - 0{,}04 \cdot \alpha]\ (\cdot)\ [10 + \alpha;\ 14 - 3 \cdot \alpha]$$

the result of which would obviously be different from the result arrived at before. The lack of adaptation of this expression is given by the fact that in order to make each level of the sum of percentages of equity and outside capitals always equal to the unit, it is necessary to consider the lower extreme of an interval in conjunction with the upper extreme of the other in the fuzzy numbers $\underset{\sim}{\gamma}$ and $(1 - \underset{\sim}{\gamma})$. Therefore the normal sum of the products is not applicable to this problem.

In fact, from a purely theoretical approach, the normal sum of the products would give rise to:

$$[a_1(\alpha) \cdot x_1(\alpha) + b_1(\alpha) \cdot y_1(\alpha); a_2(\alpha) \cdot x_2(\alpha) + b_2(\alpha) \cdot y_2(\alpha)]$$

This expression is not coherent with the need of the sum of the "weights" in the weighting (percentages of equity and outside capitals) give rise to the unit.

12.4 Mechanisms that Justify the Constancy of Capital Cost

A very generalised tendency sustains that there exists no relation of dependence between the variations of the ratio of indebtedness and capital cost, taking into account the fact that, always according to this hypothesis, the cost of outside capitals is also independent from the degree of indebtedness of the business. In other words, a financial institution may or may not grant a credit (for example), but will not charge more for it by the mere fact that it considered the requesting entity to be excessively indebted. An objection has been made to this position that states when a business requires outside capital it resorts, in the first place, to the entity lending at the best price and only when not being able to get this will resort to financial institutions with higher prices.

Without entering into this discussion, which on the other hand has been the subject of numerous analyses, we propose making a valid explanation in a context of uncertainty, of the reasons that cause the independence of the financial structure. The following hypotheses can be taken for this:

1. The cost of the debts $\underset{\sim}{\mathbf{i}}_d$ as well as the capital cost $\underset{\sim}{\mathbf{i}}_c$ does not vary from one period to another although the proportion between outside capital and equity may well vary. It will be assumed that the ratio of indebtedness increases and, therefore $\underset{\sim}{\mathbf{r}}_d^{(a)} < \underset{\sim}{\mathbf{r}}_d^{(b)}$.
2. Also in this case the assumption is maintained that the cost of equity is higher than the cost of the debts, therefore $\underset{\sim}{\mathbf{i}}_a > \underset{\sim}{\mathbf{i}}_d$.

In the equation that defines capital cost we will have, obviously, that $i_c = k_1$, $i_d = k_2$, but on the other hand, i_a will be a function of $\gamma^{(b)}, i_a(\gamma^{(b)})$.

The result in this case will be in the transfer from one moment to another, rather than in the field of certainty:

$$i_c^{(b)} = \gamma^{(b)} \cdot i_d + (1 - \gamma^{(b)}) \cdot i_a(\gamma^{(b)}) = i_c^{(a)}$$

on assuming capital cost as invariable.

And finally:

$$i_a(\gamma^{(b)}) = \frac{i_c^{(b)}}{1 - \gamma^{(b)}} - \frac{\gamma^{(b)}}{1 - \gamma^{(b)}} \cdot i_d$$

In order to pass over to a sphere of uncertainty we are going to continue using the estimates from the previous example. For this we take as capital cost (invariable):

$$\underset{\sim}{\mathbf{i}}_c^{(b)} = \underset{\sim}{\mathbf{i}}_c^{(a)} = (13{,}25; 15{,}56; 18{,}20)$$

which, expressed in α-cuts, will be:

$$\underset{\sim}{\mathbf{i}}_c^{(b)}(\alpha) = [13{,}25 + 2{,}31\alpha; 18{,}20 - 2{,}64\alpha]$$

and as the cost of the debts, also constant:

$$\underset{\sim}{\mathbf{i}}_d^{(b)} = \underset{\sim}{\mathbf{i}}_d^{(a)} = (10; 11; 14)$$

which, expressed in α-cuts will be:

$$\underset{\sim}{\mathbf{i}}_d^{(b)}(\alpha) = \underset{\sim}{\mathbf{i}}_d^{(a)}(\alpha) = [10 + \alpha; 14 - 3\alpha]$$

The ratio of indebtedness, as under the previous heading, becomes:

$$\underset{\sim}{\mathbf{r}}_d^{(a)} = \frac{(0{,}30; 0{,}3\widehat{3}; 0{,}35)}{(0{,}65; 0{,}6\widehat{6}; 0{,}70)}$$

as:

$$\underset{\sim}{\mathbf{r}}_d^{(b)} = \frac{(0{,}36; 0{,}40; 0{,}42)}{(0{,}58; 0{,}60; 0{,}64)}$$

and in α-cuts, from being:

$$r_d^{(a)}(\alpha) = \frac{[0{,}30 + 0{,}0\widehat{3}\alpha; 0{,}35 - 0{,}0\widehat{1}\alpha]}{[0{,}70 - 0{,}0\widehat{3}\alpha; 0{,}65 + 0{,}0\widehat{1}\alpha]}$$

to become:

$$r_d^{(b)}(\alpha) = \frac{[0{,}36 + 0{,}04\alpha; 0{,}42 - 0{,}02\alpha]}{[0{,}64 - 0{,}04\alpha; 0{,}58 + 0{,}02\alpha]}$$

Capital cost $i_c^{(b)}(\alpha) = [13{,}25 + 2{,}31\alpha; 18{,}20 - 2{,}64\alpha]$ will have as the lower extreme:

$$(0{,}36 + 0{,}04\alpha)(10{+}\alpha) + (0{,}64 - 0{,}04\alpha)[p_1 + \alpha(p_2 - p_1)]$$
$$\wedge(0{,}42 - 0{,}02\alpha)(10{+}\alpha) + (0{,}58 + 0{,}02\alpha)[p_1 + \alpha(p_2 - p_1)]$$

and as the upper extreme:

$$(0{,}36 + 0{,}04\alpha)(14 - 3\alpha) + (0{,}64 - 0{,}04\alpha)[p_3 + \alpha(p_3 - p_2)]$$
$$\vee(0{,}42 - 0{,}02\alpha)(14 - 3\alpha) + (0{,}58 + 0{,}02\alpha)[p_3 + \alpha(p_3 - p_2)]$$

where:

$$i_a^{(b)}(\alpha) = [p_1 + \alpha(p_2 - p_1); p_3 + \alpha(p_3 - p_2)]$$

In the first place we proceed to an analysis of the lower extreme when $\alpha = 0$. We have:

$$0{,}36 \cdot 10 + 0{,}64 \cdot p_1 \wedge 0{,}42 \cdot 10 + 0{,}58 \cdot p_1 \,\square\, 3{,}6 + 0{,}64 \cdot p_1 \wedge 4{,}2 + 0{,}58 \cdot p_1$$

The graphical representation of these equations (see Fig. 12.1) shows that up to a certain value of p_1 (in this case $p_1 = 10$) it is lower:

$$(0{,}36 + 0{,}04\alpha)(10{+}\alpha) + (0{,}64 - 0{,}04\alpha)[p_1 + \alpha(p_2 - p_1)]$$

while when $p_1 > 10$ is less:

$$(0{,}42 - 0{,}02\alpha)(10{+}\alpha) + (0{,}58 + 0{,}02\alpha)[p_1 + \alpha(p_2 - p_1)]$$

since:

$$\begin{aligned} 0{,}36 \cdot 10 + 0{,}64 \cdot p_1 &= 0{,}42 \cdot 10 + 0{,}58 \cdot p_1 \\ 0{,}06 \cdot p_1 &= 0{,}06 \\ p_1 &= 10 \end{aligned}$$

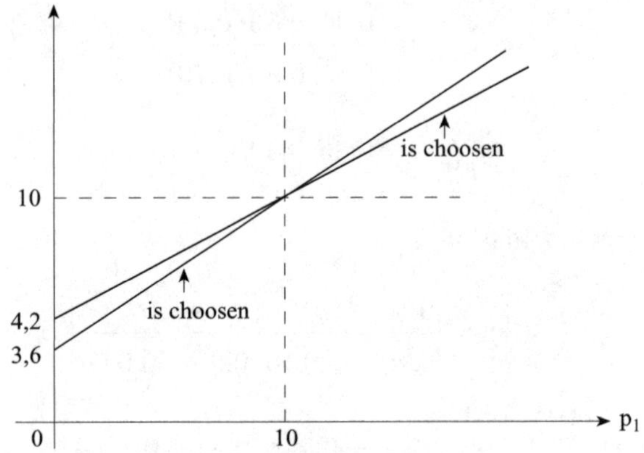

Fig. 12.1.

In this specific case, the condition that the cost of equity must be higher than the cost of outside capitals makes it necessary to choose as the lower extreme:

$$(0{,}42 - 0{,}02\alpha)(10 + \alpha) + (0{,}58 + 0{,}02\alpha)[p_1 + \alpha(p_2 - p_1)]$$

Now we continue to study the upper extreme when $\alpha = 0$. We arrive at:

$$0{,}36 \cdot 14 + 0{,}64 \cdot p_3 \vee 0{,}42 \cdot 14 + 0{,}58 \cdot p_3 \;\square\; 5{,}04 + 0{,}64 \cdot p_3 \vee 5{,}88 + 0{,}58 \cdot p_3$$

Taking into account that:

$$\begin{aligned} 5{,}04 + 0{,}64 \cdot p_3 &= 5{,}88 + 0{,}58 \cdot p_3 \\ 0{,}06 \cdot p_3 &= 0{,}84 \\ p_3 &= 14 \end{aligned}$$

Will be had when $p_3 < 14$, the following will be higher:

$$(0{,}42 - 0{,}02\alpha)(14 - 3\alpha) + (0{,}58 + 0{,}02\alpha)[p_3 + \alpha(p_3 - p_2)]$$

while when $p_3 > 14$ the following will be higher:

$$(0{,}36 + 0{,}04\alpha)(14 - 3\alpha) + (0{,}64 - 0{,}04\alpha)[p_3 + \alpha(p_3 - p_2)]$$

Figure 12.2 shows us what we have just stated.

If we take into account the hypothesis $\underset{\sim a}{\mathbf{i}} \succ \underset{\sim d}{\mathbf{i}}$ as the upper extreme the following is selected:

$$(0{,}36 + 0{,}04\alpha)(14 - 3\alpha) + (0{,}64 - 0{,}04\alpha)[p_3 + \alpha(p_3 - p_2)]$$

In order to arrive at the cost of equity we are going to operate, by starting out from the known equation:

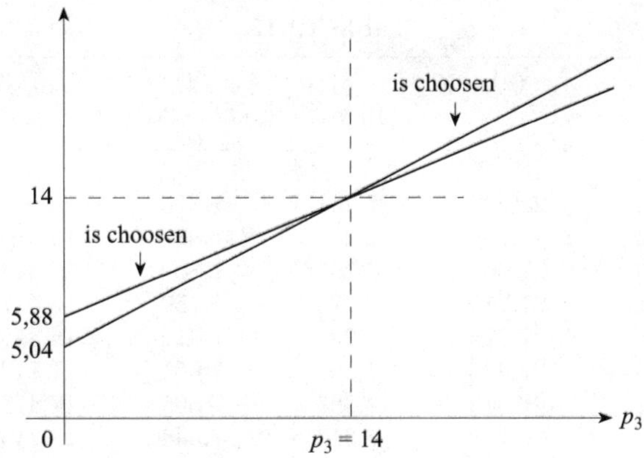

Fig. 12.2.

$$\begin{aligned}
&[13{,}25 + 2{,}31\alpha; 18{,}20 - 2{,}64\alpha] \\
&\quad = \{[p_1 + \alpha(p_2 - p_1)](0{,}58 + 0{,}02\alpha) + (0{,}42 - 0{,}02\alpha)(10 + \alpha); \\
&\quad\quad (0{,}64 - 0{,}04\alpha)[p_3 + \alpha(p_3 - p_2)] + (0{,}36 + 0{,}04\alpha)(14 - 3\alpha)\} \\
&\quad = \{[p_1 + \alpha(p_2 - p_1)](0{,}58 + 0{,}02\alpha); [p_3 + \alpha(p_3 - p_2)](0{,}64 - 0{,}04\alpha)\} \\
&\quad\quad (+)[(0{,}42 - 0{,}02\alpha)(10 + \alpha); (0{,}36 + 0{,}04\alpha)(14 - 3\alpha)]
\end{aligned}$$

From where:

$$\begin{aligned}
&[13{,}25 + 2{,}31\alpha; 18{,}20 - 2{,}64\alpha](\bar{m})[(0{,}42 - 0{,}02\alpha)(10 + \alpha)\,; \\
&\quad (0{,}36 + 0{,}04\alpha)(14 - 3\alpha)] \\
&\quad = [p_1 + \alpha(p_2 - p_1); p_3 + \alpha(p_3 - p_2)](\,\cdot\,)[0{,}58 + 0{,}02\alpha; 0{,}64 - 0{,}04\alpha]
\end{aligned}$$

Given the fact that we are resolving an equation, in order to arrive at $[p_1 + \alpha(p_2 - p_1); p_3 + \alpha(p_3 - p_2)]$ the quotient $(\dot{m})$ must be used, and therefore:

$$\begin{aligned}
&[p_1 + \alpha(p_2 - p_1); p_3 + \alpha(p_3 - p_2)] = \{[13{,}25 + 2{,}31\alpha; 18{,}20 - 2{,}64\alpha] \\
&\quad (\dot{m})[0{,}58 + 0{,}02\alpha; 0{,}64 - 0{,}04\alpha]\} \\
&\quad (\bar{m})\{[(0{,}42 - 0{,}02\alpha)(10 + \alpha); (0{,}36 + 0{,}04\alpha)(14 - 3\alpha)] \\
&\quad (\dot{m})[0{,}58 + 0{,}02\alpha; 0{,}64 - 0{,}04\alpha]
\end{aligned}$$

and finally:

$$\begin{aligned}
i_a^{(b)}(\alpha) &= [p_1 + \alpha(p_2 - p_1); p_3 + \alpha(p_3 - p_2)] \\
&= \left[\frac{13{,}25 + 2{,}31\alpha}{0{,}58 + 0{,}02\alpha}; \frac{18{,}20 - 2{,}64\alpha}{0,64 - 0{,}04\alpha}\right] \\
&\quad (\bar{m}) \left[(10 + \alpha)\frac{0{,}42 - 0{,}02\alpha}{0{,}58 + 0{,}02\alpha}; (14 - 3\alpha)\frac{0{,}36 + 0{,}04\alpha}{0{,}64 - 0{,}04\alpha}\right]
\end{aligned}$$

Table 12.1.

α	(1) $\frac{13{,}25+2{,}31\alpha}{0{,}58+0{,}02\alpha}$	(2) $\frac{18{,}20-2{,}64\alpha}{0{,}64-0{,}04\alpha}$	(3) $(10+\alpha)$ $\frac{0{,}42-0{,}02\alpha}{0{,}58+0{,}02\alpha}$	(4) $(14-3\alpha)$ $\frac{0{,}36+0{,}04\alpha}{0{,}64-0{,}04\alpha}$	(5) $(1)(\bar{m})(3)$	(6) $(2)(\bar{m})(4)$
0	22,844	28,437	7,241	7,875	15,603	20, 562
0,1	23,163	28,201	7,253	7,840	15,910	20, 361
0,2	23,479	27,962	7,265	7,802	16,214	20, 160
0,3	23,793	27,719	7,276	7,759	16,517	19,960
0, 4	24,105	27,474	7,287	7,712	16,818	19,762
0,5	24,415	27,225	7,296	7,661	17,119	19,564
0,6	24,722	26,974	7,305	7,605	17,417	19,369
0,7	25,028	26,718	7,313	7,544	17,715	19,174
0,8	25,332	26,460	7,320	7,478	18,012	18,982
0,9	25,633	26,198	7,327	7,408	18,306	18,790
1	25,933	25,933	7,333	7,333	0, 000	18,600

We now immediately express $i_a^{(b)}(\alpha)$ by means of the hendecagonal system (see Table 12.1).

If the triangular approximation is accepted as sufficiently representative, we will arrive at the fact that the cost of equity in the position will be:

$$\underset{\sim}{\mathbf{i}}_a^{(b)}(\alpha) = (15{,}603; 18{,}600; 20{,}562)$$

It can be seen that the operators $(\bar{m})$ and $(\dot{m})$ that have been used are correct as follows. For the lower extreme:

$$15{,}603 \times 0{,}58 + 10 \times 0{,}42 = 9{,}049 + 4{,}2 \cong 13{,}25$$

For the maximum presumption:

$$18{,}600 \times 0{,}60 + 11 \times 0{,}40 = 11{,}16 + 4{,}4 \cong 15{,}56$$

For the upper extreme:

$$20{,}562 \times 0{,}64 + 14 \times 0{,}36 = 13{,}159 + 5{,}04 \cong 18{,}20$$

The comparison of the cost of equity in position a, $\underset{\sim}{\mathbf{i}}_a^{(a)}$ and in position b, $\underset{\sim}{\mathbf{i}}_a^{(b)}$ brings to light the fact that as:

$$(15; 18; 20) \prec (15{,}603; 18{,}600; 20{,}562)$$

the result being that the cost equity of increases on substituting equity for outside capitals, given the advantages that are obtained as a consequence of the increased financial profitability as the purchasers of shares require a higher remuneration for their capitals in order to compensate for the increased financial risk they assume, since at least maintaining the quotation of the shares continues to be a valid objective.

13 Incidence of the Financial Structure on the Value of the Business

13.1 Classical Treatment of the Problem

In the extensive literature arising from the important matter constituted by the hypothetical existence of an optimum financial structure, a relevant place is occupied by the relation between financial structure and the "value of the business". Let us say beforehand that a new position with regard to this problem is not to be found among our objectives, based on starting out points that are different from those that excellent writers and illustrious professors have sustained. Neither do we wish to lengthen the extensive list of treatises that include, study, compile or summarise the theses sustained by the different authors. The finality that we are seeking is perhaps more modest, but has the hope that it may provide a new orientation to this problem, with the support of the modern techniques for treatment of uncertainty. Even though this contribution may be as miniscule as a grain of sand, but allow us the audacity to think that the result of our efforts has been well worth while.

In the first place let us point out that the sense that we give to the notion of "value of the business" coincides with the most accepted meaning in the books on this subject, which identifies the term with the sum of the market value of the shares and the market value of the debts. If the value of the securities of the business are represented by V (shares and liabilities), by A the market value of the shares and by D the market value of the debts we will obviously arrive at:

$$V = A + D$$

Within the two components of the value of the business, the different authors pay special attention to the procedure for arriving at the market value of the shares, A. According to which of these procedures are used the value of the shares of a business will vary. Without being exhaustive on the subject we mention the following:

1. The market value of the shares is arrived at by means of the capitalisation of the net results, that is, by the results of the exploitation less financial expenses. It is the so-called RN procedure. Taking into account that in the calculation of the net results financial expenses take a part, it is obvious then that the financial structure has an influence on the value of the business.

Anna Maria Gil-Lafuente: *Fuzzy Logic in Financial Analysis*, StudFuzz **175**, 217–240 (2005)
www.springerlink.com

Let us take a look at how the different elements intervening in the problem play their roles in the field of certainty. We shall call:

R = net results;
B = operating profit;
F = financial expenses,

Where:

$$R = B - F$$

Under the case we have admitted, the market value of the shares will be:

$$A = \frac{R}{i_a}$$

where i_a = cost of equity.
At an initial stage, 0, the value of the business will be:

$$V_0 = \frac{R}{i_a} + D = \frac{B - F}{i_a} + D = \frac{B - i_d \cdot D}{i_a} + D = \frac{B}{i_a} + D \cdot \left(1 - \frac{i_d}{i_a}\right)$$

where i_d = the cost of the debts.
We now move on to assume that the financial structure varies, for example, substituting equity by debts, with this the ratio of indebtedness will also be modified. With this calculation we want to see how the value of the business changes. We will have:

$$V_1 = \frac{B - i_d \cdot (D + \Delta D)}{i_a} + (D + \Delta D) = \frac{B}{i_a} + D \cdot \left(1 - \frac{i_d}{i_a}\right) + \Delta D \cdot \left(1 - \frac{i_d}{i_a}\right)$$

In order to arrive at any immediate conclusions the following hypotheses are adopted:

(a) The cost of the debts is lower than that of equity. Then, as $i_d < i_a$ we will have $\frac{i_d}{i_a} < 1$.
(b) The rates i_d and i_a are constants whatever the capital and the volume of debts may be. This is equivalent to saying that the costs of equity and outside capitals are independent of the ratio of indebtedness D/A.

Only then can the conclusion be reached that as the ratio of indebtedness increases the value of the business also increases.
In fact, as $\frac{i_d}{i_a} < 1$ then $D \cdot (1 - \frac{i_d}{i_a}) > 0$ will be and, therefore, $V_1 > V_0$.

2. The market value of the shares is obtained by means of the capitalisation of the operating profits to a rate equal to capital cost. This is the so called RE procedure. Underlying in this procedure is the hypothesis that the value of the business is independent from its financial structure[1], that is,

[1] Among the more important defenders of this position mention can be made of Modigliani F and Miller M (June 1958) with the well known article "The Cost of Capital, Corporate Finance, and the Theory of the Firm". (Ed) The American Economic Review.

from the ratio of indebtedness, and only indebtedness, and only depends on the capacity of the assets for generating profits.
In this case, the value of the business will be:

$$V = \frac{B}{i_c}$$

where i_c = capital cost.
Underlying in this procedure is another hypothesis: capital cost i_c remains constant whatever the structure, that is the ratio of indebtedness. In order to be able to admit this hypothesis certain considerations are necessary. If for example, the ratio of indebtedness increases as a consequence of substituting equity for outside capitals, on always assuming $i_a > i_d$ economic advantages will be obtained by substituting the most expensive capitals for others that are less expensive. If the cost of the debts i_d remains constant whatever the level of indebtedness, the constancy of capital cost i_c can only be justified if the cost of equity i_a increases. We conclude by saying that in the market place, purchasers of shares will only accept the maintenance of their quotation if they get greater profitability (increase of i_a) which compensates the higher financial risk they consume, as a consequence of a relatively higher indebtedness.

3. There are many other approaches to this subject. We only mention the fact that there exists a traditional position which sustains the existence of an optimum financial structure for which the value of the business is maximised. Thus, it is assumed, as we did for the RE hypothesis, that the differential obtained on substituting equity for outside capitals is absorbed, in a certain manner, by the increase of equity. But contrary to the RE hypothesis, until the optimum financial structure $(D/A)^{\text{opt}}$, the cost of equity i_a does not increase sufficiently in order to compensate for the differential and, therefore capital cost i_c decreases. Starting out from the optimum of the financial structure $(D/A)^{\text{opt}}$, the increase in the cost of equity i_a is higher than the differential, and therefore, capital cost i_c grows. The mechanism that moves the variation of the value of the business operates in reverse. Up to the optimum financial structure the value of the business grows. From the point of $(D/A)^{\text{opt}}$ it decreases.

If in reality an economic system of perfect competition were to exist, "market forces" would make sure of balancing the differential risks with differential yields. But in reality, for several reasons, "frictions" exist in the market which makes it imperfect. It is in this environment that Durand[2] proposes a path in which he introduces the "super-premiums of security", which are the difference between the market price of a debt and the price,

[2] Durand D (1952) "Cost of Debt Funds for Business: Trends and Problems of Measurement" in Conference and Research on Business Finance, National Bureau of Economic Research, New York, pp. 215–247.

lower, that investors subscribe as a consequence of the prestige, faith in the solvency of the business, etc.

In order to finish this section we are going to develop a numerical example which we will place in the sphere of uncertainty. With the object of simplifying the operation to the maximum and focussing our attention on the comparative aspects among the different positions, we will use estimates by means of confidence intervals.

We are going to start out from a hypothetical business, the economic structure of which is foreseen will be formed, at an initial point in time, only by equity with a volume of 10.000 monetary units.

In order to maintain a determined level for the quotation of its shares, the holders of capital require a cost of $i_a = [0{,}14; 0{,}16]$, while if resorting to outside financing, financial means could be got at a cost of $i_d = [0{,}10; 0{,}12]$. Operating profits estimated for this moment are $\underset{\sim}{\mathbf{B}} = [1.400; 1.600]$. In this highly schematic case, the value of the business is the same both for position RN as it is for position RE, since $\underset{\sim}{\mathbf{i}}_c = \underset{\sim}{\mathbf{i}}_a$ and also $\underset{\sim}{\mathbf{R}} = \underset{\sim}{\mathbf{B}}$ as no cost of debts exists. This will be:

$$V_0 = [1.400; 1.600] : [0{,}14; 0{,}16] = [8.750; 11.428]$$

with the common number that best represents it: $V_0 = 10.089$.

Now we are going to look at what happens when we resort to outside financing. For this we are going to assume that equity is changed for debts, in an amount estimated at $[2.000; 3.000]$.

(1) Value of the business according to RN. The net result is calculated:

$$\underset{\sim}{\mathbf{R}} = \underset{\sim}{\mathbf{B}} - \underset{\sim}{\mathbf{F}}$$
$$\square\ \underset{\sim}{\mathbf{R}} = [1.400; 1.600](-)[0{,}10; 0{,}12](\cdot)[2.000; 3.000]$$
$$\square\ \underset{\sim}{\mathbf{R}} = [1.400; 1.600](-)[200; 360] = [1.040; 1.400]$$

and the value of the business[3]:

$$\underset{\sim}{\mathbf{V}}_1^{RN} = [1.040; 1.500](:)[0{,}14; 0{,}16](+)[2.000; 3.000]$$
$$\square\ \underset{\sim}{\mathbf{V}}_1^{RN} = [6.500; 10.000](+)[2.000; 3.000] = [8.500; 13.000]$$

which has as its most representative common number:

$$\bar{\underset{\sim}{\mathbf{V}}}_1^{RN} = 10.750$$

(2) Value of the business according to RE. If the hypothesis is accepted of the constancy of capital cost i_c, it is obvious that indebtedness will not

[3] It is necessary to use the original formula, since $V_0 = \frac{B}{i_a} + D \cdot (1 - \frac{i_d}{i_a})$ is the result of doing determined arithmetical operations in the field of certainty which, as is known, do not always coincide with those required by uncertainty.

modify the initial value of the business V_0, since $i_a^{(1)} > i_a$ in order to compensate the lower cost of the [2.000; 3.000] which now have a lower price (specifically [0,10; 0,12].
This will be then:

$$\underset{\sim 1}{\mathbf{V}}^{RE} = \underset{\sim 0}{\mathbf{V}} = [8.750; 11.428]$$

with:

$$\underset{\sim 1}{\bar{\mathbf{V}}}^{RE} = 10.089$$

(3) Value of the business according to variable RE. We will start out now, with the same RE hypotheses but with a variation. In stead of i_c remaining constant we will assume that i_a is constant.
We have to calculate the new capital cost. For this let us remember that, taking into account the need that the percentage of equity plus the percentage of outside capitals must be equal to 1, it is not possible to continue the elemental arithmetic of confidence intervals. We will operate then as follows. On the one hand we will have:

$$\begin{aligned} &[0{,}10; 0{,}12](\cdot)\frac{2.000}{10.000}(+)[0{,}14; 0{,}16](\cdot)\frac{8.000}{10.000} \\ &= [0{,}020; 0{,}024](+)[0{,}112; 0{,}128] \\ &= [0{,}132; 0{,}152] \end{aligned}$$

And on the other:

$$\begin{aligned} &[0{,}10; 0{,}12](\cdot)\frac{3.000}{10.000}(+)[0{,}14; 0{,}16](\cdot)\frac{7.000}{10.000} \\ &= [0{,}030; 0{,}036](+)[0{,}098; 0{,}112] \\ &= [0{,}128; 0{,}148] \end{aligned}$$

For increased security, we will choose the lowest of the lower extremes and the highest of the upper extremes, with which we arrive at:

$$\underset{\sim c}{\mathbf{i}} = [0{,}128; 0{,}148]$$

We can now find the value of the business:

$$\underset{\sim 1}{\mathbf{V}}^{RN} = \frac{B}{i_c} = [1.400; 1.600](:)[0{,}128; 0{,}148] = [9.210; 12.500]$$

with a common number which represents it, $\underset{\sim 1}{\bar{\mathbf{V}}}^{RN} = 10.855$.
Value of the business according to Durand. With the same basic data let us assume now that, as a consequence of the hoped for solidity of the business and good position in the market (imperfections or "frictions" of market economy) investors are prepared to assume a debt of [2.000; 3.000] at a price of [0,08; 0,09]. In this way they pay a "super-premium" estimated at:

$$[0{,}10; 0{,}12](\cdot)[0{,}08; 0{,}09] = [0{,}01; 0{,}04]$$

Profitability for creditors will be:

$$[2.000; 3.000](\cdot)[0{,}08; 0{,}09] = [160; 270]$$

The value of the debts on capitalising the profitability obtained at the market rate of $[0{,}10; 0{,}12]$ will give rise to:

$$[160; 270](:)[0{,}10; 0{,}12] = [1.333; 2.700]$$

In this way, the "value of equity" for the shareholders will be:

$$10.000(-)[1.333; 2.700] = [7.300; 8.667]$$

Now then, the "value of the debts" for the shareholders who go out into the market will continue to be:

$$[2.000; 3.000]$$

Finally, we arrive at the fact that the "value of the business" will be:
If the debt issue is 2.000:

$$V_1^D = 8.667 + 2.000 = 10.667$$

If the debt issue is 3.000:

$$V_1^D = 7.300 + 3.000 = 10.300$$

We arrive then in short at:

$$\underset{\sim}{\mathbf{V}}_1^D = [10.300; 10.667]$$

the common number which best represents this is $\bar{\underset{\sim}{\mathbf{V}}}_1^D = [10.300; 10.667]$.

Obviously we could go on establishing new hypotheses in order to find other schemes which would give rise to different values for the business. We feel that this sample should be sufficient for providing a brief idea of the mechanisms that move the relations of causality between the financial structure and the value of the business.

We have limited ourselves to representing the estimates of the values intervening in this problem by means of confidence intervals as this is the simplest and most elementary form for representing uncertainty. But all that we have stated above can be made extensive to confidence triplets, T.F.N., fuzzy numbers, etc.

From now onwards we are going to assume, for the effect of greater generalisation, that the estimate obtained for the value of the business is expressed in a special form of fuzzy numbers.

13.2 New Approaches for Arriving at the Long-Term Value

The specialist works on this subject are occupied, as we have also been, in bringing to light how, under certain hypotheses, the most significant values move as a consequence of the activity of the business. These movements will in the end give rise to the greater or lesser appreciation they deserve from their surroundings, that is to say to the economic estimate and financial prestige they warrant. In short they will give an account of their worth.

All are aware that one of the subjects that has undergone the most debate and study in the field of financial analysis is the incidence of the financial structure on the value of the business and above all on capital cost, which has led to the controversy relative to the existence or not of an optimum financial structure.

For this we are going to commence with a statement that is as elemental as the fact that the incorporation of equity and outside capitals in a certain proportion will give rise to a determined "financial structure". For greater ease it is normal to use, as a representation of the same, the quotient between outside sources and in house sources, which is known as the "ratio of indebtedness". Therefore with these financing sources, that is with a determined "ratio of indebtedness" the acquisition takes place of a determined economic structure (fixed assets, stocks, financial assets, etc.), that is, of certain means of production by means of the mobilisation of which the primary objectives of producing, selling, collecting and obtaining profits (the nature of which we are not interested in specifying) are carried out.

This activity takes place through a succession of periods of time, throughout which these values are modified as a consequence of very divers facts linked on occasions to the situation of the general economic system and other to the internal evolution of the business itself.

The interest awakened by this subject is directed at, in all truth, on certain occasions to obtaining estimates of the value of the business (and also of capital cost) in the short-term, but it is no less truthful to state that, in general, what is required to be known is "where is" the value of the business going, that is to say, what will be the estimate of this important value in the long-term: if the business tends to go forward towards a determined value, or else will there be periodic fluctuations in its value, or is it not possible, with the available means, to foresee a long-term stable situation.

In whichever of these cases there obviously exists a similarity relative to the general scheme with which the functioning of the system can be represented (see Fig. 13.1).

Our objective will consist of studying the performance of the internal financial system, in order to determine, if this is possible, what in the future will be the estimate for the value of the business. The following possibilities exist:

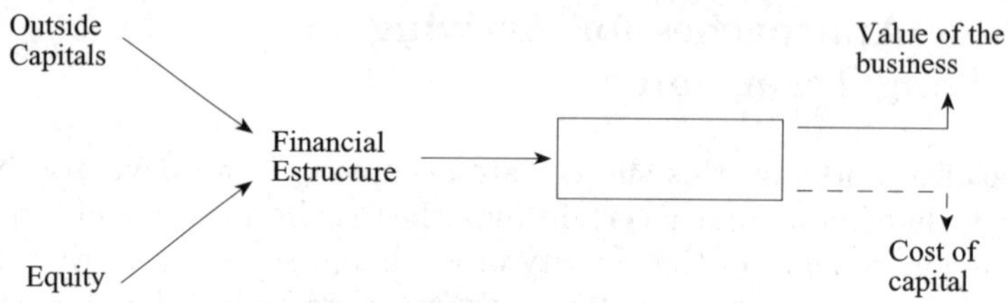

Fig. 13.1.

1. The system tends towards a certain value for the business.
2. A value for the business does not exist that tends towards the system, but a certain periodicity is seen
3. The system has a chaotic performance.

We are now going to bring to light a technique that allows for the treatment of this problem in the first of these hypotheses, that is when the system tends towards its limit. In the field of randomness it would be said that the system is ergodic: we are going to place ourselves in the field of uncertainty.

13.3 Determination of the Value in the Hypothesis of Convergence

With the object of avoiding any type of complication due to an excess of formalism, we will develop the process by means of a practical case taken from reality. Obviously its generalisation is immediate.

A group of experts have been consulted in order for them to come up with the value for a business over the long-term. Obviously the number of situations has been limited, and as we are working in a sphere of uncertainty, they have estimated these by means of the following confidence intervals:

$$E_1 = [20; 24] \quad E_2 = [25; 29] \quad E_3 = [30; 34] \quad E_4 = [35; 39]$$
$$E_5 = [40; 44] \quad E_6 = [45; 49] \quad E_7 = [50; 54]$$

Therefore it would be said that system S will be formed by situations E_i, $i = 1; 2; \cdots; 7$:

$$S = \{E_1, E_2, \ldots, E_7\}$$

On the other hand, in accordance with the opinion of the experts and the corresponding operations of fuzzy arithmetic, a normal fuzzy sub-set has been arrived at that expresses, by means of valuations, that is, subjective numerical estimates, the possibility that the system may be found a an initial moment 0, in each one of the possible situations E_i, $i = 1; 2; \cdots; 7$. We are presenting what in the field of randomness is called a situation vector. In this case let us assume that it will be:

$\underset{\sim}{q}_0 =$

E_1	E_2	E_3	E_4	E_5	E_6	E_7
0,3	0,4	0,8	1	0,9	0,2	0,1

As is known to each of the values of the function characteristic of membership $M(x)$ we can correspond with an expression in a semantic scale by means of the hendecagonal system[4].

Having arrived at these estimates we then proceed to draw up, also by the experts, a "normal fuzzy relation of transition". This is a subjective matrix which expresses the possibility that system S passes over to situation E_i at moment t, to a situation E_j at moment $t+1$, in which, $i, j = 1,2,\ldots,7$, $t = 0,1,2,\ldots$.

Therefore, if the experts have said that there is a possibility of passing from one value of the business $E_1 = [20; 24]$ in one period, to $E_1 = [20; 24]$ in the following period (that is, no variation), it will be 0.5; that it passes from $E_1 = [20; 24]$ to $E_2 = [25; 29]$ it will be 0.8; that it passes from to $E_1 = [20; 24]$ to $E_3 = [30; 34]$ it will be 1; ...; that it passes from $E_7 = [50; 54]$ to $E_7 = [50; 54]$ (that is it remains the same) it will be 1, it is possible to construct a fuzzy relation of transition $\underset{\sim}{N}$ as follows:

$\underset{\sim}{N} =$

	E_1	E_2	E_3	E_4	E_5	E_6	E_7
E_1	0,5	0,8	1	0,6	0,1	0	0
E_2	0	0,7	1	0,9	0,4	0,1	0
E_3	0	0,3	0,8	1	0,2	0	0
E_4	0	0	0,6	0,7	1	0,2	0
E_5	0	0	0	0,9	1	0,5	0
E_6	0	0	0	0,6	0,9	1	0,3
E_7	0	0	0	0	0,8	0,9	1

It will be seen that this is dealing with a business with expansionist objectives, since hypothetically starting out from a determined situation in one period, the maximum presumption always exists that in the following period

[4] In this respect a selection can be made of any of the scales proposed in Kaufmann A and Gil Aluja J (1992) Técnicas de gestión de empresas. Previsiones, decisiones y estrategias. (Ed) Pirámide, Madrid, by doing, if suitable, the most convenient variation.

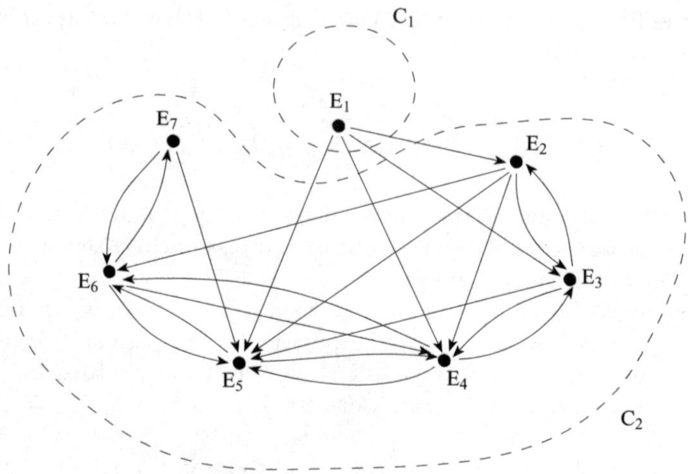

Fig. 13.2.

the increased presumption will be to increase its value or in any event maintain it. Neither will the fact escape the reader that we started out from a "stationary" fuzzy relation, in the sense that we maintained the subjective matrix for all the periods that could be considered. In this way, the passage from situation E_i to another E_j, $i, j = 1,2,\ldots,7$, has the same valuation whichever period is considered. It is also possible to use an "evolutionary" fuzzy relation, but in this case the possibilities for treatment would be limited to the first of the processes we are about to develop.

The graphical representation shown in Fig. 13.2 can be associated to the fuzzy relation $\underset{\sim}{\mathbf{N}}$.

Taking into account that this graphic representation is not heavily connected[5] (when leaving E_1 one cannot return to E_1) we must proceed to breakdown this matrix into classes of equivalence C_i, $i = 1,2,\ldots,$ which is equivalent to breaking down a non heavily connected graphic representation into heavily connected sub-graphs. For this we will use the Malgrange[6] method, which consists in taking into consideration any situation, for example E_5, in an arbitrary manner. The transitive closing is found for E_5, which we shall call $\hat{\Gamma}(\underset{\sim}{\mathbf{E}}_5)$ and also the inverse transitive closing of the same situation E_5, which we designate $\hat{\Gamma}^{-1}(\underset{\sim}{\mathbf{E}}_5)$. The intersection $\hat{\Gamma}(\underset{\sim}{\mathbf{E}}_5) \cap \hat{\Gamma}^{-1}(\underset{\sim}{\mathbf{E}}_5)$ will give rise to all the situations that are members of the same class of

[5] We should remember that in a heavily connected graphical representation one can always pass over from one vertex to another, that is, that there always exists at least one path that leads from any one vertex to another.

[6] Malgrange Yves (1967) Descompositions d'un graphe en sous-graphes fortement connexes maximaux, Cie. Machines Bull, quoted by Kaufmann A and Gil Aluja J (1991) Nuevas técnicas para la dirección estratégica, University of Barcelona, p. 30.

equivalence as E_5. For our objective, we are going to consider the Boolean matrix[7] associated to $\underset{\sim}{N}$:

	E_1	E_2	E_3	E_4	E_5	E_6	E_7	$\hat{\Gamma}(\underset{\sim}{E}_5)$
E_1	1	1	1	1	1			X
E_2		1	1	1	1	1		3
E_3		1	1	1	1			2
E_4			1	1	1	1		1
E_5				1	1	1		0
E_6				1	1	1	1	1
E_7					1	1	1	2
$\hat{\Gamma}^{-1}(\underset{\sim}{E}_5)$	1	1	1	1	0	1	1	

The intersection will give rise to

$$\begin{aligned}\hat{\Gamma}(\underset{\sim}{E}_5) \cap \hat{\Gamma}^{-1}(\underset{\sim}{E}_5) &= \{E_2, E_3, E_4, E_5, E_6, E_7\} \cap \{E_1, E_2, E_3, E_4, E_5, E_6, E_7\} \\ &= \{E_2, E_3, E_4, E_5, E_6, E_7\}\end{aligned}$$

We arrive at the fact that with this situations E_2, E_3, E_4, E_5, E_6 and E_7 form the same class of equivalence and the corresponding sub-graph will be heavily connected.

The process will continue to eliminate from the associated Boolean matrix the rows and columns we have found, i.e. E_2, E_3, E_4, E_5 and E_7, with which in this very simple case, only E_1 remains, which would give rise to a matrix of 1×1, with which another class of equivalence will have been arrived at formed only by E_1. In the more complex case, several situation would remain forming a square matrix of the order of $r > 1$, which would permit continuing with the procedure by arbitrarily choosing a situation from those remaining, in order to arrive at the transitive closing and the inverse transitive closing, doing the intersection in order to find the situations that form the new equivalency class and so on successively. But let us return to our case. There exist, then, two classes of equivalency:

$$(C_1) = \{E_1\}$$
$$C_2 = \{E_2, E_3, E_4, E_5, E_6, E_7\}$$

[7] As is known, in order to arrive at a Bolean matriz associated to a fuzzy relation it is sufficient to substitute the values that are not nil in each box with one.

Fig. 13.3.

Taking into account that the sub-indices of the C_1 have been assigned in an arbitrary manner we are going to order them, that is, we are going to arrive at the ordinal function. In this specific case[8] a simple glance at the classes graph will be sufficient to see that:

$$C_1 \prec C_2$$

The corresponding class graph will be represented by Fig. 13.3 and the normal broken down fuzzy relation of transition $\underset{\sim}{\mathbf{N}}$ and ordered in equivalence classes would be:

	E_1	E_2	E_3	E_4	E_5	E_6	E_7
E_1	C_1						
E_2							
E_3							
E_4				C_2			
E_5							
E_6							
E_7							

The order, in this case, coincides with the order initially assigned for each one of the situations. If this were not so, an adequate permutation would have to be done between the rows and also between the columns, with the object of placing successively together the situations that are a member of the same equivalency class. Once the normal fuzzy relation of transition has been reconstructed in the new way (in the random it is said in a normal way), which in this case coincides with the original form, we proceed with the opportune calculations.

[8] Simple algorithms exist that permit arriving at the ordinal function of a graph. See, for example, Kaufmann A and Gil Aluja J Nuevas técnicas para la dirección estratégica, op. cit., pp. 37–40.

In this respect, and in order to find an estimate of the value of the business in a future moment of time, which could be considered as "situation on the limit" several paths can be followed. We are going to consider two of these, which in the end give rise to coinciding results.

The first consists in successively arriving at the estimate of the situation of the system at moment $t = 1$, $[\underset{\sim}{\mathbf{q}}_1]$, starting out from the fuzzy sub-set that shows the situation of the system at moment $t = 0$ by means of the fuzzy sub-set $[\underset{\sim}{\mathbf{q}}_0]$ expressed in the form of a situation vector, which is composed, by means of the max-min operator with the normal fuzzy relation of transition[9]. In the same way $[\underset{\sim}{\mathbf{q}}_2]$, $[\underset{\sim}{\mathbf{q}}_3], \ldots,$ are arrived at until such time as convergence occurs.

This will take place when $[\underset{\sim}{\mathbf{q}}_r] = [\underset{\sim}{\mathbf{q}}_{r+1}]$, $0 < r < n$. What we will have is:

$$[\underset{\sim}{\mathbf{q}}_1] = [\underset{\sim}{\mathbf{q}}_0] \circ [\underset{\sim}{\mathbf{N}}]$$
$$[\underset{\sim}{\mathbf{q}}_2] = [\underset{\sim}{\mathbf{q}}_1] \circ [\underset{\sim}{\mathbf{N}}]$$
$$\cdots$$
$$[\underset{\sim}{\mathbf{q}}_r] = [\underset{\sim}{\mathbf{q}}_{r-1}] \circ [\underset{\sim}{\mathbf{N}}]$$

where $\circ$ represents the composition or max-min convolution.

Let us take a look at the calculations for our practical case:

E_1	E_2	E_3	E_4	E_5	E_6	E_7
0,3	0,4	0,8	1	0,9	0,2	0,1

$\circ$

t=0

	E_1	E_2	E_3	E_4	E_5	E_6	E_7
E_1	0,5	0,8	1	0,6	0,1	0	0
E_2	0	0,7	1	0,9	0,4	0,1	0
E_3	0	0,3	0,8	1	0,2	0	0
E_4	0	0	0,6	0,7	1	0,2	0
E_5	0	0	0	0,9	1	0,5	0
E_6	0	0	0	0,6	0,9	1	0,3
E_7	0	0	0	0	0,8	0,9	1

$=$

E_1	E_2	E_3	E_4	E_5	E_6	E_7
0,3	0,4	0,8	0,9	1	0,5	0,2

t=1

[9] It will be seen that, in the case of uncertainty, the process to be followed coincides with the process for randomness, but changing the sum-product composition operator by the max-min convolution.

Fuzzy situation vector at moment $t = 1$:

E_1	E_2	E_3	E_4	E_5	E_6	E_7
0,3	0,4	0,8	0,9	1	0,5	0,2

t=1 ∘

	E_1	E_2	E_3	E_4	E_5	E_6	E_7
E_1	0,5	0,8	1	0,6	0,1	0	0
E_2	0	0,7	1	0,9	0,4	0,1	0
E_3	0	0,3	0,8	1	0,2	0	0
E_4	0	0	0,6	0,7	1	0,2	0
E_5	0	0	0	0,9	1	0,5	0
E_6	0	0	0	0,6	0,9	1	0,3
E_7	0	0	0	0	0,8	0,9	1

=

E_1	E_2	E_3	E_4	E_5	E_6	E_7
0,3	0,4	0,8	0,9	1	0,5	0,3

t=2

Fuzzy situation vector at moment $t = 2$:

E_1	E_2	E_3	E_4	E_5	E_6	E_7
0,3	0,4	0,8	0,9	1	0,5	0,3

t=1 ∘

	E_1	E_2	E_3	E_4	E_5	E_6	E_7
E_1	0,5	0,8	1	0,6	0,1	0	0
E_2	0	0,7	1	0,9	0,4	0,1	0
E_3	0	0,3	0,8	1	0,2	0	0
E_4	0	0	0,6	0,7	1	0,2	0
E_5	0	0	0	0,9	1	0,5	0
E_6	0	0	0	0,6	0,9	1	0,3
E_7	0	0	0	0	0,8	0,9	1

=

E_1	E_2	E_3	E_4	E_5	E_6	E_7
0,3	0,4	0,8	0,9	1	0,5	0,3

t=3

Fuzzy situation vector at moment $t = 3$. We stop here because $[q_2] = [q_3]$.

The same result would be arrived at by following another process, as a consequence of certain properties of normal fuzzy relations and the max-min composition operator. This new procedure will consist of composing the normal fuzzy relation of transition 1,2,...,r times until $[\underset{\sim}{\mathbf{N}}]^r = [\underset{\sim}{\mathbf{N}}]^{r+1}$, $0 \leq r \leq n$ occurs. We arrive at:

$[\underset{\sim}{\mathbf{N}}] =$

↱	E_1	E_2	E_3	E_4	E_5	E_6	E_7
E_1	0,5	0,8	1	0,6	0,1	0	0
E_2	0	0,7	1	0,9	0,4	0,1	0
E_3	0	0,3	0,8	1	0,2	0	0
E_4	0	0	0,6	0,7	1	0,2	0
E_5	0	0	0	0,9	1	0,5	0
E_6	0	0	0	0,6	0,9	1	0,3
E_7	0	0	0	0	0,8	0,9	1

$[\underset{\sim}{\mathbf{N}}]^2 =$

↱	E_1	E_2	E_3	E_4	E_5	E_6	E_7
E_1	0,5	0,8	1	0,6	0,2	0	0
E_2	0	0,7	0,8	1	0,9	0,4	0,1
E_3	0	0,3	0,8	0,8	1	0,2	0
E_4	0	0,3	0,6	0,9	1	0,5	0,2
E_5	0	0	0,6	0,9	1	0,5	0,3
E_6	0	0	0,6	0,9	0,9	1	0,3
E_7	0	0	0	0,8	0,9	0,9	1

$[\underset{\sim}{\mathbf{N}}]^3 =$

↱	E_1	E_2	E_3	E_4	E_5	E_6	E_7
E_1	0,5	0,7	0,8	0,8	1	0,5	0,2
E_2	0	0,7	0,8	0,9	1	0,5	0,3
E_3	0	0,3	0,8	0,9	1	0,5	0,2
E_4	0	0,3	0,6	0,9	1	0,5	0,3
E_5	0	0,3	0,6	0,9	1	0,5	0,3
E_6	0	0,3	0,6	0,9	0,9	1	0,3
E_7	0	0	0,6	0,9	0,9	0,9	1

$[\underset{\sim}{\mathbf{N}}]^4 =$

↱	E_1	E_2	E_3	E_4	E_5	E_6	E_7
E_1	0,5	0,7	0,8	0,9	1	0,5	0,3
E_2	0	0,7	0,8	0,9	1	0,5	0,3
E_3	0	0,3	0,8	0,9	1	0,5	0,3
E_4	0	0,3	0,6	0,9	1	0,5	0,3
E_5	0	0,3	0,6	0,9	1	0,5	0,3
E_6	0	0,3	0,6	0,9	0,9	1	0,3
E_7	0	0,3	0,6	0,9	0,9	0,9	1

$[\underset{\sim}{\mathbf{N}}]^5 =$

↱	E_1	E_2	E_3	E_4	E_5	E_6	E_7
E_1	0,5	0,7	0,8	0,9	1	0,5	0,3
E_2	0	0,7	0,8	0,9	1	0,5	0,3
E_3	0	0,3	0,8	0,9	1	0,5	0,3
E_4	0	0,3	0,6	0,9	1	0,5	0,3
E_5	0	0,3	0,6	0,9	1	0,5	0,3
E_6	0	0,3	0,6	0,9	0,9	1	0,3
E_7	0	0,3	0,6	0,9	0,9	0,9	1

We stop at this point as $[\underset{\sim}{\mathbf{N}}]^4 = [\underset{\sim}{\mathbf{N}}]^5$.

As we are all well aware, in the sphere of uncertainty the fuzzy relation does not always provide the highest value of each one of the elements[10]. For this reason the transitive closing is generally considered in stead of the limit relation. In this case it will be:

$$[\hat{\underset{\sim}{\mathbf{N}}}] = [\underset{\sim}{\mathbf{N}}] \cup [\underset{\sim}{\mathbf{N}}]^2 \cup [\underset{\sim}{\mathbf{N}}]^3 \cup [\underset{\sim}{\mathbf{N}}]^4 \cup [\underset{\sim}{\mathbf{N}}]^5$$

[10] The limit fuzzy relation of transition only coincides with the transitive closing if the initial fuzzy relation is reflexive.

	E_1	E_2	E_3	E_4	E_5	E_6	E_7
E_1	0,5	0,8	1	1	1	0,5	0,3
E_2	0	0,7	1	1	1	0,5	0,3
E_3	0	0,3	0,8	1	1	0,5	0,3
E_4	0	0,3	0,6	0,9	1	0,5	0,3
E_5	0	0,3	0,6	0,9	1	0,5	0,3
E_6	0	0,3	0,6	0,9	0,9	1	0,3
E_7	0	0,3	0,6	0,9	0,9	0,9	1

In order to arrive at the fuzzy situation vector as from moment $t = 4$ it is sufficient to do:

$$[\underset{\sim}{q}_0] \circ [\hat{\underset{\sim}{N}}] = [\underset{\sim}{q}_n], \quad n \geq 4$$

which in this case will be:

E_1	E_2	E_3	E_4	E_5	E_6	E_7
0,3	0,4	0,8	1	0,9	0,2	0,1

$\circ$

t=0

	E_1	E_2	E_3	E_4	E_5	E_6	E_7
E_1	0,5	0,8	1	1	1	0,5	0,3
E_2	0	0,7	1	1	1	0,5	0,3
E_3	0	0,3	0,8	1	1	0,5	0,3
E_4	0	0,3	0,6	0,9	1	0,5	0,3
E_5	0	0,3	0,6	0,9	1	0,5	0,3
E_6	0	0,3	0,6	0,9	0,9	1	0,3
E_7	0	0,3	0,6	0,9	0,9	0,9	1

=

E_1	E_2	E_3	E_4	E_5	E_6	E_7
0,3	0,4	0,8	0,9	1	0,5	0,3

t=n

It will be seen that this result coincides with the result obtained when following the process of temporal sequences that was described before.

On the other hand the matrix of transitive closing expresses by itself the possibilities that the value of the business can be found in any of the situations, if there is the maximum presumption that at the initial moment 0 it were to have been in a determined situation. The reasoning is the following: if the value of the business is at moment 0 equal to $E_1 = [20; 24]$, the maximum presumption exist throughout the period that it will be in $\{E_3; E_4; E_5\} = \{[30; 34]; [35; 39]; [40; 44]\}$; a presumption of 0,8 that it will be found in $E_2 = 0{,}8$; a presumption of 0,5 that it will be found in $\{E_1; E_6\} = \{[20; 24]; [45; 49]\}$ and a presumption of 0,3 that it will be found in $E_7 = [50; 54]$. All of this can be seen in the first row of $[\hat{\underset{\sim}{N}}]$; the same can be said for, if starting out from E_2 (with which we would have the second row), of $E_3, \ldots, E_7$ (last row).

This reasoning can be expressed by composing a vector in which there is a single 1 in the box corresponding to the initial situation in which it is assumed the system and matrix $[\hat{\underset{\sim}{N}}]$ are to be found.

In this way, if it is assumed that the value of the business at the initial moment is $E_1 = [20; 24]$, we have:

E_1	E_2	E_3	E_4	E_5	E_6	E_7		E_1	E_2	E_3	E_4	E_5	E_6	E_7
1							$\circ\ [\hat{\underset{\sim}{N}}] =$	0,5	0,8	1	1	1	0,5	0,3

which is none other than the first row of matrix $[\hat{\underset{\sim}{N}}]$.

If initially we start out from row $E_1 = [25; 29]$ it will be:

E_1	E_2	E_3	E_4	E_5	E_6	E_7		E_1	E_2	E_3	E_4	E_5	E_6	E_7
	1						$\circ\ [\hat{\underset{\sim}{N}}] =$	0	0,7	1	1	1	0,5	0,3

second row of matrix $[\hat{\underset{\sim}{N}}]$.

An so on successively for the initial positions $E_3 = [30; 34]$, $E_4 = [35; 39], \ldots$

E_1	E_2	E_3	E_4	E_5	E_6	E_7		E_1	E_2	E_3	E_4	E_5	E_6	E_7
		1					$\circ\ [\hat{\underset{\sim}{N}}] =$	0	0,3	0,8	1	1	0,5	0,3

E_1	E_2	E_3	E_4	E_5	E_6	E_7		E_1	E_2	E_3	E_4	E_5	E_6	E_7
			1				$\circ\ [\hat{\underset{\sim}{N}}] =$	0	0,3	0,6	0,9	1	0,5	0,3

E_1	E_2	E_3	E_4	E_5	E_6	E_7		E_1	E_2	E_3	E_4	E_5	E_6	E_7
				1			$\circ\ [\hat{\underset{\sim}{N}}] =$	0	0,3	0,6	0,9	1	0,5	0,3

E_1	E_2	E_3	E_4	E_5	E_6	E_7		E_1	E_2	E_3	E_4	E_5	E_6	E_7
					1		$\circ\ [\hat{\underset{\sim}{N}}] =$	0	0,3	0,6	0,9	0,9	1	0,3

E_1	E_2	E_3	E_4	E_5	E_6	E_7		E_1	E_2	E_3	E_4	E_5	E_6	E_7
						1	$\circ\ [\hat{\underset{\sim}{N}}] =$	0	0,3	0,6	0,9	0,9	0,9	1

Rows 3, 4, 5, 6 and 7 of the matrix of transitive closing $[\hat{\underset{\sim}{N}}]$ have been arrived at successively.

It has been clearly shown then that matrix $[\hat{\underset{\sim}{N}}]$ is sufficient for showing the reasoning; if initially the value were to be in E_i presumptions $\mathcal{M}_{E_j}$ would exist that in the long-term the value would be found in $E_j, i, j = 1,2,\ldots,7$.

The reader will have noticed that the development of this practical case includes several peculiarities which we have been careful to point out, as well as at least to describe the path for its generalisation. In spite of any eventual inconveniences that this might mean, we have preferred to follow the estimates arising from an actual application arrived at from the opinion of the experts linked to an important promotion-construction company from the Catalan region, rather than carry out a purely speculative exercise. We might add that the results obtained deserved the acquiescence of the governing body of said company, who held the same general ideas, tinged with certain nuances in their personal point of view.

In this case the circumstance has arisen that the normal fuzzy relation of transition converged toward a limit position (it would be said, if we were to be in the stochastic field, that the matrix is ergodic). But it does not always happen in this way, and the case could have been that this convergence towards a limit position did not exist, and that therefore it could not be said, as in this case, that in the future the maximum presumption exists that the value of the business will be located in interval $E_5 = [40; 44]$, or either that starting out, for example, from a value at the initial moment of $E_3 = [30; 34]$, the greatest presumption that in the future it would be found in $\{E_4; E_5\} = [35; 44]$. We would find ourselves then faced with a system with periodicity, or even with a system with a chaotic performance.

Both in one or the other case, it is possible in the current situation of science to carry out the correct treatment. We have limited ourselves here to provide a solution, we hope sufficiently adequate, to the first of the listed cases.

13.4 Development of the Scheme in the Event of Periodicity

By means of the description and development of an actual case the circumstances have been brought to light, conveniently specified, that the "system" has a limit, and that this has been arrived at by using certain techniques extracted from pseudo-Markovian analysis. But it could happen that even when starting out from the same referential:

$$S = \{E_1; E_2; \cdots; E_7\}$$

and from the same initial situation fuzzy vector:

	E_1	E_2	E_3	E_4	E_5	E_6	E_7
$[\underset{\sim}{q}_0] =$	0,3	0,4	0,8	1	0,9	0,2	0,1

the convergence in the limit were not to occur, as a consequence of the particular form of the graph associated to the normal fuzzy relation of transition. This phenomenon may appear in those businesses the objective of which is subjected to the very marked fluctuations of the economic situation, to the comings and goings of fashion, or simply because their products have a seasonal sales period. A fact should be pointed out here, although obvious, that the moments in time to be considered $T = 0,1,2,\ldots$, do not necessarily have to coincide with calendar years or accounting periods. With the object of developing this new case we are going to consider a hypothetical business that wishes to know its value in the future. For this we start out from the same subjective estimate of the value at moment 0 and we establish by means of the corresponding opinion of the experts the following normal fuzzy relation of transition the associated graph of which is as shown in Fig. 13.4.

$[\hat{\underset{\sim}{N}}] =$

	E_1	E_2	E_3	E_4	E_5	E_6	E_7
E_1		1					
E_2			1				
E_3	0,1			0,8	1		
E_4					1		
E_5						1	
E_6							1
E_7					1		

As we did for the previous case, we are going to carry out the successive max-min compositions from $[\hat{\underset{\sim}{N}}]$ in order to obtain $[\hat{\underset{\sim}{N}}]^2$, $[\hat{\underset{\sim}{N}}]^3,\ldots$, until convergence or a periodicity arises.

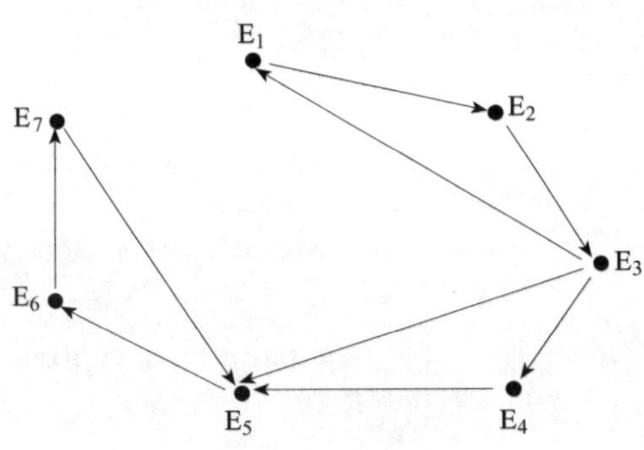

Fig. 13.4.

We will have:

$[\underset{\sim}{N}] =$

	E_1	E_2	E_3	E_4	E_5	E_6	E_7
E_1		1					
E_2			1				
E_3	0,1			0,8	1		
E_4					1		
E_5						1	
E_6							1
E_7					1		

$[\underset{\sim}{N}]^2 =$

	E_1	E_2	E_3	E_4	E_5	E_6	E_7
E_1			1				
E_2	0,1			0,8	1		
E_3		0,1			0,8	1	
E_4						1	
E_5							1
E_6					1		
E_7						1	

$[\underset{\sim}{N}]^3 =$

	E_1	E_2	E_3	E_4	E_5	E_6	E_7
E_1	0,1			0,8	1		
E_2		0,1			0,8	1	
E_3			0,1			0,8	1
E_4							1
E_5					1		
E_6						1	
E_7							1

$[\underset{\sim}{N}]^4 =$

	E_1	E_2	E_3	E_4	E_5	E_6	E_7
E_1		0,1			0,8	1	
E_2			0,1			0,8	1
E_3	0,1			0,1	1		0,8
E_4					1		1
E_5						1	
E_6							1
E_7					1		

$[\underset{\sim}{N}]^5 =$

	E_1	E_2	E_3	E_4	E_5	E_6	E_7
E_1			0,1			0,8	1
E_2	0,1			0,1	1		0,8
E_3		0,1			0,8	1	
E_4						1	
E_5							1
E_6					1		
E_7						1	

$[\underset{\sim}{N}]^6 =$

	E_1	E_2	E_3	E_4	E_5	E_6	E_7
E_1	0,1			0,1	1		
E_2		0,1			0,8	1	
E_3			0,1			0,8	1
E_4							1
E_5					1		
E_6						1	
E_7							1

$[\underset{\sim}{N}]^7 =$

	E_1	E_2	E_3	E_4	E_5	E_6	E_7
E_1		0,1			0,8	1	
E_2			0,1			0,8	1
E_3	0,1			0,1	1		0,8
E_4					1		
E_5						1	
E_6							1
E_7					1		

It can be seen that $[\hat{\underset{\sim}{N}}]^7 = [\hat{\underset{\sim}{N}}]^4$, therefore if we continued with the max-min composition we would arrive at:

$$[\hat{\underset{\sim}{N}}]^8 = [\hat{\underset{\sim}{N}}]^5$$
$$[\hat{\underset{\sim}{N}}]^9 = [\hat{\underset{\sim}{N}}]^6$$

$$[\hat{\underset{\sim}{\mathbf{N}}}]^{10} = [\hat{\underset{\sim}{\mathbf{N}}}]^{7}$$

$$\ldots$$

reproducing again the same fuzzy relations. Therefore, what takes place is a periodicity and, in a strict sense, there is no convergence towards a limit position.

Having reached this point the important question arises to see if it is possible to become aware of, in view of a normal fuzzy relation, when this is periodic and to determine what is its period. For this we are going to propose a calculation procedure the principal advantage of which is its simplicity and ease of use.

1. Boolean matrices $[B_j]$ are obtained which are associated to the fuzzy relations $[\hat{\underset{\sim}{\mathbf{N}}}]_j$. For this it will be sufficient to substitute the valuations of each element of the matrix that are different from 0 by 1.
2. We then find the different lines for each $[B_j]$ which, as we know, are given by the Boolean sum of the elements of the principal diagonal, that is:

$$tr[B_j] = \sum_{i=1}^{7} \mathcal{M}_{ii}$$

   where $\sum$ means Boolean sum.
   In this didactic example the lines will be:

$$\begin{array}{lll} tr[B_1] = 0\,, & tr[B_2] = 0\,, & tr[B_3] = 1 \\ tr[B_4] = 0\,, & tr[B_5] = 0\,, & tr[B_6] = 1 \\ tr[B_7] = 0\,, & \ldots & \end{array}$$

3. It will be seen that the lines with a value of 1 are those corresponding to B_3 and B_6. The sub-indices 3 and 6 are considered and we arrive at the maximum common divisor. In this case the m.c.d. of $\{3; 6\}$ is obviously 3. If there not to be a m.c.d. higher than 1 the fuzzy relation would be periodic, but as $3 > 1$ it can be concluded that the normal fuzzy relation of transition is periodic and has the period 3.

Finally let us see which are the sub-sets of these situations that give rise to the periodicity. For this we resort to the following procedure[11]. Select any vertex from the graph (Fig. 13.4) for example E_1. From this vertex the arcs are counted that are necessary for reaching any other, always seeking the path that is equal to or higher than the period obtained (in this case 3). Thus, in order to reach E_2 the following has to be done $(E_1; E_2) \rightarrow (E_2; E_3) \rightarrow (E_3; E_1) \rightarrow (E_1; E_2)$. There are 4 arcs, but as the period is 3, the difference

[11] We overlook its justification. For those interested in it reference can be made to Kaufmann A and Gil Aluja J (1991) Nuevas técnicas para la dirección estratégica (in Spanish), University of Barcelona, pp. 52–53.

between 4 and 3 is 1. The sub-set of which E_2 is a member will have the sub-index 1 and we will call it G_1. We do not accept the path $(E_1; E_2)$ since it only has one arc and $1 < 3$.

Let us take a look at another vertex, for example E_5. In order to reach this vertex two paths can be followed: $(E_1; E_2) \to (E_2; E_3) \to (E_3; E_4) \to (E_4; E_5)$ and also $(E_1; E_2) \to (E_2; E_3) \to (E_3; E_5)$. We will choose the shortest, provided it is equal to or higher than the period (in this case 3). Since there are three arcs it will be $3 - 3 = 0$. We assign a 0 sub-index to G, therefore E_5 is a member of the sub-set of situations G_0. If we do the same for all the vertices we arrive at:

$$\begin{array}{ll} E_1 = 3 \to G_0 & E_5 = 3 \to G_0 \\ E_2 = 4 \to G_1 & E_6 = 4 \to G_1 \\ E_3 = 5 \to G_2 & E_7 = 5 \to G_2 \\ E_4 = 3 \to G_0 & \end{array}$$

In this way, the previous graph can be represented as in Fig. 13.5.

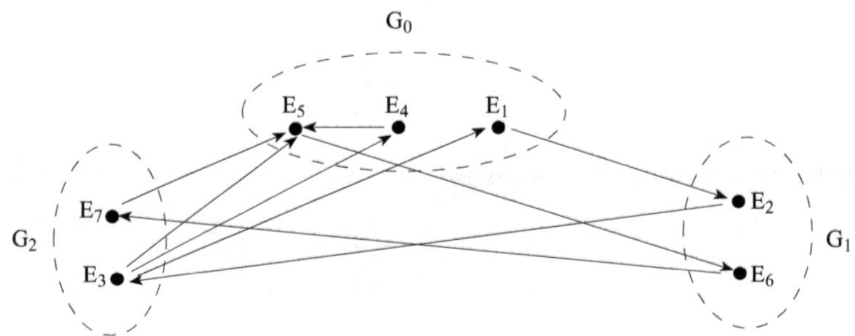

Fig. 13.5.

If we only consider the sub-sets of situations G_0, G_1 and G_2 that form the periodicity, we arrive at Fig. 13.6.

In order to find the estimate of the future value of the business it will suffice to compose the fuzzy initial situation vector $[\underset{\sim}{q}_0]$ with fuzzy relations $[\underset{\sim}{N}]^4$, $[\underset{\sim}{N}]^5$, $[\underset{\sim}{N}]^6$ by means of the max-min operator. In this way we arrive at:

$[\underset{\sim}{q}_0] \circ [\underset{\sim}{N}]^4 =$

E_1	E_2	E_3	E_4	E_5	E_6	E_7
0,1	0,1	0,1	0,1	1	0,9	0,8

$[\underset{\sim}{q}_0] \circ [\underset{\sim}{N}]^5 =$

E_1	E_2	E_3	E_4	E_5	E_6	E_7
0,1	0,1	0,1	0,1	0,8	1	0,9

$[\underset{\sim}{q}_0] \circ [\underset{\sim}{N}]^6 =$

E_1	E_2	E_3	E_4	E_5	E_6	E_7
0,1	0,1	0,1	0,1	0,9	0,8	1

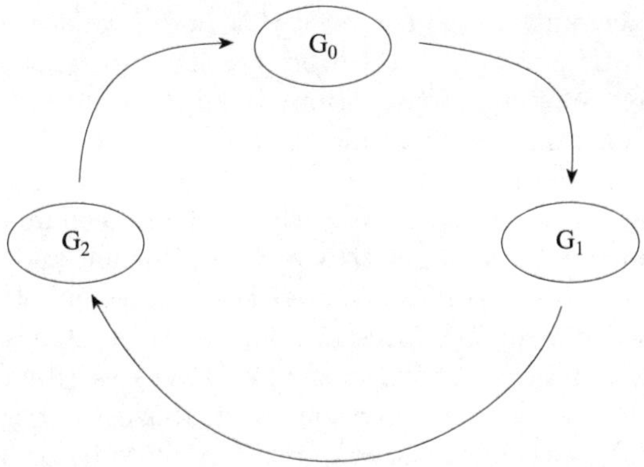

Fig. 13.6.

Taking into account that $[\underset{\sim}{\mathbf{N}}]^7 = [\underset{\sim}{\mathbf{N}}]^4$, $[\underset{\sim}{\mathbf{N}}]^8 = [\underset{\sim}{\mathbf{N}}]^5$, $[\underset{\sim}{\mathbf{N}}]^9 = [\underset{\sim}{\mathbf{N}}]^6$ the same fuzzy situations vectors will be reproduced at all moments after the sixth.

Arriving at these vectors allows us to make certain conclusion for each specific case. In the case we are studying, it will be seen that the greatest presumptions take place for situations $E_5 = [40; 44]$, $E_6 = [45; 49]$ and $E_7 = [50; 54]$, which is equivalent to stating that, in the future, the value of the business will move within the interval $[40; 54]$ with movements taking place every three periods. If we started out from a different fuzzy situation vector and a different normal fuzzy relation of transition, the considerations would obviously be different.

What has been brought to light in this second alternative is that although a convergence towards a limit has taken place, it is possible to make valid estimates on the value of the business for future periods. We feel then that the proposed scheme could be used for the type of business the characteristics of which would allow us to think that in the future periodic fluctuations in its value may occur. In any event, the path that has been proposed allows us to determine ex ante if the value of the business tends towards a certain value or, on the contrary, it is estimated that any type of periodicity may occur.

13.5 Attempts for Solution of Other Cases

The hypothetical case remains to be studied where the performance of the value of the business does not allow for any type of stability. The analysis of the range of possibilities for which the techniques we have described are not valid permits us to state that current scientific status only approaches, in a very timid fashion, certain attempts for solution the acceptance of which does not always enjoy general consensus. Among the better known we can point out the set of techniques arising from the fuzzy theory of chaos.

Taking into account the objective of this book, we are not going to give a detailed description of any of the works written on this subject in which we have had a very modest participation, given their eminently experimental nature[12]. Let us say, merely as an indication, that the new approaches start out from the assumption that both the estimates and the opinions of the experts are of an uncertain nature and the control mechanisms are based on a set of rules, concepts and values that are both formal and/or fuzzy. In this way, a determinist system is conceived with a fuzzy control reaction.

The works of Professor Teodorescu, done between 1990 and 1992 bring to light that a fuzzy reaction system, such as we have described, can become, in certain cases, a chaotic system, if the control strategy is not sufficiently stable. The mathematical basis that he uses stems from a model by Verhulst, who, as is known, established the basic hypothesis that there exist, on the one hand, a natural tendency to growth of the value under study that is proportional to the extension of its objective and, therefore on the other hand, that factors exist that are adverse to the growth which tend to the reduction of the value in question.

We will finish this section, even though we are fully aware that there remain many aspects to be studied. We hope that in the not too distant future these black holes may see the light. Science and scientists will never stop from surprising us.

[12] We refer to the following contributions to the International 92 Geneva Conference "Signals & Systems" (17–19 June 1992): Teodorescu N H and Gil Aluja J, "Phénomènes économiques chaotiques de croissance", and Gil Aluja J, Teodorescu N H and Gil Lafuente A M, "A chaotic fuzzy model for management problems", as well as Gil Lafuente A M, Gil Aluja J, Teodorescu, M M and Tacu A P "Chaotic Fuzzy Models in Economy", in the 2nd International Conference on Fuzzy Logic Systems in Neural Networks, 17–22 July, 1992, Lizuka (Japan), and Gil Aluja J, Gil Lafuente A M and Teodorescu N H, "Periodicity and Chaos in Economic Fuzzy Forecasting", in International Fuzzy Systems Symposium IS-KIT'92, 12–15 July 1992 (Japan).

14 Sequential Study of the Modifications in the Ratio of Indebtedness

14.1 Prior Considerations

The ratio of indebtedness is an element that permits expressing the financial structure of a business in a simplified manner, by means of the quotient of two aggregated values from liabilities that represent the sources of outside financing and in-house financing.

One of the most important problems that arise in financial analysis is the establishment of the incidence of the variations of the financial structure on capital cost and the value of the business.

There are important studies[1] that include works written by several authors, among these Modigliani and Miller, Durand, etc., which bring to light, with analytical formulation and numerical examples, what occurs with capital cost and the value of the business when the ratio of indebtedness varies as a consequence of increasing (or decreasing) debts and decreasing (or increasing) equity, under certain restrictive hypotheses. With this mechanism certain conclusions are reached, one of the these that is repeated the most states (always under certain conditions) that capital cost decreases as the ratio of indebtedness increases. Obviously if we modify the initial hypotheses then the conclusions also vary. In this way a varied (and limited) range of situations is brought to light that give rise to the drawing up of models that are known, some by the name of their author or authors, other by the principal initial hypothesis.

Obviously the cases on which these models are supported are not fully covered by the published works and that it is possible to draw up new schemes based on different hypotheses. The path exists, therefore, that is open to further research which, without a doubt, will provide interesting results.

Nevertheless we feel that in the sphere of study of the incidence of the variations of the financial structure on the two values we have mentioned, there exists a vacuum that we have seen to be filled with the models that

[1] Mention can be made of Mao James GT (1974) Análisis financiero (Ed) El Ateneo, Buenos Aires, Chap. 11, and Suárez Suárez Andrés S, (1985) Decisiones óptimas de inversion y financiación en la empresa. (Ed) Pirámide, Madrid, Chaps. 37 and 38.

Anna Maria Gil-Lafuente: *Fuzzy Logic in Financial Analysis*, StudFuzz **175**, 241–255 (2005)
www.springerlink.com

have been published to date and which, in our opinion, constitute a prior step to the establishment of the oft repeated relation of incidences.

In fact, when we start out from, for example, the fact that the ratio of indebtedness increases in a determined amount as a consequence of the change over from equity to outside capitals, what is wanted is to show a modification in the financial structure on moving from one moment in time to another. In order to able to see the possible situations which could be reached in a future moment in time variations in different amounts are assumed, with which different capital costs and different valuations of the business are arrived at. All of this for the study of a sequence that goes from the current moment to a moment in the future (it could be assumed that a certain period or the accounting period transpires). If with this process it is desired to move on to a second future period, the data from each one of the results of the first period will have to be included and the reasoning repeated the same number of times in order to move on from period one to period two, and so on successively.

Whichever way we look at this it is obvious that the scheme lacks operational ease and even we feel coherence. The reason for this is simple and is due to the fact that in none of the studies that have been carried out has the problem of determining the possible financial structures of each one of the moments of time that form the economic horizon been considered as basic.

14.2 General Approach to the Problem

We propose to draw up a model of a sequential nature that will allow us to estimate each one of the possibilities that the ratio of indebtedness (and as a more general case any type of ratio) attains, in each of the future moments, a determined position, that is to say that it have a determined value.

For this let us assume a ratio of indebtedness such as:

$$r_d = \frac{\text{Outside capitals}}{\text{Equity}}$$

has at the current time (that we will call moment zero), a value a and that it could reach at each of the future moments $n = 1,2,3,\ldots,r$, certain positions (values) $x \in \{a,b,c,d\}$. It can be considered then that these "situations" form a set:

$$E = \{a,b,c,d\}$$

From moment 1, each one of the possible situations a,b,c,d can give way to a situation in a future moment or any other, provided that it is included in E. And, evidently, the possibility of arriving at one or the other does not necessarily have to be the same. This reasoning can be shown by means of Fig. 14.1.

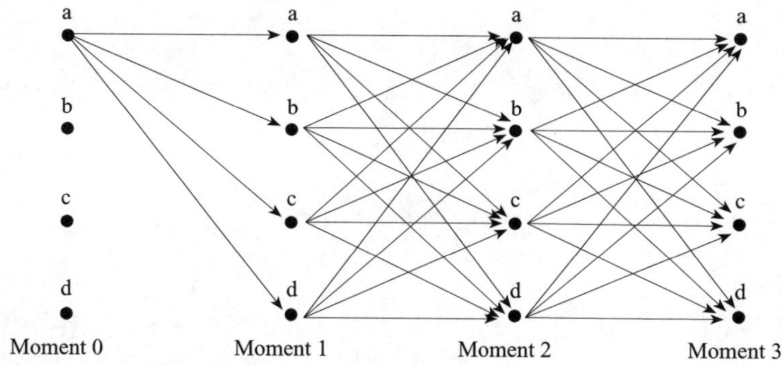

Fig. 14.1.

Having reached this point it becomes necessary to underline certain aspects of the process.

The first of these refers to the nature of the transition, or passage from one situation, relative to a moment in time, to another, corresponding to the following period. In many fields of human activity in which phenomena occur with characteristics such as they permit being repeated a great number of times, it is possible to use the techniques of randomness, in such a way that one could talk of the probability of transition or passage from a situation a at moment 1, to a situation a at moment 2, P_{aa}; probability of transition or passage from a in 1 to b in 2, $P_{ab}; \ldots$, and so on successively. In this way we would have a stochastic matrix such as:

[P] =

	a	b	c	d
a	P_{aa}	P_{ab}	P_{ac}	P_{ad}
b	P_{ba}	P_{bb}	P_{bc}	P_{bd}
c	P_{ca}	P_{cb}	P_{cc}	P_{cd}
d	P_{da}	P_{db}	P_{dc}	P_{dd}

where obviously, the sum of the probabilities of each row must be equal to the unit, must be complied with.

To this matrix the graph shown in Fig. 14.2 can be associated.

It would appear to be obvious to point out that when any of these probabilities were to be equal to zero the corresponding arrow can be eliminated.

Nevertheless, in the field of social sciences and in the same in financial phenomena, it is difficult to believe that the past provides sufficient information in order to be able to establish this type of probability. Evidently if this were to be possible we would not doubt for a single instant following a stochastic path.

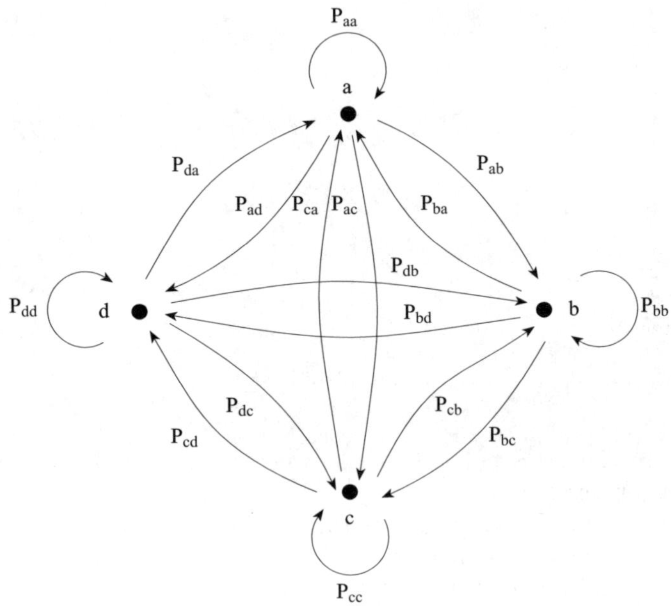

Fig. 14.2.

But, as we have so often repeated, current reality, characterised by very rapid changes, makes it difficult for allowing the phenomena to be repeated in an unchanged environment. This is the main reason why we resort to certain information, that is more pliable, supplied by the opinion of experts, the basis of which is not to be found in past "data" but in their knowledge and experience. The data provided will not in reality be measurements but "valuations", that is to say, subjective numerical estimates. As we all know, these valuations must comply with $v_{ij} \in [0;1]$. In this case we will have, in stead of a stochastic matrix, a "fuzzy relation". If in each row the existence of a 1 is established, that is that in each row there exists a maximum presumption, we are then in the presence of a "normal fuzzy relation" which will be symbolised by $[\underset{\sim}{\mathbf{N}}]$ with the same form as $[P]$, but in place of the p_{ij} we will have the v_{ij}, and in stead of the condition that the sum of the elements of each row be equal to the unit, we will have the condition that there exist, at least, a valuation equal to 1 in each row.

Now we can start out from a normal fuzzy relation $[\underset{\sim}{\mathbf{N}}]$ which will express, according to the opinion of the experts, the greater or lesser sensation (feeling) (according to whether they assign valuations closer to zero than to one) that the transition will occur of one situation of moment r to another situation of moment $r+1$, always within the possible situations $\{a, b, c, d\}$.

The second aspect to consider refers to the permanence or not of the same matrix of normal fuzzy relations, for all the period of time $n = 1,2,3,\ldots,r$. In other words, if there always exists that same sensation of change throughout time or, on the contrary, this will vary when we are dealing with different moments. In the first case, the linking is called "stationary" and in the second this give rise to "non stationary" chains, in which the elements of the matrix of fuzzy relations (in its normal occurrence), depends on moment of time n. This could be written as:

$[\underset{\sim}{N}(n)] =$

	a	b	c	d
a	$v_{aa}(n)$	$v_{ab}(n)$	$v_{ac}(n)$	$v_{ad}(n)$
b	$v_{ba}(n)$	$v_{bb}(n)$	$v_{bc}(n)$	$v_{bd}(n)$
c	$v_{ca}(n)$	$v_{cb}(n)$	$v_{cc}(n)$	$v_{cd}(n)$
d	$v_{da}(n)$	$v_{db}(n)$	$v_{dc}(n)$	$v_{dd}(n)$

In a first approximation, we are going to start out from the assumption that reality permits the consideration of a stationary chain.

Thus, we are going to carry out the treatment of this problem based on two hypotheses.

1. No valid probabilities are known for considering the passage from one period to another, so that valuations are resorted to, establishing a matrix of normal fuzzy relations.
2. We have the conviction that the passage from one period to another will not modify the opinion of the experts relative to the possibilities of transition of one situation to another.

We should remember, before continuing with the process, that the transitive closing of a fuzzy relation is given by:

$$[\hat{\underset{\sim}{N}}] = [\underset{\sim}{N}] \cup [\underset{\sim}{N}]^2 \cup [\underset{\sim}{N}]^3 \cup \ldots$$

and when $[\underset{\sim}{N}]$ is reflexive:

$$([\underset{\sim}{N}]^{r+1} = [\underset{\sim}{N}]^r) \rightarrow ([\underset{\sim}{N}]^n = [\hat{\underset{\sim}{N}}])$$

As can be seen, the convergence in uncertainty does not use the same formulae as in the field of randomness, where, when a limit exists for the Markov chains (ergodic chain), we have:

$$[M] = \lim_{r \to \infty} [M]^r$$

and in the event that $[M]$ were to be "completely ergodic" all the rows of $[M^*]$ would be identical and would provide the permanent vector of the situation.

14.3 Sequential Development of the Model

We now move on to develop a first scheme by means of an example with the object of avoiding a formulation that could hide the thread of our reasoning.

The financial manager of a business feels that the possible ratios of indebtedness that can reasonably be assumed are:

$$a = \frac{0{,}3\widehat{3}}{0{,}6\widehat{6}} = 0{,}5 \qquad b = \frac{0{,}40}{0{,}60} = 0{,}6\widehat{6}$$

$$c = \frac{0{,}45}{0{,}55} = 0{,}81 \qquad d = \frac{0{,}5}{0{,}5} = 1$$

The information available at this time leads to thinking that at the end of the current accounting period a very comfortable financial situation will be reached in such a way that he estimates with the greatest presumption that the business will be at ratio of value a (that is $\mu_a = 1$); with a presumption of 0,7 at b; with a presumption of 0,4, at c; with a presumption of 0,1, at d.

We therefore arrive at:

$[\mathrm{v}_0]$ =

a	b	c	d
1	0,7	0,4	0,1

On the other hand, a group of experts is asked to express an opinion relative to the possibilities existing that there will a passage from one ratio to another (from one financial structure to a different one) on passing from a period $r = n$ to another period $r = n + 1$. In this way we arrive at the following normal fuzzy relation (also called "uncertain matrix"):

$[\underset{\sim}{\mathrm{N}}]$ =

	a	b	c	d
a	0,7	1	0,3	
b		0,8	1	
c		0,7	1	0,4
d		0,2	1	0,6

The associated graph, in this case, is that shown in Fig. 14.3.

In order to obtain the corresponding vectors for each moment $n = 1, 2, 3, \ldots, r$, the following formula will be used:

$$[V_{n+1}] = [V_n] * [\underset{\sim}{\mathrm{N}}]$$

where * represents the max-min composition operator (rows by columns).

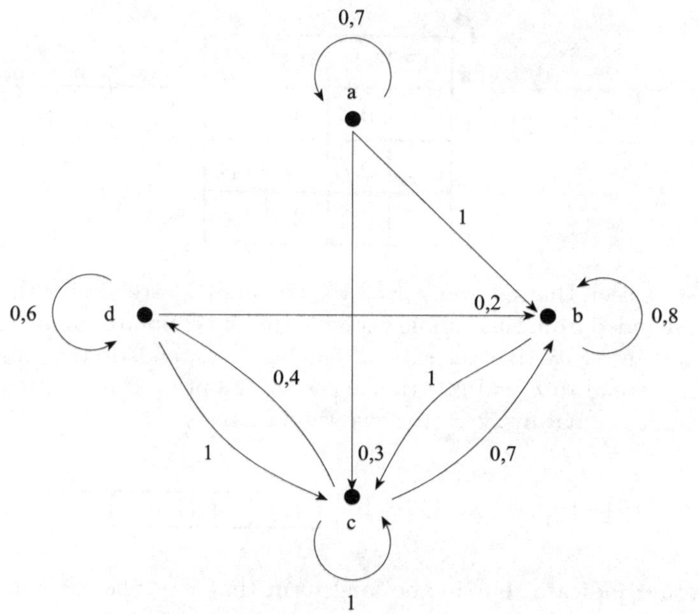

Fig. 14.3.

We then arrive at successively:

n=0:

a	b	c	d
1	0,7	0,4	0,1

*

	a	b	c	d
a	0,7	1	0,3	
b		0,8	1	
c		0,7	1	0,4
d		0,2	1	0,6

=

a	b	c	d
0,7	1	0,4	0,4

n=1

n=1:

a	b	c	d
0,7	1	0,4	0,4

*

	a	b	c	d
a	0,7	1	0,3	
b		0,8	1	
c		0,7	1	0,4
d		0,2	1	0,6

=

a	b	c	d
0,7	0,8	1	0,4

n=2

a	b	c	d
0,7	0,8	1	0,4

n=2

$*$

	a	b	c	d
a	0,7	1	0,3	
b		0,8	1	
c		0,7	1	0,4
d		0,2	1	0,6

$=$

a	b	c	d
0,7	0,8	1	0,4

n=3

It will be seen that as from $n = 3$ the vectors $[v_n]$ are always the same. It can be concluded from this simple example that an estimate can be made with the greatest intensity that the ratio of indebtedness tends to be designated by the letter c equal to $r_d = 0{,}81$. In the case of adopting the transitive closing as the "limit normal fuzzy vector" we would have:

$[\hat{\underset{\sim}{v}}] = [\underset{\sim}{v}_1] \cup [\underset{\sim}{v}_2] \cup [\underset{\sim}{v}_3] =$

a	b	c	d
0,7	1	1	0,4

which would indicate that in the long term that it is the amounts of ratio $b = 0{,}6\widehat{6}$ and $c = 0{,}81$ which it is felt with the greatest intensity will occur. In any event, a tendency exists to higher indebtedness and, therefore to a financial structure with a greater weight of outside financing.

The same result is reached by carrying out the max-min composition (also called convolution) of the normal fuzzy matrix $[\underset{\sim}{N}]$, that is to say, obtaining successively $[\underset{\sim}{N}]^2, [\underset{\sim}{N}]^3, \ldots$, and the transitive closing $[\hat{\underset{\sim}{N}}]$. In fact we arrive at:

$[\underset{\sim}{N}] =$

	a	b	c	d
a	0,7	1	0,3	
b		0,8	1	
c		0,7	1	0,4
d		0,2	1	0,6

$[\underset{\sim}{N}]^2 = [\underset{\sim}{N}] * [\underset{\sim}{N}]$

$=$

	a	b	c	d
a	0,7	1	0,3	
b		0,8	1	
c		0,7	1	0,4
d		0,2	1	0,6

$*$

	a	b	c	d
a	0,7	1	0,3	
b		0,8	1	
c		0,7	1	0,4
d		0,2	1	0,6

$=$

	a	b	c	d
a	0,7	0,8	1	0,3
b		0,8	1	0,4
c		0,7	1	0,4
d		0,7	1	0,6

$[\underset{\sim}{N}]^3 = [\underset{\sim}{N}]^2 * [\underset{\sim}{N}]$

	a	b	c	d
a	0,7	0,8	1	0,3
b		0,8	1	0,4
c		0,7	1	0,4
d		0,7	1	0,6

$*$

	a	b	c	d
a	0,7	1	0,3	
b		0,8	1	
c		0,7	1	0,4
d		0,2	1	0,6

$=$

	a	b	c	d
a	0,7	0,8	1	0,4
b		0,8	1	0,4
c		0,7	1	0,4
d		0,7	1	0,6

$[\underset{\sim}{N}]^4 = [\underset{\sim}{N}]^3 * [\underset{\sim}{N}]$

	a	b	c	d
a	0,7	0,8	1	0,4
b		0,8	1	0,4
c		0,7	1	0,4
d		0,7	1	0,6

$*$

	a	b	c	d
a	0,7	1	0,3	
b		0,8	1	
c		0,7	1	0,4
d		0,2	1	0,6

$=$

	a	b	c	d
a	0,7	0,8	1	0,4
b		0,8	1	0,4
c		0,7	1	0,4
d		0,7	1	0,6

We stop the process at this point taking into account that:

$$[\underset{\sim}{N}]^4 = [\underset{\sim}{N}]^3$$

By calculating $[v_0] * [\underset{\sim}{N}]^3$ we will obtain:

a	b	c	d
1	0,7	0,4	0,1

$*$

	a	b	c	d
a	0,7	0,8	1	0,4
b		0,8	1	0,4
c		0,7	1	0,4
d		0,7	1	0,6

$=$

a	b	c	d
0,7	0,8	1	0,4

result which coincides with the result we found in the previously described process.

We now move on to calculate the transitive closing:

$$[\hat{\underset{\sim}{N}}] = [\underset{\sim}{N}] \cup [\underset{\sim}{N}]^2 \cup [\underset{\sim}{N}]^3 =$$

	a	b	c	d
a	0,7	1	1	0,4
b		0,8	1	0,4
c		0,7	1	0,4
d		0,7	1	0,6

with which $[v_0] * [\hat{\underset{\sim}{N}}]$ will be:

a	b	c	d
1	0,7	0,4	0,1

$*$

	a	b	c	d
a	0,7	1	1	0,4
b		0,8	1	0,4
c		0,7	1	0,4
d		0,7	1	0,6

$=$

a	b	c	d
0,7	1	1	0,4

Once again we have arrived at the result obtained with the previous process when the transitive closing of the limit normal fuzzy vector was used.

14.4 The Solution by Means of Dynamic Programming

Another path can be taken to arrive at the objective set based on the fact that operators $\wedge$ (min) and $\vee$ (max) possess associative, distributive and monotonous properties. The new process that we are proposing is supported by the principle of dynamic programming.

For its presentation we are going to construct a sequential graph, shown in Fig. 14.4. Given that the normal fuzzy relation $[\underset{\sim}{N}]$ is stationary, the same arcs as those that will be ascribed to the same valuations and will be repeated in each sequence. As a mere example, and with the object of not having to repeat the calculations, we only consider the first row of matrix $[\underset{\sim}{N}]$ on the understanding that the procedure is the same for all the remaining rows.

Let us assume then that an attempt is made to obtain row a of matrix $[\hat{\underset{\sim}{N}}]$ In order to obtain each one of the elements of row a of the fuzzy relation $[\underset{\sim}{N}]$ we operate as follows:

$a : (a \rightarrow a) = 0{,}7$	we place	0,7	in	a
$b : (a \rightarrow b) = 1$	we place	1	in	b
$c : (a \rightarrow c) = 0{,}3$	we place	0,3	in	c
$d : (a \rightarrow d) = 0$	we place	0	in	d

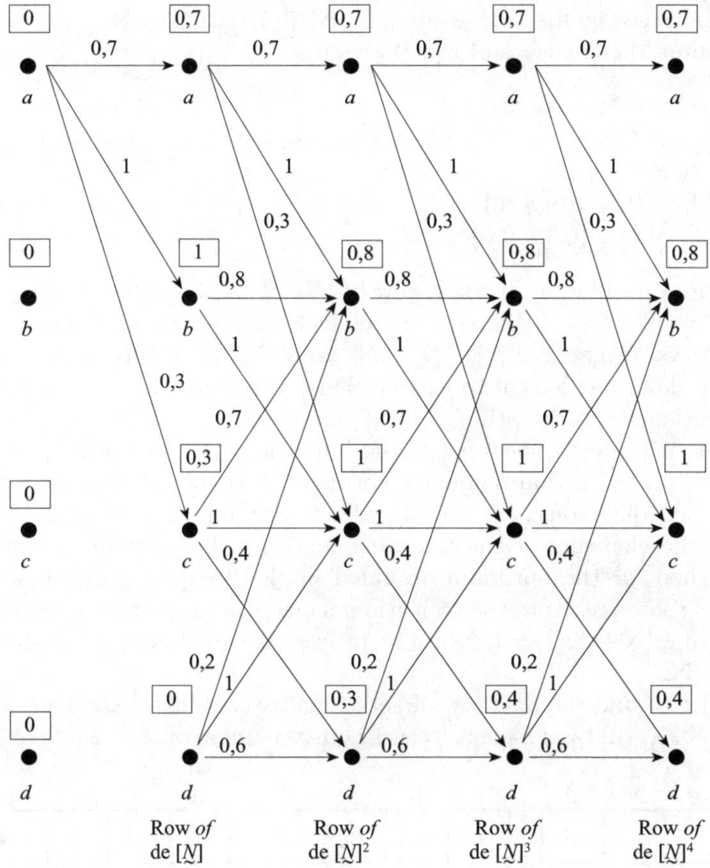

Fig. 14.4.

In this way, we obtain as the first row of the normal fuzzy matrix. $[\underset{\sim}{\mathbf{N}}]$, $(0{,}7; 1; 0{,}3; 0)$.

Obtaining the elements of row a in $[\underset{\sim}{\mathbf{N}}]^2$:

In a : $(0{,}7 \wedge 0{,}7) \vee (0 \wedge 1) \vee (0 \wedge 0{,}3) \vee (0 \wedge 0) = 0{,}7$	we place 0,7 in a
In b : $(0{,}7 \wedge 1) \vee (1 \wedge 0{,}8) \vee (0{,}3 \wedge 0{,}7) \vee (0 \wedge 0{,}4) = 0{,}8$	we place 0,8 in b
In c : $(0{,}7 \wedge 0{,}3) \vee (1 \wedge 1) \vee (0{,}3 \wedge 1) \vee (0 \wedge 1) = 1$	we place 1 in c
In d : $(0{,}7 \wedge 0) \vee (1 \wedge 0) \vee (0{,}3 \wedge 0{,}4) \vee (0{,}6 \wedge 0) = 0{,}3$	we place 0,3 in d

Here we have as the first row of matrix $([\underset{\sim}{\mathbf{N}}]^2, (0{,}7; 0{,}8; 1; 0{,}3)$.

Obtaining the elements of row a of $[\underset{\sim}{\mathbf{N}}]^3$

In a: $(0{,}7 \wedge 0{,}7) \vee (0{,}8 \wedge 0) \vee (0 \wedge 1) \vee (0{,}3 \wedge 0) = 0{,}7$	we place 0,7 in a
In b: $(0{,}7 \wedge 1) \vee (0{,}8 \wedge 0{,}8) \vee (1 \wedge 0{,}7) \vee (0{,}3 \wedge 0{,}2) = 0{,}8$	we place 0,8 in b
In c: $(0{,}7 \wedge 0{,}3) \vee (0{,}8 \wedge 1) \vee (1 \wedge 1) \vee (0{,}3 \wedge 1) = 1$	we place 1 in c
In d: $(0{,}7 \wedge 0) \vee (0{,}8 \wedge 0) \vee (1 \wedge 0{,}4) \vee (0{,}3 \wedge 0{,}6) = 0{,}4$	we place 0,4 in d

Here we have as the first row of matrix $[\underset{\sim}{\mathbf{N}}]^3$, $(0{,}7; 0{,}8; 1; 0{,}4)$.

Obtaining the elements of row $[\underset{\sim}{\mathbf{N}}]^4$:

In a : $(0{,}7 \wedge 0{,}7) = 0{,}7$ — we place 0,7 in a

In b : $(0{,}7 \wedge 1) \vee (0{,}8 \wedge 0{,}8) \vee (1 \wedge 0{,}7) \vee (0{,}4 \wedge 0{,}2) = 0{,}8$ — we place 0,8 in b

In c : $(0{,}7 \wedge 0{,}3) \vee (0{,}8 \wedge 1) \vee (1 \wedge 1) \vee (0{,}3 \wedge 1) = 1$ — we place 1 in c

In d : $(1 \wedge 0{,}4) \vee (0{,}4 \wedge 0{,}6) = 0{,}4$ — we place 0,4 in d

Therefore we have as the first row of $[N]^4$, $(0{,}7; 0{,}8; 1; 0{,}4)$.

If the results that have been successively arrived at are compared with the first rows of matrices $[\underset{\sim}{\mathbf{N}}]$, $[\underset{\sim}{\mathbf{N}}]^2$, $[\underset{\sim}{\mathbf{N}}]^3$ and $[\underset{\sim}{\mathbf{N}}]^4$, it will be seen that they are identical, as was logical to expect. This then is another approach based, as we mentioned, on the principle of dynamic programming.

In fact, it is easy to see one of the limits assigned to the elements of a, b, c and d (placed within a square) corresponding to matrices $[\underset{\sim}{\mathbf{N}}]$, $[\underset{\sim}{\mathbf{N}}]^2, \ldots$, are positions that could be considered as optimum. Therefore in order to arrive at the elements of the following matrices the max-min composition is established for the optimum obtained in the previous position with the valuations that constitute the transition or passage from one position $[\underset{\sim}{\mathbf{N}}]^n$ to the following $[\underset{\sim}{\mathbf{N}}]^{n+1}$, $n = 1, 2, 3$. The process is ended here, obviously when $[\underset{\sim}{\mathbf{N}}]^{n+1} = [\underset{\sim}{\mathbf{N}}]^n$.

In order to find the first row of the transitive closing of the normal fuzzy relation $[\hat{\underset{\sim}{\mathbf{N}}}]$, it will be sufficient to carry out the union of the results obtained:

a	b	c	d
0,7	1	0,3	0

row a of $[\underset{\sim}{\mathbf{N}}]$

$\cup$

a	b	c	d
0,7	0,8	1	0,3

row a of $[\underset{\sim}{\mathbf{N}}]^2$

$\cup$

a	b	c	d
0,7	0,8	1	0,4

row a of $[\underset{\sim}{\mathbf{N}}]^3$

=

a	b	c	d
0,7	1	1	0,4

row a of $[\hat{\underset{\sim}{\mathbf{N}}}]$

To obtain the remaining rows of the transitive closing $[\hat{\underset{\sim}{\mathbf{N}}}]$ the process is repeated after row a commencing sequences b, c, and d in the first column of Fig. 14.4.

The result is the same by any of the paths we have described relative to the presumptions of variation in the ratio of indebtedness. In this sense, estimated with the maximum intensity that the passage from moment 0 to moment 1 the following change will take place:

$$\text{Moment } 0 \rightarrow \text{Moment } 1$$

$$r_d^{(1)} = \frac{0{,}40}{0{,}60} = 0{,}6\widehat{6} \qquad r_d^{(2)} = \frac{0{,}45}{0{,}55} = 0{,}81$$

No change is expected in the transition from moment 2 to moment 3 and following. This index becomes established therefore in the long term relative to 45% of outside capital and 55% equity.

14.5 Incorporation of Uncertain Ratios

It is convenient to point out at this point that uncertainty has taken place through the feelings of the experts relative to the amounts that could be reached for the ratio of indebtedness in certain future moments $n = 1, 2, 3, \ldots,$. However, the figures that will be reached by the ratio of indebtedness under consideration have been considered as "certain" for each position a, b, c, d of our example (which could represent "situations" of the ratio due to different financial policies).

We now move on to a more general case in which this ratio can be estimated with certainty within each one of the four financial policies of our example, but on the other hand the experts are capable of expressing their respective amounts by means of fuzzy numbers. In their opinions they have maintained as the maximum presumption the crisp number give before. The information received is as follows:

$$\text{Position } a : \underset{\sim}{\mathbf{r}}_d^{(a)} = \frac{(0{,}30; 0{,}3\widehat{3}; 0{,}35)}{(0{,}65; 0{,}6\widehat{6}; 0{,}70)}\,.$$

$$\text{Position } b : \underset{\sim}{\mathbf{r}}_d^{(b)} = \frac{(0{,}36; 0{,}40; 0{,}42)}{(0{,}58; 0{,}60; 0{,}64)}\,.$$

$$\text{Position } c : \underset{\sim}{\mathbf{r}}_d^{(c)} = \frac{(0{,}43; 0{,}45; 0{,}49)}{(0{,}51; 0{,}55; 0{,}57)}\,.$$

$$\text{Position } d : \underset{\sim}{\mathbf{r}}_d^{(d)} = \frac{(0{,}50; 0{,}50; 0{,}55)}{(0{,}45; 0{,}50; 0{,}50)}\,.$$

Let us recall that, given the fact that the chosen ratio of indebtedness is the quotient between outside capitals (D) and equity (A), we can also write:

$$r_d = \frac{D}{A} = \frac{D}{D+A} \cdot \frac{A}{D+A} = \frac{\gamma}{1-\gamma} \quad \text{been} \quad \gamma = \frac{D}{D+A}$$

In this way, the result will be that the percentage of outside capital over the total will, for example, be in position a:

$$\underset{\sim}{\gamma} = (0{,}30; 0{,}3\widehat{3}; 0{,}35)$$

and expressed in α-cuts:

$$\gamma_\alpha = [0{,}30 + 0{,}0\widehat{3}\alpha; 0{,}35 - 0{,}01\widehat{6}\alpha]$$

The percentage of equity over the overall financing will be, therefore:

$$\begin{aligned} 1-\gamma_\alpha &= 1(-)[0{,}30+0{,}0\widehat{3}\alpha;0{,}35-0{,}01\widehat{6}\alpha] \\ &= [1-(0{,}35-0{,}01\widehat{6}\alpha);1-(0{,}30+0{,}0\widehat{3}\alpha)] \\ &= [0{,}65+0{,}01\widehat{6}\alpha;0{,}70-0{,}0\widehat{3}\alpha] \end{aligned}$$

which can be expressed in the ternary form as follows:

$$1-\underset{\sim}{\gamma} = (0{,}65;0{,}6\widehat{6};0{,}70)$$

which is nothing more than the denominator of the fraction $r_d^{(a)}(\alpha)$, which represents the ratio at position[2] a.

If the ratio is represented in the form of α-cuts e will arrive at, always for position a:

$$r_d^{(a)}(\alpha) = \frac{[0{,}30+0{,}0\widehat{3}\alpha;0{,}35-0{,}01\widehat{6}\alpha]}{[0{,}65+0{,}01\widehat{6}\alpha;0{,}70-0{,}0\widehat{3}\alpha]} = \left[\frac{0{,}30+0{,}0\widehat{3}\alpha}{0{,}70-0{,}0\widehat{3}\alpha};\frac{0{,}35-0{,}01\widehat{6}\alpha}{0{,}65+0{,}01\widehat{6}\alpha}\right]$$

With the object of visualising the resulting (non triangular) fuzzy number we have expressed this ratio by means of the hendecagonal system (see Table 14.1).

This fuzzy number brings to light the fact that, in any case, the ratio of indebtedness of position a will be no less than 0,428 nor any more than 0,538 (with nil presumption) and is presumed with the maximum intensity ($a = 1$) will be 0,5.

If the triangular approximation were to be admitted as sufficient, the T.F.N. $\underset{\sim}{\mathbf{r}}_d^{(a)} = (0{,}428;0{,}500;0{,}538)$ would be accepted.

Continuing with this same procedure we will arrive at the triangular approximations for positions b, c, d. In order not to tire our reader unnecessarily we have obviated the corresponding calculations and passed on directly to the results. These are respectively:

$$\begin{aligned} \underset{\sim}{\mathbf{r}}_d^{(a)} &= (0{,}428;0{,}500;0{,}538) \\ \underset{\sim}{\mathbf{r}}_d^{(b)} &= (0{,}562;0{,}66\widehat{6};0{,}724) \\ \underset{\sim}{\mathbf{r}}_d^{(c)} &= (0{,}754;0{,}818;0{,}960) \\ \underset{\sim}{\mathbf{r}}_d^{(d)} &= (1;1;1{,}22\widehat{2}) \end{aligned}$$

The sequential methods we have described for estimating the greater or lesser presumption existing that in each of moment $n = 1, 2, 3, \ldots,$ the ratio

[2] It can be easily seen that we have taken a T.F.N. both for the numerator and for the denominator of this quotient.

Table 14.1.

α	$\frac{0{,}30+0{,}0\overset{\frown}{3}\,\alpha}{0{,}70-0{,}0\overset{\frown}{3}\,\alpha}$	$\frac{0{,}35-0{,}01\overset{\frown}{6}\,\alpha}{0{,}65-0{,}01\overset{\frown}{6}\,\alpha}$
0	0,428	0,538
0,1	0,434	0,534
0,2	0,440	0,530
0,3	0,449	0,527
0,4	0,455	0,523
0,5	0,463	0,519
0,6	0,470	0,516
0,7	0,477	0,512
0,8	0,484	0,508
0,9	0,491	0,505
1	0,500	0,500

in a determined position of $\{a, b, c, d\}$, have been reasoned as follows based on the example we developed.

At moment 1 the maximum presumption is that the ratio will be no less than 0,562 nor greater than 0,724, but the figure, with the greatest intensity is presumed to be $0{,}66\overset{\frown}{6}$ (position b). But in period 2 and following it is estimated with the most conviction that the ratio will be no less than 0,754 nor greater than 0,960, and if just one figure had to be given then this would be 0,818 (position c).

With the joint use of sequential techniques on the one hand, and of uncertain ratios on the other, we have arrived at two levels of uncertainty; the level relative to the uncertain transition and the level that refers to the uncertain quantification of the ratio.

15 Determination of the Possible Strategies for Reducing Indebtedness in Business

15.1 Approach to the Problem

The current economic-financial context in which business activity evolves is characterised by the existence of high interest rates, difficulties in obtaining external financing and instability in internal situations caused by high rates of indebtedness. All of which constitutes a constant relative to the financial structures of the same, as well as to the environment in which normal operations are carried out.

It should come as no surprise that faced with this situation increasingly more frequent attempts are being made by those financially responsible for the businesses, to modify, and in the majority of cases, reduce their current levels of indebtedness.

But putting these decisions into practice implicitly carries a number of problems, the nature of which is varied, but on the other hand underlying in all of them is the uncertainty of the information relative to the future and even, on many occasions, the lack of information that is susceptible to being represented numerically. For this we have considered the possibility of presenting a model for decision, the basis of which is to be found in non numerical mathematics.

In this sense, we are focusing all out attention on an attempt to list all the possible paths that can be followed through several periods of time in order to reach the set objective, that consists in modifying or in the event reducing the rate of indebtedness, represented, for example, by the ratio corresponding to the quotient between short, medium and long term debts and total liabilities.

We have been unable to find an approach of this nature in the specialised literature, since this question was traditionally tackled by techniques that only permitted a partial view of this problem, as there did not appear to be anything with total immersion in an overall process for the same.

In fact, traditionally placed at the disposal of any interested professional is a range of instruments that permit getting to know the level of indebtedness of a business, Perhaps the most used is the ratio method. A ratio is nothing more than a quotient between two values which, in the case we are interested in, belong to the financial structure of the balance sheet, and the result of which can be a number or a numerical estimate representing a determined situation of the business. By means of these ratios relative to the financial

Anna Maria Gil-Lafuente: *Fuzzy Logic in Financial Analysis*, StudFuzz **175**, 257–277 (2005)
www.springerlink.com

structure what is attempted is to support or rectify the financing position. But on many occasions we feel that, by acting in this way, an excessive strain is put on this important instrument. In fact, what these ratios[1] do is show, on the one hand, is the stability of the financing and, on the other, the financial autonomy reflected by the relation existing between equity and debts. With regard to the stability of financing, it is customary to carry out the corresponding analysis by means of two ratios, both complementary to each other relative to the unit:

$$R_1 = \frac{\text{Permanent capitals}}{\text{Total liabilities}}$$
$$R_2 = \frac{\text{Short-term debts}}{\text{Total liabilities}}$$

With regard to financial autonomy in general two ratios can be considered, also complementary to each other relative to the unit:

$$R_3 = \frac{\text{Equity}}{\text{Total liabilities}}$$
$$R_4 = \frac{\text{Debts}}{\text{Total liabilities}}$$

Finally, if the autonomy of the business is analysed relative to the expiry dates of its resources, it is possible to analyse the performance of the capitals, in a more significant manner, with the following ratios rather than with the previous ones:

$$R_5 = \frac{\text{Equity}}{\text{Permanent capitals}}$$
$$R_6 = \frac{\text{Medium and long-term debts}}{\text{Permanent capitals}}$$

For our part, we are going to propose a very different approach, which is not limited to the use of the ratios, but that these constitute the starting out point.

The work we now present is an approach to a combinatory model based on the method of the P-Latin[2] composition, the object of which is to permit listing all those strategies that the business can follow with the purpose of arriving at a modification, and in the event reduction, of the levels of indebtedness in order to arrive at a position of comfortable solvency.

---

[1] Cohen E (1990) Analyse financière, 2nd Edition. Collection Gestion Económica., Series: Politique Générale, finance et marketing. Paris, p. 173.

[2] Kaufmann A and Gil Aluja J (1991) Nuevas técnicas para la dirección estratégica, University Press. Barcelona, p. 190.

15.2 Sequential Movements of the Index of Indebtedness

Normally a business, at any given time, finds itself in a specific situation of indebtedness, given for example, by means of a ratio that describes the quotient between total debts and the liabilities of the businesses. Let us assume that this ratio is in a initial position of 70%.

But it is felt that in the future this situation may evolve to other situations. In short, we are faced with a system, characterised by a set of possible situations $E = \{E_1, E_2, \ldots, E_n\}$.

In order to simplify this, we will limit ourselves, for example to 5 situations: $E_1 = a$, $E_2 = b$, $E_3 = c$, $E_4 = d$, $E_5 = e$.

Evidently, each situation corresponds to a specific situation of indebtedness. Thus:

$$
\begin{aligned}
E_1 &= a = 70\% \\
E_2 &= b = 60\% \\
E_3 &= c = 50\% \\
E_4 &= d = 40\% \\
E_5 &= e = 30\%
\end{aligned}
$$

Since we are faced with a context of uncertainty, the passage of one situation E_i, to another situation $E_j, i, j = 1, 2, 3, 4, 5$ cannot be known in certain terms neither can they be expressed by means of probabilities. We are obliged therefore to resort to valuations, that is to subjective numerical estimates.

In this way the "possibilities of transition or passage" from one situation to another will be estimated by valuations in $[0; 1]$ by means of a fuzzy relation $[\underset{\sim}{R}]$. Obviously this matrix, in an evolving economic system, will be different for each one of the moments in time of the economic horizon, but with the object of not making this too difficult it can be assumed that, in a first approximation, there is no variation over the course of time.

This fuzzy relation, which we assume is known, is as follows:

$[\underset{\sim}{R}] =$

	a	b	c	d	e
a	0,4	0,8	0,7	0,5	0
b	0,6	0,6	0,9	0,2	0,1
c	0,5	0,6	0,8	0,9	0,7
d	0,7	0,5	0,6	0,6	0,8
e	0,3	0,4	0,5	0,8	0,9

We are going to study this fuzzy relation, by means of α-cuts, commencing by the highest level, which in this case is $\alpha \geq 0{,}9$. We then arrive at the following Boolean matrix:

$[R_{0,9}] =$

	a	b	c	d	e
a					
b			1		
c				1	
d					
e					1

For $\alpha \geq 0{,}8$:

$[R_{0,8}] =$

	a	b	c	d	e
a		1			
b			1		
c			1	1	
d					1
e				1	1

For $\alpha \geq 0{,}7$:

$[R_{0,7}] =$

	a	b	c	d	e
a		1	1		
b			1		
c			1	1	1
d	1				1
e				1	1

We are going to pause at this level, which we can consider is acceptable for those responsible for the financial area of the business.

From the fuzzy relation at level $\alpha \geq 0{,}7$, that is of the Boolean matrix $[R_{0,7}]$, we are going to construct a Latin matrix $[E]$. For this each of the 1 is substituted by Latin letters which define the row and the column of which they are members. Therefore as there exists a 1 in row a column b in the corresponding box we place (ab). Taking into account the fact that there exists a 1 in the box relative to row a and column c we now place (ac) in the corresponding box. And so on successively till we arrive at:

$[E]^1$ =

	a	b	c	d	e
a		ab	ac		
b			bc		
c			cc	cd	ce
d	da				de
e				ed	ee

Here we now place a superscript 1 as this is the first Latin matrix drawn up with the initial data, that is with arcs, the length of which are therefore equal to the unit.

This matrix then shows the possibilities there are for the business to pass from a determined level of indebtedness to another. As is sufficiently well known, every matrix, such as $[E]^1$ can be presented by means of its associated graph. In this case this will be:

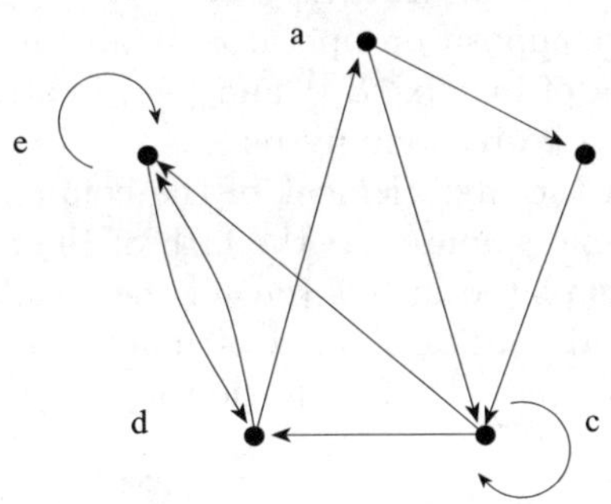

On the other hand, from the graph associated to $[E]^1$ the length of each one of the "paths" can be established along which the business will pass from one situation of indebtedness to another.

We should remember that a "path" is formed by a succession of vertices and arcs in which the first represents the end and beginning of each one of seconds. Therefore the "length" of a path will be given by the number of arcs existing between the initial vertex and the final vertex.

In our example there appears to be no problem whatsoever in assuming the length of each one of the paths as representing the number of periods, years that the business will require for modifying (reducing) its level of indebtedness.

Starting out then from a determined vertex of the graph (specific situation of indebtedness) it will be possible to arrive at another vertex (another situation of indebtedness different or equal to the previous one) by means of a path the length of which is $1, 2, 3, \ldots$, that is to say, in a period of one, two, three, ..., years. Obviously the vertices that comprise the path will indicate the intermediary situations of indebtedness through which the business must pass in order to arrive at the "objective situation".

It can be said then that the set of possible paths that will be obtained from the process that we develop below should allow for arriving at the set objective.

15.3 Possibilities of Arriving at the Objective in Two Periods

With the object of having an overall vision of the possibilities for modifying the situation of indebtedness, we are going to proceed to obtain the possible paths of different lengths (first length 2, then length 3, an so on successively). For this we resort to the "composition" of Latin matrix $[E]^1$, that is the matrix that represents the paths with length 1, with this same matrix "amputated" from its initial $[E']^1$. The results will give us a matrix listing all the paths of length 2. We call this $[E]^2$.

In order to use this composition operator, it is sufficient to take row a of matrix $[E]^1$ and row a of matrix $[E']^1$ and join, in strict order, the Latin letters that appear in the corresponding boxes, always considering the first element of the row with the first element of the column, the second of the row with the second of the column, ..., the fifth of the row with the fifth of the column. The result is placed in box (aa) of the resulting matrix.

We then move on and consider row a of matrix $[E]^1$ and column b of matrix $[E']^1$ and so on successively for all the rows of $[E]^1$. The result is:

$[E]^1 =$

	a	b	c	d	e
a		ab	ac		
b			bc		
c			cc	cd	ce
d	da				de
e				ed	ee

$[E']^1 =$

	a	b	c	d	e
a		b	c		
b			c		
c			c	d	e
d	a				e
e				d	e

$[E]^2 =$

	a	b	c	d	e
a			abc acc	acd	ace
b			bcc	bcd	bce
c	cda		ccc	ccd ced	cce cde cee
d		dab	dac	ded	dee
e	eda			eed	ede eee

The resulting matrix $[E]^2$ indicates all the possible paths that can be followed of length 2, that is, successions of three vertices and two arcs that join them. The first letter of each box indicates the starting out point, that is the hypothetical situation of indebtedness at the initial moment 0: the second letter, the situation of indebtedness at moment 1: and the last letter, the situation of indebtedness for period 2, considering this first phase as final. It will be seen that the passage from one or even two periods, can give rise to the fact that the situation of indebtedness does not vary as happens with box (*cc*).

As an example we will analyse some of the paths obtained from matrix $[E]^2$, such as path (*ace*). Its representation in the graph would be as follows:

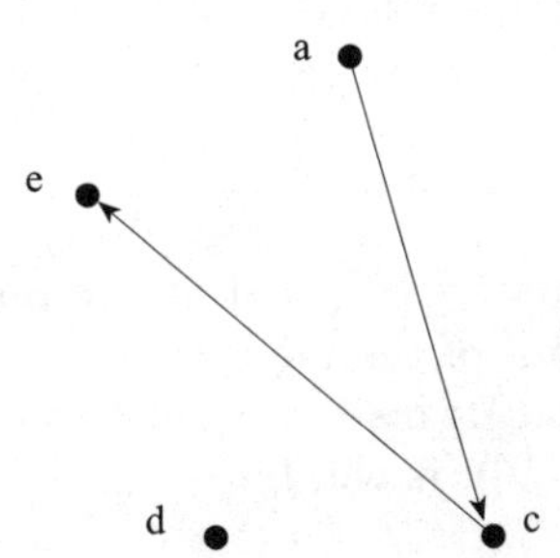

In the terms we have used in this work this path can be adopted if the business is in a situation of indebtedness of $a = 70\%$ and wishes to reduce this rate from 70% to 30% in 2 years. For this, during the first year and following certain conditions the intermediary objective must be set of passing from an indebtedness of 70% to 50%. And during the second year from this rate of 50% to 30% adopting the corresponding strategies for this.

At this point in time we cannot state what real possibilities exist of arriving at this final objective. For this we would have to resort to the initial matrix of transition or passage $[\underset{\sim}{\mathbf{R}}]$. In it will be seen that the possibility of passing from a to c are 0,7, and of passing from c to e are equally 0,7.

The question that arises is, in short, to know what is the overall possibility that the business has of passing from a level of indebtedness of 70% to another level of 30%. For this we propose resorting to the use of inferences or implications.

15.4 Establishment of Valuations by means of Inferences

One of the more important aspects of inferences in the multivalent environment is that, on the contrary to binary logic, if one proposal implies another, that is, $P \rightarrow Q$, the valuation of $P \rightarrow Q$, that is $v(P \rightarrow Q)$ does not necessarily have to equal 1, but it can take values included in $[0; 1]$ in the same way that $v(P)$ and $v(Q)$ can[3].

From among the many inferences that can be used in the multivalent field we are going to select, even though only as an example, the Lukaciewicz inference.

The Lukaciewicz inference is expressed as follows:

$$v(P \rightarrow Q) = 1 \wedge (\overline{v(P)} + v(Q))$$

The simple inference graph we reproduce below shows the only path existing for passing from a to e, that is from a ratio of indebtedness of 70% to one of 30%.

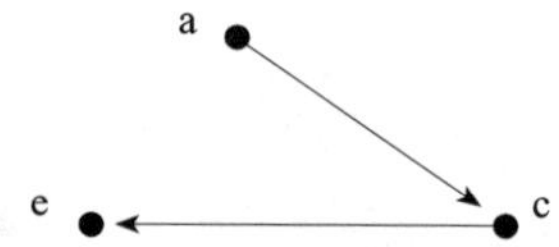

It can be assumed that matrix $[\underset{\sim}{\mathbf{R}}]$, that is, the possibilities of transition, supplies us with the valuations of the inferences.

We see in the first place the transition from a to c. If we find ourselves at a (ratio of indebtedness of 70%), it will be:

$$v(a) = 1$$

and as:

[3] Gil Lafuente A M (1993) Fundamentos del análisis financiero. (Ed) Ariel, Barcelona, p. 153.

$$v(a \to c) = 0{,}7$$

with the Lukaciewicz inference it will be:

$$0{,}7 = 1 \wedge (0 + v(c))$$

therefore:

$$v(c) = 0{,}7$$

We now take a look at the transition from c to e. As:

$$v(c) = 0{,}7$$
$$v(c \to e) = 0{,}7$$

we will have:

$$0{,}7 = 1 \wedge (0{,}3 + v(e))$$

and, finally:

$$v(e) = 0{,}4$$

If we admit the established hypotheses, we arrive at the fact that the transition from a ratio of indebtedness of 70% to another of 30% can be considered and this situation can be admitted with a reduced presumption of 0,4.

Taking into account the low level of presumption, that is the low valuation obtained, it would appear to be recommendable, with our example, to resort to a longer period of time. Obviously this is a case which we have shown for teaching effects.

15.5 Arriving at the Objective in the Medium and Long Term

In the majority of cases, to proceed to a profound modification of a position of indebtedness so that later it remains stable, requires a passage of time that is greater than the two periods that were previously set. Also, when a sufficiently extensive period of time is available it is more feasible that the hoped for results are attained, since the margin for manoeuvre and, therefore, movement is much greater. In this sense, as will be seen immediately, the paths that can be followed are increasingly greater, as a longer period of time is granted.

The technique to be used in order to pass over from the paths with a length of 2 to those of a length of 3 does not add any additional problems. Suffice it to do the convolution or composition of matrix $[E]^2$ with $[E']^1$ and we obtain matrix $[E]^3$ which shows paths of length 3. This process will be repeated successively if we want to find the matrices that show paths of length $4, 5, \ldots$.

Now we compose the Latin matrix $[E]^3$. We find:

$[E]^2 =$

	a	b	c	d	e
a			abc acc	acd	ace
b			bcc	bcd	bce
c	cda		ccc	ccd ced	cce cde cee
d		dab	dac	ded	dee
e	eda			eed	ede eee

$[E']^1 =$

	a	b	c	d	e
a		b	c		
b			c		
c			c	d	e
d	a				e
e				d	e

$[E]^3 =$

	a	b	c	d	e
a	acda		abcc accc	abcd accd aced	abce acce acde acee
b	bcda		bccc	bccd bced	bcce bcde bcee
c	ccda ceda	cdab	cdac cccc	cccd cced cded ceed	ccce ccde cede ccee cdee ceee
d	deda		dabc dacc	dacd deed	dace dede deee
e	eeda	adab	edac	eded eeed	eede edee eeee

On this occasion, the same as occurred with matrix $[E]^2$ but now with greater clarity, we can see the repetition of one or several vertices on one and the same path. Perhaps we should make a brief pause here and comment on this aspect.

From a methodological point of view, or better said of procedure, the repetition of a vertex occurs as a consequence of property P required of the composition, which in this case is that the "set of vertices formed by the composition be one path", nothing further. If, for example, the property required would have been obtaining a "Hamiltonian path" (path without repetition of vertices), obviously by a request of principle only a single listing of each vertex comprising it would have appeared in the set[4].

[4] Due to its interest we will look at this aspect further on.

From a financial point of view, we have adopted this hypothesis as the starting out point, in this phase of the work, taking into account that on may occasions it may be necessary to remain for more than one period in one and the same situation of indebtedness (in time of depression this may be an intermediary objective) and even to vary it in the contrary sense to the sought after objective. It is like taking a few steps backwards in order to get up speed and later jump even further.

We feel then that the scheme approached in this way acquired a greater wealth of elements, without prejudice to treating other variations that may prove very useful in other specific cases.

Now we immediately move on to look at, as was done before, the paths that go from a to e, that is from a ratio of indebtedness of 70% to one of 30%. It will be seen that there are now 4 paths: ($abce$), ($acce$), ($acde$) and ($acee$). Nevertheless, in the latter a special aspect appears that needs to be pointed out. It is the fact that the repetition of a vertex takes place "at the end" of the process. This means that the objective has been reached in period 2 and that the permanent situation can be maintained, on the condition, of course, that the necessary steps are taken to ensure that it is complied with. This always occurs when in the final vertex a loop occurs, as in this case.

We are going to represent these paths by means of a graph expressed by arrow diagrams:

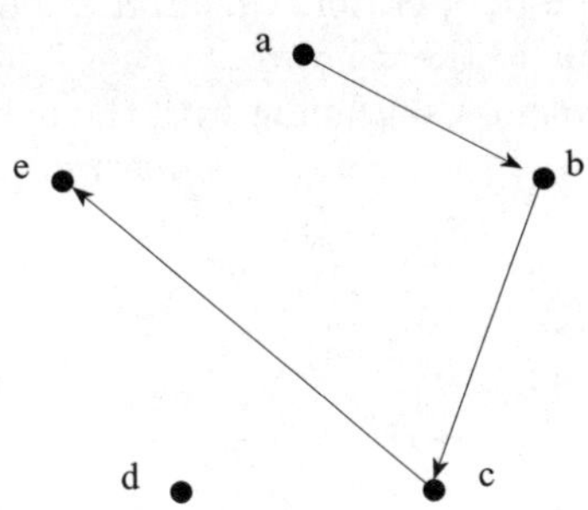

Path A

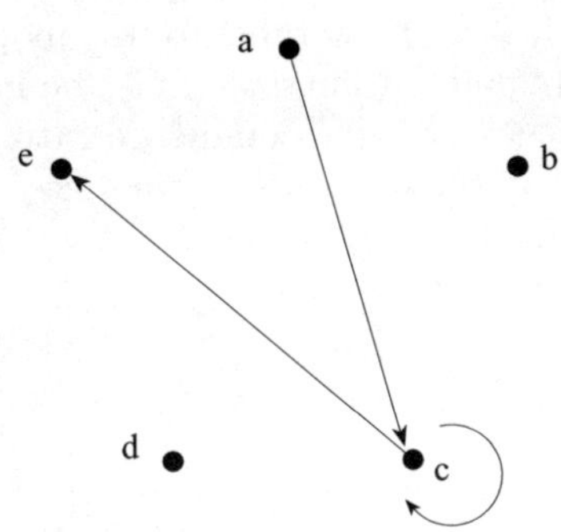

Path B

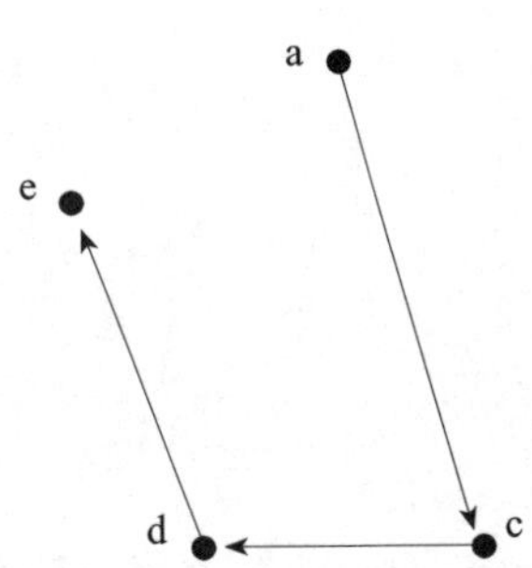

Path C

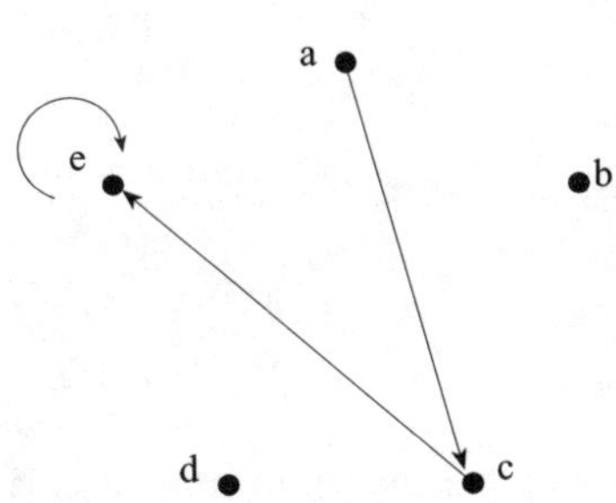

Path D

If we draw the previous paths in a single graph we arrive at the following:

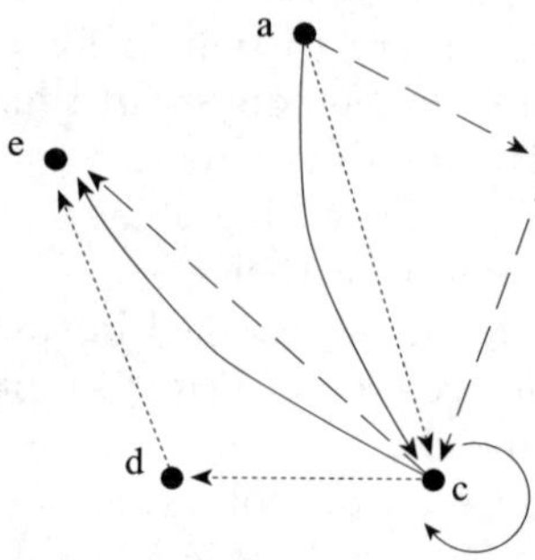

where:

······➤ way A

— —➤ way B

——➤ way C

Now we know the paths to follow in order to pass from one situation of indebtedness to another, it will be necessary to select which if these to follow. We now also propose using inferences. Obviously, it is possible that in certain cases it is more suitable to use another type of operator, which, if the desire is to remain within [0; 1] must be looked for in the τ-norms.

We are going to commence with path ($abce$). Continuing with the hypothesis that the elements of the fuzzy relation $[\underset{\sim}{\mathbf{R}}]$ express the valuation of the inferences, we will have as data:

$$v(a) = 1$$
$$v(a \to b) = 0{,}8$$
$$v(b \to c) = 0{,}9$$
$$v(c \to e) = 0{,}7$$

From the Lukaciewicz inference:

$$0{,}8 = 1 \wedge (0 + v(b))$$
$$v(b) = 0{,}8$$
$$0{,}9 = 1 \wedge (0{,}2 + v(c))$$
$$v(c) = 0{,}7$$
$$0{,}7 = 1 \wedge (0{,}3 + v_1(e))$$
$$v_1(e) = 0{,}4$$

which will be the valuation of arriving at a ratio of indebtedness equal to 30% from path ($abce$) which we have designated path 1.

We move on to path 2 ($acce$). We have as data:

$$v(a) = 1$$
$$v(a \to c) = 0{,}7$$
$$v(c \to c) = 0{,}8$$
$$v(c \to e) = 0{,}7$$

We do the corresponding calculations:

$$0{,}7 = 1 \wedge (0 + v_I(c))$$
$$v_I(c) = 0{,}7$$
$$0{,}8 = 1 \wedge (0{,}3 + v_{II}(c))$$
$$v_{II}(c) = 0{,}5$$
$$0{,}7 = 1 \wedge (0{,}5 + v_2(e))$$
$$v_2(e) = 0{,}2$$

that constitutes the valuation of the objective as from path ($acce$) designated as path 2.

We move on to path 3, ($acde$). The data is:

$$v(a) = 1$$
$$v(a \to c) = 0{,}7$$
$$v(c \to d) = 0{,}9$$
$$v(d \to e) = 0{,}8$$

We arrive at:

$$0{,}7 = 1 \wedge (0 + v(c))$$
$$v(c) = 0{,}7$$
$$0{,}9 = 1 \wedge (0{,}3 + v(d))$$
$$v(d) = 0{,}6$$
$$0{,}8 = 1 \wedge (0{,}4 + v_3(e))$$
$$v_3(e) = 0{,}4$$

By means of path ($acde$) designated path 3, we arrive at a valuation of 0,4.

We eliminate path 4, ($acee$) as it is, as we have said before, an objective already arrived at for two periods.

Finally, given that the three paths considered are alternatives, we take as the operator for the decision, the T-conorm ($\vee$):

$$v_1(e)(\vee)v_2(e)(\vee)v_3(e) = 0{,}4$$

which brings to light the possibility of following two paths, 1 and 3 with the same valuation 0,4.

In order to select one of these, we should resort to elements not considered when drawing up the data, such as financial cost, production repercussions, incidence on sales, etc.

We will now make a brief reference to the first paths we pointed out previously. Therefore, by taking path (*abce*) an option has been made for a progressive but continued decrease of the indebtedness rate of 70% (initial situation) to 60%. In the second period, this rate of 60% becomes 50%, and finally, during the third period it goes from 50% to 30%, level that represents the final objective set by the business.

Briefly analysing another of the possible paths, for example (*acce*), we can see that in this case the strategy carried out by the business has consisted in a more accelerated reduction of the rate of indebtedness in the first period since during the first year, the business moves from a rate if indebtedness of 70% (initial situation) to a level of 50%. During the second period an option is made to remain at this level of indebtedness for the effects of consolidating its financial situation. Finally, during the third period, the business once again proceeds to a further reduction in its rate of indebtedness passing from 50% to 30% and in this way complying with the initial objective.

Among the paths that appear in matrix $[E]^3$ there are some that warrant special comment. These are paths (*cded*), (*eded*), (*cede*) and (*dede*). In these the final vertex also appears along the interior of the path, which indicates that the objective has been reached in less time than the time set for attaining the same and that, following the financial policy after abandoning the objective, it is once again aimed at. If complementary measures are not taken, this cyclical process will continue indefinitely, if the sub-graph formed by the corresponding arcs and vertices, being strongly connected, will form a class of terminal equivalency (once entered into it cannot be left). In this case it can be seen that this occurs in the sub-graph formed by vertices $\{e, d\}$ and the arcs that join them:

The financial consequences of such situations are obvious, and the steps to be taken are within the orthodoxy of financial management.

Finally we will take a look, with regard to this paragraph, at the path formed by vertices (*acda*). This path has the peculiarity that its initial extreme coincides with its final extreme. This then is a Hamiltonian circuit, represented below in the following graph:

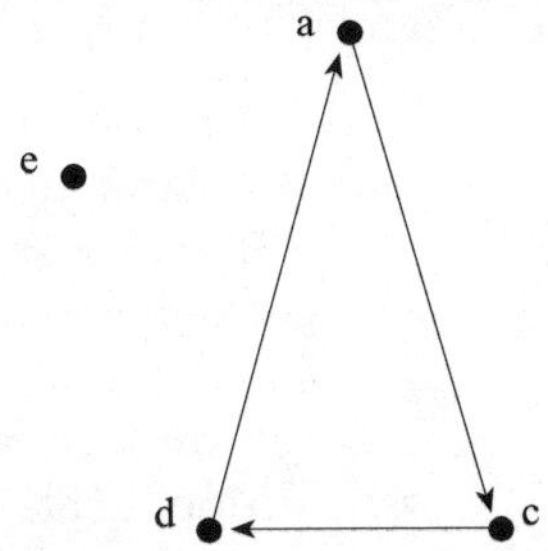

This path brings up a very special case in which the business, situated initially at a rate of indebtedness of 70% during the first year manages to reduce this level to 50%. Throughout the second year this could descend to a rate of 40%, but during the third year the proportion of debts relative to total liabilities would return a become situated at the initial levels, these being 70% of indebtedness.

In this case what we have is a "transitory" variation in the objectives, and this policy is normally adopted when modifications, that are important but not permanent, are foreseen for the economic situation.

Having made these considerations, an analysis may be useful of the possible strategies that a business could follow with the object of reducing its level of indebtedness starting out from a determined situation. Normally different objectives would be considered for which several options exist.

In this context we will start out from an initial position in which the level of indebtedness is 70%. From this situation and given the fact that in our example the graph is strongly connected[5] one can arrive at, by means of a more or less lengthy path, to any other position of indebtedness.

Given that, up to this point, we have only studied the paths of length 3, we will limit ourselves to these periods. Thus, in the following table and as an example, we will show the various paths that the business can follow.

Having made these considerations, we now move on to arrive at the possible paths of length 4, that is, the exhaustive list of the transitions from an initial financial situation to another final position when between these 4 periods were to have gone by. For this it will be sufficient to compose matrix $[E]^3$ with matrix $[E']^1$ in the same way as we did before:

We are avoiding at this point making any possible analysis since any considerations that could be made either coincide or constitute an extension of those we have already made.

[5] A graph is strongly connected when, starting out from any one of the elements of the same, it is possible to return to the starting point (vertex) by means of the different inter-connections that form the arcs.

	Indebtedness			
Strategy	Inicial Situation	Final Objective	No of Years	Characteristics of the Strategy
1	70%	30%	2	– Drastic reduction: 1st year from 70% 50%, 2nd year 50% to 30%. – Requires solid financial structure. – Having reached the objective this can be maintained or indebtedness can increase again to 40%.
2	70%	30%	3	– More gentle reduction: 1st and 2nd year from 70% to 60% to 50%; 3rd year from 50% to 30%. – Having reached the objective, and based on the solidity of the financial structure, maintain the rate at 30% or else increase the same to 40%.
3	70%	30%	4	– Gradual reduction: from 70% to 60% to 50% to 40% to 30% annually. – Another possibility: carry out strategy (2) but stopping in one period at situation 50% of indebtedness in order to consolidate financial structure.
4	70%	40%	2	– Reduction from 70% to 50% during the first year (greater effort); second year from 50% to 40% (less effort). – Having reached the objective, if the business has managed to maintain a solid financial structure it can continue reducing its level of indebtedness. – To the contrary, if it does not maintain its solidity it could see the rate of indebtedness increase to the initial levels.
5	70%	40%	3	– Gradual reduction from 70% to 60% to 50% and to 40%, respectively per annum. – Having reached the objective of 40% consider the same as for strategy 4. – Another possibility is to apply strategy 4 stopping during one period at indebtedness level 50% in order to consolidate the process before continuing on to the final objective.

6	70%	50%	1	– Reduction in one period from 70% to 50%. – Having reached the objective remain there in order to consolidate this financial situation. – Another possibility is to continue the indebtedness reduction process.
7	70%	50%	2	– Gradual reduction of indebtedness from 70% to 60% and to 50% respectively per annum. – Having reached the objective it is possible to carry out the consideration as per strategy 6.
8	70%	60%	1	– Reduction of indebtedness in one period from 70% to 60%. – Having reached this objective, the tendency will be to continue the process of reducing indebtedness.

The process stops here although it would be possible to continue to arrive at matrices that list paths of lengths greater than 4, by means of new compositions, in the same way as we have done up to now.

Perhaps we should point out in the event of finding matrices listing paths of a length greater than 4 that, as is could be in no other way, in all of them one or several situations or positions would be repeated, that is that the path would pass more than once by some of the vertices that make up the corresponding matrix. This evidence arises since we are faced with a set of 5 possible situations that can only form Hamiltonian paths (that is, without repetition) of length 4 (4 arcs that are joined through five vertices).

15.6 Direct Financial Policies

Very often the idea occurs of taking actions included in financial policies that are aimed in a single direction, that is, that the decisions include either successive reductions of the indebtedness or in the event successive increases of the same. But these decisions do not contemplate either maintenance or a backward movement (or an advance in the event) of the indebtedness policy. Faced with this situation, and in spite of the fact that the previous matrices list all the possible solutions, that is, all the paths that permit passing from one situation of financial indebtedness to another, they do not foresee a solution to this particular problem.

Fortunately we have sufficient techniques[6] available in order to give ample satisfaction in this area. And for this it is sufficient to adapt the scheme that we have developed up to this point, in such a way that Latin matrices can be

[6] Kaufmann A and Gil Aluja J (1991) Nuevas técnicas para la dirección estratégica. (Ed) Barcelona University, p. 193.

	a	b	c	d	e
a	acda		abcc accc	abcd accd aced	abce acce acde acee
b	bcda		bccc	bccd bced	bcce bcde bcee
c	ccda ceda	cdab	cdac cccc	cccd cced cded ceed	ccce ccde cede ccee cdee ceee
d	deda		dabc dacc	dacd deed	dace dede deee
e	eeda	adab	edac	eded eeed	eede edee eeee

$[E]^3$

	a	b	c	d	e
a		b	c		
b			c		
c			c	d	e
d	a				e
e				d	e

$[E']^1$

=

	a	b	c	d	e
a	abcda accda aceda	acdab	acdac abccc acccc	abcdd acccd abced acced acded aceed	abcce accce abcde accde acede abcee accee cadee aceee
b	bccda bceda	bcdab	bcdac bcccc	bcccd bcced bcded bceed	bccce bccde bcede bcee bcdee bceee
c	cccda cceda cdeda ceeda	ccdab cedab	ccdac cedac cdabc cdacc cccc	cdadc ccccd ccced ccded ceded cceed cdeed ceeed	edace cccce cccde cccde cdede ceede cccee ccdee cedee cceee cdeee ceeee
d	dacda deeda	dedab	dedac dabccd accc	dabcd daccd daced deded deeed	dabce dacce dacde deede dacee dedee deeee
e	ededa eeeda	eedab	eedac edabc edacc	edacd eeded edeed eeeed	edace edede eeede eedee edeee eeeee

$[E]^4$

arrived at that show paths of lengths 2, 3 and 4 "without any repetition" of vertices, that is to say, without the business having to pass more than once nor be detained for more than one period in the same financial situation.

For this we will do the compositions of the Latin matrices by changing the required property, that is the P, which in this case will be "path without repetition", that is to say, that on arriving at the successive results of the composition, the paths (of length 2, 3 and 4) will be rejected that do not comply with the required property P, and therefore do contain the repetition of an element.

Very quickly we arrive at the Latin matrices that list the paths of length 2, 3 and 4:

	a	b	c	d	e
a		ab	ac		
b			bc		
c			cc	cd	ce
d	da				de
e				ed	ee

$[E]^1$

	a	b	c	d	e
a		b	c		
b			c		
c			c	d	e
d	a				e
e				d	e

$[E']^1$

=

	a	b	c	d	e
a			abc	acd	ace
b				bcd	bce
c	cda			ced	cde
d		dab	dac		
e	eda				

$[E]^2$

	a	b	c	d	e
a			abc	acd	ace
b				bcd	bce
c	cda			ced	cde
d		dab	dac		
e	eda				

$[E]^2$

	a	b	c	d	e
a		b	c		
b			c		
c			c	d	e
d	a				e
e				d	e

$[E']^1$

=

	a	b	c	d	e
a				abcd aced	abce acde
b	bcda			bced	bcde
c	ceda	cdab			
d			dabc		dace
e		edab	edac		

$[E]^3$

	a	b	c	d	e
a				abcd aced	abce acde
b	bcda			bced	bcde
c	ceda	cdab			
d			dabc		dace
e		edab	edac		

$[E]^3$

	a	b	c	d	e
a		b	c		
b			c		
c			c	d	e
d	a				e
e				d	e

$[E']^1$

=

	a	b	c	d	e
a				abced	abcde
b	bceda				
c		cedab			
d					dabce
e			edabc		

$[E]^4$

As we have pointed out before, it is not possible to arrive at Latin matrices of order 5, $[E]^5$, and following without some repetition occurring, therefore it is obvious that the process must end here. If this were not so, we would arrive at a matrix $[E]^5$, the result of doing the P-Latin composition of matrix $[E]^4$, with the amputated initial matrix $[E']^1$, which would be void.

A brief analysis of matrix $[E]^4$ brings to light the fact that all the possible paths include the same vertices $\{a, b, c, d, e\}$ ordered in a different manner, in accordance, as is logical, with the direction of the arcs that join them. This is so as a consequence that in the graph that has served as our example there is at least one circuit $\{a, b, c, e, d, a\}$ which gives rise to 5 different paths. But as there exists another circuit (de) by means of it a sixth path is born.

In short, then, the fact of providing different paths with no repetition makes it possible to present to those responsible for finances in businesses who have to take the decisions, a whole range of paths in the form of possible policies, by means of which it is ensured that the business does not pass more than once neither does it become detained in one and the same situation of financial indebtedness.

In this case, on following the most direct path time is gained, although economic or strategic optimisation of the chosen path is not necessarily ensured.

15.7 Final Considerations

After analysing the results obtained by means of the method based on the P-Latin composition, it is possible to verify compliance of the objective that was set. We have managed to list all the strategies that the business could carry out with the object of modifying a position of indebtedness existing at any given moment.

It is obvious that a model such as we have proposed, based fundamentally on the use of non numerical mechanisms, also allows the use of numerical data.

Taking into account that we are immersed in a context of uncertainty in which information relative to the future is characterised, on many occasions, by the impossibility of being represented in a crisp manner, it becomes necessary to resort, as happens repeatedly, to the theory of fuzzy sub-sets.

Throughout the development of the model the need arises for taking decisions. This is one of the elements that define the essence of economic performance, which requires, for it solution, resorting to operators of diverse nature but which, given the context in which the scheme has been thought over, its results must remain within$[0; 1]$. It is well known that for this two operators exist, τ-norms and inferences. In this case we have opted for inferences, because it would appear that with them a good adaptation to reality is arrived at for the formal scheme. This does not mean that for other decisions a T-norm or T-co-norm may be more adequate.

All of this, accompanied by the incorporation of the P-Latin composition method for determining the possible paths to be followed for modifying the indebtedness of a business has permitted, in our opinion, taking a very positive step forward towards the taking of decisions by those responsible for finances, at the same time constituting one more piece of evidence of the fact that we are faced by a process of change in the paradigms of the theory of decision.

Nevertheless we do not assume that this model will give full satisfaction to the problems that are brought up by the changes in the situation of indebtedness, and in spite of the fact that with its development a more overall vision is attained of the possible policies to be taken on, the model can be completed by the explicit incorporation of other elements implied in the process such as financial cost, repercussions in the production process, incidences in sales, etc.

We will leave this subject here open to the path so that future works may contribute new elements for widening the horizon of this important matter.

16 Sequential Strategies for Raising Financial Means

16.1 Approach to the Problem

One of the basic aspects of financial activity in business is to be found in the process for raising financial means. These can be very varied and of diverse natures. From a general point of view it can be considered that the businessman, in order to continue developing the production process requires a determined amount of monetary means for attending to cash requirements over time.

Businesses that are immersed in an economic system with continued expansion frequently see how their monetary flows derived from invoicing are not always sufficient, so that they often find it necessary to resort to financial means initially from external sources.

Indeed while a part of these are susceptible to being contributed by the owners of the business and in this way becoming equity, and another part can possibly be obtained from the excess derived from sales of products or services, there is a more or less important residue that must come from resources obtained from financial entities.

Obviously, in a context in which events and phenomena take place at speeds that were unsuspected up to this point in time, the need arises for placing at the disposal of businesses a set of techniques that permit their executives choose those financial products that are best suited to their requirements.

In financial literature, there are a multitude of elements that, incorporated into certain models of a determinist or random nature, attempt to provide a satisfactory solution to this matter. Nevertheless, we feel that in an eminently changing environment it is necessary to use another type of instrument for the treatment of business financial problems in a context of uncertainty.

From among the many alternatives provided by the mathematics of uncertainty we are going to attempt to contribute a valid solution through a well-known concept: the concept of logical inferences. For this, we are going to present a model by means of an example, which will be analysed from this perspective, which will permit us to get a better visualisation and understanding of the technique.

We are going to assume that a business requires, in order to reach certain general objectives, that a determined volume of financial means is considered:

Anna Maria Gil-Lafuente: *Fuzzy Logic in Financial Analysis*, StudFuzz **175**, 279–301 (2005)
www.springerlink.com

increase in the quotation of the shares, increase in sales (attain a greater presence in the market) and increase its profit levels.

In order to reach these general objectives or ends it is necessary to arrive at certain intermediary objectives, which are listed below:

Extend industrial building
Increase the commercial distribution network
Improve the quality control service
Increase investments in advertising
Improve the supply system
Purchase production equipment
Improve the liquidity situation

In order to arrive at these objectives the people responsible for the finances of the business analyse the possibilities of using the following financial means:

Contributions from shareholders by means of a capital increase
Increase of the margin between sales price and unitary variable cost
Launching an issue of debentures convertible into shares
Obtaining a long-term credit from a financial entity
Increase in payment terms to suppliers
Obtaining a short-term credit from a financial entity

16.2 The Objective by Means of Inference Chains

In the context we have mentioned above and for the effects of getting to know the possibilities for reaching these objectives, we will analyse the problem by means of inference chains. In the financial sphere the notion of inference acquires a special relevance, since it brings to light the connections existing between the values that make up the possible financial means that are susceptible to being used and the management elements that will permit the business to reach its objectives. In fact, everyone is aware of the relation existing between the contributions made by shareholders and the possible extension of a plant, since the former will provide a level of financing that will be decisive for attaining the latter. Likewise the relation existing between the acquisition of a long-term credit and an increase in the distribution network is brought to light. In this case, the long-term financing takes on the function of reinforcing the existing network for distribution of the products of the business in this way making it possible to take on any future possible extensions and improvements in the same, in order to provide clients with a better service. In this way we could analyse, one by one, the relations existing between financial means such as sales margin, the launching of a new debenture issue, increase payment terms to suppliers, short-term credits, etc., over basic elements of financial management such as are quality control, improvements in supply costs, renewal of production equipment, and

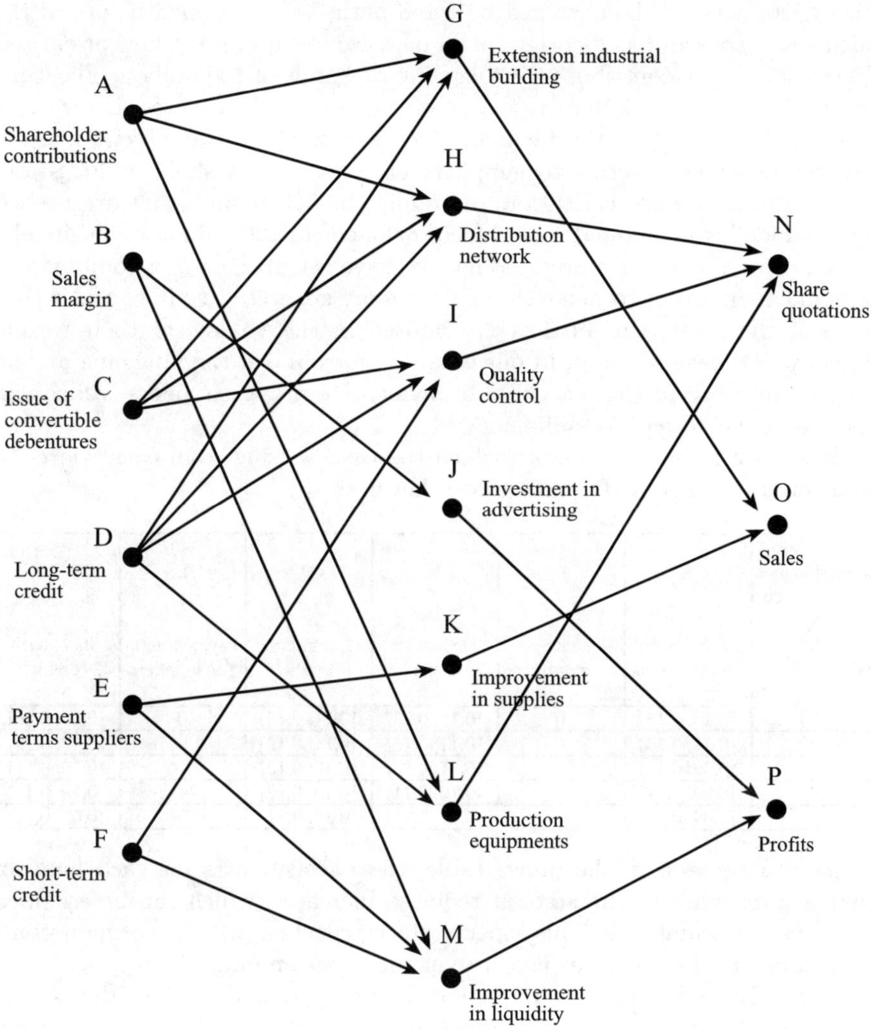

Fig. 16.1.

the improvement of the level of liquidity, which in short are the elements that determine the possible securing of the objectives set by the business.

Precisely in order to bring to light this framework of relations between financial elements, it is possible to construct an inference chain in which the links existing between the different values are established. These are shown in Fig. 16.1 below:

We now move on to study the financial support on which the construction of the graph is based. For this we start out with a hypothetical business which, at a given moment, needs to raise a certain volume of financial means.

These products can be obtained by the 6 paths we mentioned before. If the raising of resources has its origin in an increase in capital, the new or current shareholders may contribute the financing by bands of 100 million. The same occurs if they proceed from the issue of convertible debentures. Likewise, if it is decided to obtain the financing by increasing the commercial margin, increasing payment terms to suppliers or requesting a short or long-term credit from a finance institution, obtaining the 500 million that are needed can be done by a combination of blocks or bands of 100 million indivisible by the different financing sources. What we have set up here is a combinatory component. In order to make the scheme more general, we will consider that some of the sources of finance are limited in the volume of their raising capacity. We have selected, in our case, as an example, the financing arising from an increase in the margin and payment terms to suppliers, which can only be in each case 100 million.

Below we present a table in which the costs arising from each source of financing by bands of 100 million are shown:

Finance sources / Millions	A Shareholder contributions	B Sales margin	C Debenture issue	D Long-term credit	E Payment term for suppliers	F Short-term credit
100	[0,13; 0,15]	0	[0,11; 0,13]	[0,11; 0,14]	0	[0,14; 0,16]
200	[0,11; 0,13]	--	[0,11; 0,13]	[0,11; 0,12]	--	[0,12; 0,13]
300	[0,09; 0,11]	--	[0,11; 0,12]	[0,10; 0,12]	--	[0,10; 0,11]
400	[0,09; 0,11]	--	[0,10; 0,11]	[0,10; 0,12]	--	[0,10; 0,11]
500	[0,09; 0,10]	--	[0,09; 0,11]	[0,10; 0,11]	--	[0,09; 0,11]

As can be seen in the above table, the unitary costs for each band of financing decrease as the amount required increases, which can be assumed as the most normal case. This aspect, which reflects a principle of monotony, is fundamental for the later use of dynamic programming.

It will be seen that for sources of financing B and E only the cost is shown (in this case nil) for the first 100 million. This is due to the already mentioned fact that these two sources of financing only reach a volume of 100 million each.

Having analysed the cost per band of each one of the financing means, we must proceed to study all the possible combinations that make it possible for the total cost of the required volume to be minimal. It is obvious that the business will, in the first instance, take those financial means that are less onerous (zero cost), that is 100 million from element B (increase sales margin) and 100 million from element E (increase in payment terms for suppliers). With this there may be a determined financial deficit, in order to reach the desired level of resources. For the effects of getting to know the cost this would mean for each one of the alternatives and each one of the bands there are up to 500 million the following table is constructed:

It can be seen that if only 100 million is required the cheapest source of financing of the four is C. If we consider 200 million the lowest cost, if only using a single source of financing, could be either element C or element D. Nevertheless the possibility could be considered that these 200 million were to be obtained in a combined way from two or more different financial sources. This also crops up when considering raising 300, 400 and 500 million.

Finance sources / Millions	A Shareholder contributions		C Debenture issue		D Long-term credit		F Short-term credit	
	Partial	Accumulated	Partial	Accumulated	Partial	Accumulated	Partial	Accumulated
100	[13, 15]	[13, 15]	[11, 13]	[11, 13]	[11, 14]	[11, 14]	[14, 16]	[14, 16]
200	[11, 13]	[24, 28]	[11, 13]	[22, 26]	[11, 12]	[22, 26]	[12, 13]	[26, 29]
300	[9, 11]	[33, 39]	[11, 12]	[33, 38]	[10, 12]	[32, 38]	[10, 11]	[36, 40]
400	[9, 11]	[42, 50]	[10, 11]	[43, 49]	[10, 12]	[42, 50]	[10, 11]	[46, 51]
500	[9, 10]	[51, 60]	[9, 11]	[52, 60]	[10, 11]	[52, 61]	[9, 11]	[55, 62]

To resolve this question is, with this example, relatively simple. But if what is desired is a general solution to problems of this nature, it will be necessary to resort to a model we propose of a sequential nature.

16.3 Incidence of the Level of Financing for Each Objective Over the Source

With the object of making an approach to the proposed model that allows its use in as wide a spectrum of real cases, it would appear to be convenient to make special mention of the required level of financing for each intermediary objective. In accordance with the established hypotheses, we should remember that the financial means necessary for reaching each objective, are susceptible to being raised by means of successive bands of 100 million each. Taking this into consideration we are going to establish the following connection between sources and intermediary objectives:

For attaining element G (Expansion of the industrial building) 500 million monetary units are required, which can be raised by contributions from the shareholders and/or issuing convertible debentures and/or by obtaining a long-term credit (fixed assets, that is, long-term assets financed by permanent capitals).

With the object of attaining element H (expansion of the distribution network) 300 million monetary units are required which can be raised via shareholder contributions and/or issuing convertible debentures and/or obtaining a long-term credit (medium term assets financed by permanent capitals).

For the effects of attaining element I (increase quality controls) also required are 300 million monetary units, which can be obtained through shareholder contributions and/or the issue of convertible debentures, and/or by means of obtaining a long-term credit (medium/long-term financed by short or long-term liabilities).

With the object of reaching objective J (increased investment in advertising) 100 million monetary units are required, which can be obtained from the increase in the sales margin (short-term assets financed by current liabilities).

In order to reach element K (improve supplies) 100 million monetary units are required which can only be raised from an increase in payment terms to suppliers (short-term assets financed by short-term liabilities).

To attain element L (renewal of production equipment) 300 monetary units are required which can be obtained from shareholder contributions and/or issuing convertible debentures and/or long-term credit (fixed assets financed by permanent capitals).

Finally for reaching element M (improve the liquidity situation) means obtaining 200 million monetary units, which can come from an increase in sales margin and/or increasing payment terms to suppliers and/or obtaining a short-term credit (liquid assets financed by short-term liabilities).

Before commencing calculations relative to the sequences that comprise the process, we are going to bring to light, by means of an arrow form graph, the possible relations existing between the sources of financing and each one of the intermediary objectives.

Since there are two sources of financing that are limited to 100 million monetary units, we will have to make adjustments to the graph we showed previously.

In fact, a simple and superficial glance at the previous scheme and taking into account the established requirements and connections will allow us to see certain aspects, which will give rise to a modification of the previous graph in order to rationalise it, taking into account the data that has now been contributed. In fact if we direct our attention to vertices J and K, it can be seen that these objectives are requirements of 100 million monetary units each one for which it is only suitable that they be satisfied by a single source, B for J and E for K. This fact (which is also very general in business financial activities) contributes certain conditions (limitations): the freedom of assigning sources to requirements. In this case (evidently simple but easily generalised) there is no possibility of selection for financing objective J and objective K. They must be financed by source E which in this way exhausts its possibilities and therefore cannot take part in financing M (arc (E, M) must be eliminated and by source B which likewise exhausts its funds and cannot take part in financing M (arc (B, M) must be eliminated). Now then when we eliminate these arcs the result is that, objective M can only be attained from source F, but as this was not considered unlimited, arc (F, I) must be maintained.

Having made the necessary assignments of objectives J and K, it can be seen from the previous graph, that the financing of the remaining objectives can be done from more than one financial means. The question that now arises, therefore, consists in obtaining the combination of sources of finance that minimises the cost of their use for financing each one of the intermediary objectives. The instrument that is going to allow us to reach this objective is based on a sequential model the contents of which we present in a very general manner. For this we designate with X the amount relative to the financial means required and $x_i, i = 1, 2, \ldots, n$, will represent the monetary bands stemming from each one of the sources of financing, in such a way that the following will be complied with:

$$X = \sum_{i=1}^{n} x_i + \sum_{j=1}^{m} x_j + \cdots + \sum_{y=1}^{z} x_y$$

As we have already commented financing of a determined monetary mass means at most times, a determined cost, although there are certain cases, such as the this one, in which certain finance sources (two in the example) exist with nil cost although with a restricted amount (100 and 100 million respectively).

What remains now, for the combinatory analysis, is to obtain the minimum cost for the respective amounts required for the remaining intermediary objectives and remaining sources of finance. For this those sources of finance with zero cost must be deducted.

With the object of resolving this problem we are going to resort to the principle of the Bellman[1] optimisation on which dynamic programming is based. We feel that by means of this important instrument it will be possible to arrive at the combination of sources of finance that will allow us to minimise the cost arising from obtaining and using the same.

We will call $C_A(x_i)$, $C_B(x_j)$, $C_C(x_h)$ and $C_D(x_l)$ the cost representing financial means A, B, C and D respectively. If we designate:

$C_{AB}(X)$: the optimum cost for obtaining X with the combination of financing sources A and B.

$C_{ABC}(X)$: optimisation with financing sources A, B and C.

and so on successively, we arrive at:

$$C_{AB}(X) = \underset{x_i+x_j=X}{MIN} \left(C_A(x_i) + C_B(x_j)\right)$$

$$C_{ABC}(X) = \underset{x_i+x_j=X}{MIN} \left(C_{AB}(x_i) + C_C(x_j)\right)$$

$$\ldots$$

[1] Bellman and Dreyfus (1965) La programation dinamique et ses aplications. (Ed) Dunod Editeurs. Paris.

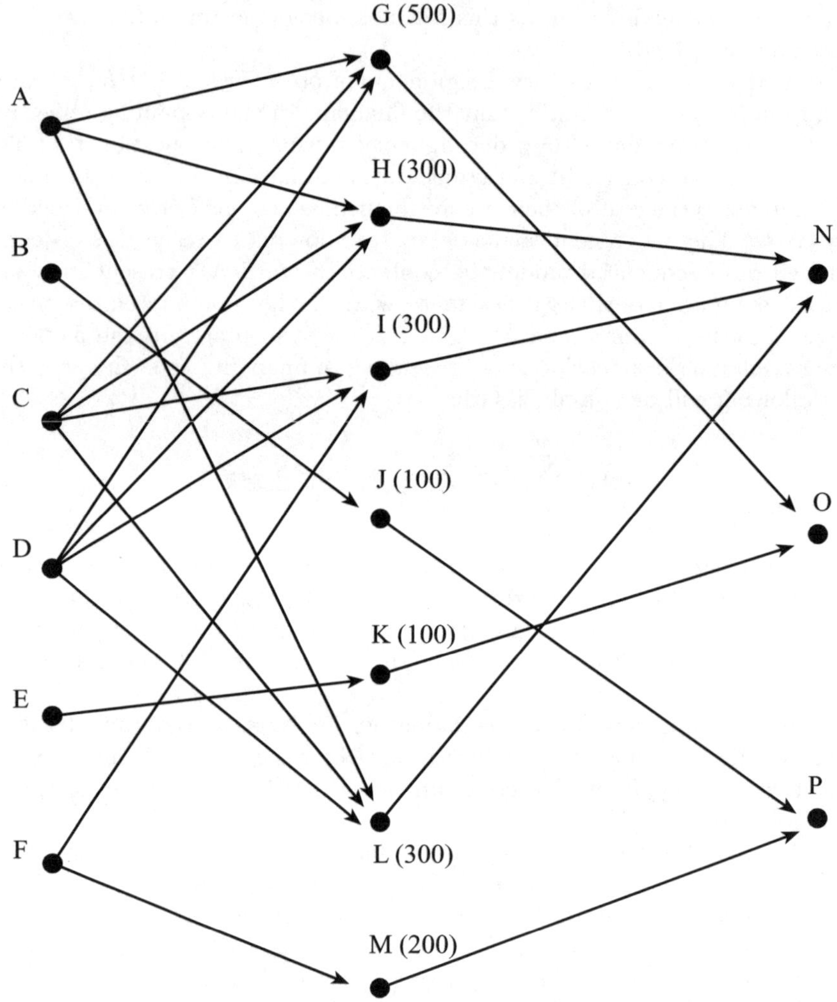

Fig. 16.2.

$$C_{ABC\ldots Z}(X) = \underset{x_i+x_j=X}{MIN}\left(C_{ABC\ldots Y}(x_i) + C_Z(x_j)\right)$$

In this way we can see that the principle of dynamic programming is complied with according to which "an optimum trajectory can only be formed by optimum sub-trajectories"[2].

In our example, the financial requirements of element G are 500 million in possible bands of 100, therefore it will be necessary, in the first place, to

[2] Kaufmann A and Gil Aluja J (1991) Nuevas técnicas para la dirección estratégica. (Ed) Universidad de Barcelona. Barcelona p. 191.

find the minimum cost (optimum) for obtaining 100 million, considering only financing sources A and C. We arrive at:

$$\begin{aligned} C_{AC}(100) &= \text{MIN}\left\{(C_A(0) + C_C(100)), (C_A(100) + C_C(0))\right\} \\ &= \text{MIN}\left\{(0\,(+)\,[11,13]), ([13,15]\,(+)\,0)\right\} = \underline{\underline{[11,13]}} \end{aligned}$$

The optimisation corresponds to 100 million from finance source C.

We then find the optimisation of the cost for obtaining 200 million:

$$\begin{aligned} &C_{AC}(200) \\ &= \text{MIN}\{(C_A(0) + C_C(200)), (C_A(100) + C_C(100)), (C_A(200) + C_C(0))\} \\ &= \text{MIN}\{(0(+)[22,26]),([13,15](+)[11,13]), ([24,28](+)0)\} = \underline{\underline{[22,26]}} \end{aligned}$$

corresponding to 200 million from C.

We then find the minimum cost for 300 million:

$$\begin{aligned} C_{AC}(300) &= \text{MIN}\left\{\begin{matrix} (C_A(0) + C_C(300)), (C_A(100) + C_C(200)) \\ (C_A(200) + C_C(100)), (C_A(300) + C_C(0)) \end{matrix}\right\} \\ &= \text{MIN}\left\{\begin{matrix} (0\,(+)\,[33,38]), ([13,15]\;(+)\,[22,26]) \\ ([24,28]\,(+)\;[11,13]), ([33,39]\,(+)\,0) \end{matrix}\right\} = \underline{\underline{[33,38]}} \end{aligned}$$

which corresponds to 300 million from C.

The minimum cost corresponding to 400 million is:

$$\begin{aligned} C_{AC}(400) &= \text{MIN}\left\{\begin{matrix} (C_A(0) + C_C(400)), (C_A(100) + C_C(300)), \\ (C_A(200) + C_C(200))\,(C_A(300) + C_C(100)), \\ (C_A(400) + C_C(0)) \end{matrix}\right\} \\ &= \text{MIN}\left\{\begin{matrix} (0\,(+)\,[43,49]), ([13,15]\;(+)\,[33,38]), \\ ([24,28]\,(+)\,[22,26]), ([33,39]\,(+)\,[11,13]), \\ ([42,50]\,(+)\,0) \end{matrix}\right\} \\ &= \left\{\underline{\underline{[43,49]}}, \underline{\underline{[42,50]}}\right\} \end{aligned}$$

In this case it can be seen that the confidence intervals [43,49] (representing the 400 million of C) and [42,50] (corresponding to 400 million from A) should be compared given that neither one of the two is clearly less than the other. It is necessary, therefore, to establish some sort of additional criterion that will allow us to decide which of the two should be considered as the least. The first of these criteria would be to see in certainty which of the two is the smaller. Thus, if we make the entropy fall:

$$\underset{\sim}{C}_C(400) = \frac{[43,\ 49]}{2} = 46 \qquad \underset{\sim}{C}_A(400) = \frac{[42,\ 50]}{2} = 46$$

we can see that the results continue to coincide.

A second criterion, which we could consider to take the confidence interval containing the least degree of uncertainty. Uncertainty is measured as the distance existing between the lower extreme and the upper extreme of a confidence interval. Therefore:

$$\text{for}: C_C(400) : (49 - 43) = \underline{\underline{6}}$$
$$\text{for}: C_A(400) : (50 - 42) = \underline{\underline{8}}$$

According to this criterion we will take as the lowest the cost arising from obtaining 400 million from finance source C.

Finally we find the optimum cost for 500 million:

$$\begin{aligned} C_{AC}(500) &= \text{MIN}\left\{\begin{array}{l}(C_A(0) + C_C(500)),\ (C_A(100) + C_C(400)),\\ (C_A(200) + C_C(300)),(C_A(300) + C_C(200)),\\ (C_A(400) + C_C(100)),(C_A(500) + C_C(0))\end{array}\right\} \\ &= \text{MIN}\left\{\begin{array}{l}(0\,(+)\,[52,60]),([13,15]\ (+)\,[43,49]),\\ ([24,28]\,(+)\,[32,38]),([33,39]\,(+)\ [22,26]),\\ ([42,50]\,(+)\,[11,13]),([51,60]\,(+)\,0)\end{array}\right\} \\ &= \underline{\underline{[51,60]}} \end{aligned}$$

which represent obtaining the 500 million from A.

In this way we have arrived at the optimum raising of financial means taking into account only finance sources A and C.

In order to continue with the optimisation process it will be necessary to include a new financing source (in this case we introduce D). If we represent the optimums we have obtained in the previous phase in a table jointly with the cost arising from D we have:

Financing Sources

Millions	$C_{AC}(X)$	$C_D(x)$
100	[11,13]	[11,14]
200	[22,26]	[22,26]
300	[33,38]	[32,38]
400	[43,49]	[42,50]
500	[51,60]	[52,61]

Now in the same way as before we proceed to obtain the respective minimum costs for 100, 200, 300, 400 and 500 million.

$$\begin{aligned} C_{ACD}(100) &= \text{MIN}\{(C_{AC}(0) + C_D(100)),(C_{AC}(100) + C_D(0))\} \\ &= \text{MIN}\{(0\,(+)\,[11,14]),([11,13]\,(+)\,0)\} = \underline{\underline{[11,13]}} \end{aligned}$$

corresponding to 100 million of C.

$$
\begin{aligned}
C_{ACD}(200) &= \text{MIN}\{(C_{AC}(0)+C_D(200)),(C_{AC}(100)+C_D(100)),\\
&\quad (C_{AC}(200)+C_D(0))\}\\
&= \text{MIN}\{(0(+)[22{,}26]),([11,13](+)[11{,}14]),([22{,}26](+)0)\}\\
&= \underline{\underline{[22{,}26]}}
\end{aligned}
$$

result corresponding to two alternatives:

200 million from finance source D
200 million from finance source C

$$
\begin{aligned}
C_{ACD}(300) &= \text{MIN}\left\{\begin{matrix}(C_{AC}(0)+C_D(300)),(C_{AC}(100)+C_D(200))\\(C_{AC}(200)+C_D(100)),(C_{AC}(300)+C_D(0))\end{matrix}\right\}\\
&= \text{MIN}\left\{\begin{matrix}(0(+)[32{,}38]),([11,13]\ (+)\ [22{,}26])\\([22{,}26]\ (+)\ [11{,}14]),([33,38]\ (+)\ 0)\end{matrix}\right\}\\
&= \underline{\underline{[32{,}38]}}
\end{aligned}
$$

representing 300 million from source D

$$
\begin{aligned}
C_{ACD}(400) &= \text{MIN}\left\{\begin{matrix}(C_{AC}(0)+C_D(400)),(C_{AC}(100)+C_D(300)),\\(C_{AC}(200)+C_D(200))\,(C_{AC}(300)+C_D(100)),\\(C_{AC}(400)+C_D(0))\end{matrix}\right\}\\
&= \text{MIN}\left\{\begin{matrix}(0(+)[42{,}50]),([11,13]\ (+)\ [32{,}38]),\\([22{,}26]\ (+)\ [22{,}26]),([33,38]\ (+)\ [11,14]),\\([43,49]\ (+)\ 0)\end{matrix}\right\}\\
&= \left\{\underline{\underline{[42{,}50]}},\underline{\underline{[43{,}49]}}\right\}
\end{aligned}
$$

Here the same considerations are made as in the case of minimising CAC (400), taking as the lowest cost interval [43, 49], which represents 400 million from C.

$$
\begin{aligned}
C_{ACD}(500) &= \text{MIN}\left\{\begin{matrix}(C_{AC}(0)+C_D(500)),(C_{AC}(100)+C_D(400)),\\(C_{AC}(200)+C_D(300)),(C_{AC}(300)+C_D(200)),\\(C_{AC}(400)+C_D(100)),(C_{AC}(500)+C_D(0))\end{matrix}\right\}\\
&= \text{MIN}\left\{\begin{matrix}(0(+)[52{,}61]),([11,13]\ (+)\ [42{,}50]),\\([22{,}26]\ (+)\ [32{,}38]),([33,38]\ (+)\ [22{,}26]),\\([43,49]\ (+)\ [11,14]),([51,60]\ (+)\ 0)\end{matrix}\right\}\\
&= \underline{\underline{[51,60]}}
\end{aligned}
$$

Represents obtaining 500 million from A.

We can see, then, that the minimum cost for raising financing of 500 million that are necessary for reaching objective G obliges us to resort to the total 500 million available from finance source A.

We now move on to arriving at the minimum cost when we need to finance objective H with 300 million. The possible sources of finance, as occurs with element G are A, C, and D. In this case we are once again going to take the calculations we found before, with which we will arrive at the conclusion that the minimum cost for a volume of 300 million is established in [32, 38] which represents obtaining the 300 million from finance source D.

We now proceed to minimise the cost for financing element I. For this we are going to use the same process we used before but the amount in this case is 300 million.

Our interest now is centred on the optimisation of the cost arising from financing sources C, D and F. For this we construct the following table from which we will minimise cost for C and D:

Financing Sources

Millions	$C_C(X)$	$C_D(X)$	$C_{CD}(x)$ Optimum Cost
100	[11,13]	[11,14]	[11,13]
200	[22,26]	[22,26]	[22,26]
300	[33,38]	[32,38]	[32,38]

We will arrive at:

$C_{CD}(100) =$ [11,13], corresponding to 100 million from C
$C_{CD}(200) =$ [22,26], corresponding to 200 million from either C or D
$C_{CD}(300) =$ [32,38], corresponding to 300 million from D.

Having found the optimum cost for financing sources C and D, we proceed to bring F into play:

Financing Sources

Millions	$C_{CD}(X)$	$C_F(X)$	$C_{CDF}(x)$ Optimum Cost
100	[11,13]	[14,16]	[11,13]
200	[22,26]	[26,29]	[22,26]
300	[32,38]	[36,40]	[32,38]

In this way:

$C_{CDF}(100) =$ [11,13], corresponding to 100 million from C
$C_{CDF}(200) =$ [22,26], corresponding to 200 million from either C or D
$C_{CDF}(300) =$ [32,38], corresponding to 300 million from D.

Just as we pointed out before, financing for intermediary objectives J and K, which each require 100 million, proceeds, respectively, from finance sources B and E the cost of which is precisely zero.

For financing 300 million for objective L we can resort to financing sources A, C and D. By following the calculations we did before, obtaining 300 million will have a minimum cost estimated in [32, 38] for financing 300 million from finance source D.

Finally as has already been stated, the cost derived from obtaining 200 million from the only possible source F, for objective M represents an amount valued in [26, 29].

16.4 Analysis of the Data Obtained

If we look at the data we found before the fact can be verified that optimisation of the cost arising from the possible use of several sources of financing, is always resolved by taking all the monetary units from a single source. This is due to two reasons:

1. The scheme has been approached in a decreasing monotonous manner, that is to say as successive bands are taken from one and the same source of financing, the cost of each of them decreases.
2. The structure under which the scheme has been developed is based on a minimisation of costs. Indeed, on considering several possible sources of finance we have always selected the one or the ones with the minimum cost.

Under these conditions (decreasing monotony and minimisation) it is possible to state that optimisation of the costs will always consist in taking all the monetary units from a single source.

The conclusions that are reached with this approach could be substantially modified if one of the two stated conditions (decreasing monotony and minimisation) were to be changed. A different structure of the problem could have been based on the fact that a business were to require a determined volume of financing and for this, it had to resort to several financial entities. In this case the business would commence by taking the bands, in monetary units that were to be least costly. On having used up all the possible bands with the least cost, it would then have to resort to those that were more expensive until its total requirements for finance were to be covered.

In this situation the business would find itself with the need to minimise the costs of the financial sources but, on the contrary to the example we developed, the cost arising from the different financing bands obtained from the different financial entities would be structured under the case of an increasing monotony.

The conclusions arising from the application of this approach will give rise to several possibilities:

A first case could be that the total amounts of monetary units required for the financing in the business were to come from one and the same financial entity.

Another possible solution would be that the different bands of financing at a minimum cost, were to come from different financial entities.

In order to make the results of the example more general for the problem of minimisation when the costs are structured under the case of decreasing monotony, below we have developed the corresponding theoretical model.

We designate by:

X: total amount to be financed
$x_i, i = 1, 2, \ldots, n$: monetary bands of equal amounts from each source of financing
$A, B, C, \ldots, Z$: sources of financing
$C_{A,B,\ldots,Z}(x_i)$: cost of financing bands from finance sources $A, B, \ldots, Z$.
$C_{A,B,\ldots,Z}(X)$: optimum of the total amount to be financed.

If we base ourselves on the decreasing monotony of the costs for each of the sources of finance we arrive at:

	Financing Sources				
Monetary Bands	A	B	C	$\ldots$	Z
x_1	$C_A(x_1)$	$C_B(x_1)$	$C_C(x_1)$	$\ldots$	$C_Z(x_1)$
x_2	$C_A(x_2)$	$C_B(x_2)$	$C_C(x_2)$	$\ldots$	$C_Z(x_2)$
x_3	$C_A(x_3)$	$C_B(x_3)$	$C_C(x_3)$	$\ldots$	$C_Z(x_3)$
.	.	.	.	.	.
.	.	.	.	.	.
.	.	.	.	.	.
X_n	$C_A(x_n)$	$C_B(x_n)$	$C_C(x_n)$	$\ldots$	$C_Z(x_n)$

If we start out from the fact that the scheme is developed under the case of decreasing monotony, the following will occur:

$$\begin{gathered} C_A(x_1) \geq C_A(x_2) \geq C_A(x_3) \geq \ldots \geq C_A(x_n) \\ C_B(x_1) \geq C_B(x_2) \geq C_B(x_3) \geq \ldots \geq C_B(x_n) \\ C_C(x_1) \geq C_C(x_2) \geq C_C(x_3) \geq \ldots \geq C_C(x_n) \\ \ldots \\ C_Z(x_1) \geq C_Z(x_2) \geq C_Z(x_3) \geq \ldots \geq C_Z(x_n) \end{gathered}$$

That is to say, each one of the additional bands corresponding to a same source of finance will have a cost equal to or less than the cost of the immediately previous band.

In this way it can be seen that, as more and more bands are taken up from one and the same source of finance, the average cost decreases in general terms. Only in the exceptional case in which the costs of each one of the bands were to remain constant, then the average cost corresponding to the different bands acquired would be maintained.

In general terms this will be expressed as follows:

$$
\begin{aligned}
C_A(x_1) &\geq \frac{C_A(x_1)\,(+)\,C_A(x_2)}{2} \geq \frac{C_A(x_1)\,(+)\,C_A(x_2)\,(+)\,C_A(x_3)}{3} \\
&\geq \cdots \geq \frac{C_A(x_1)\,(+)\,C_A(x_2)\,(+)\,C_A(x_3)\,(+)\cdots(+)\,C_A(x_n)}{n} \\
C_B(x_1) &\geq \frac{C_B(x_1)\,(+)\,C_B(x_2)}{2} \geq \frac{C_B(x_1)\,(+)\,C_B(x_2)\,(+)\,C_B(x_3)}{3} \\
&\geq \cdots \geq \frac{C_B(x_1)\,(+)\,C_B(x_2)\,(+)\,C_B(x_3)\,(+)\cdots(+)\,C_B(x_n)}{n} \\
&\cdots \\
C_Z(x_1) &\geq \frac{C_Z(x_1)\,(+)\,C_Z(x_2)}{2} \geq \frac{C_Z(x_1)\,(+)\,C_Z(x_2)\,(+)\,C_Z(x_3)}{3} \\
&\geq \cdots \geq \frac{C_Z(x_1)\,(+)\,C_Z(x_2)\,(+)\,C_Z(x_3)\,(+)\cdots(+)\,C_Z(x_n)}{n}
\end{aligned}
$$

Having analysed the matter of decreasing monotony relative to the bands from each one of the sources of finance, we should now proceed to study the values, which will give rise to the minimisation of the costs.

Indeed, if we take as the start out point the fact that to each successive band from one and the same source of finance a constantly decreasing cost will correspond, the conclusion can be reached that the total amount to be financed will be formed by successive monetary bands all from the same source of financing.

In fact, if in a first stage the lowest cost is required for a determined amount $\times 1$ monetary units, it will be sufficient to find the minimum cost represented by the use of this amount from among sources of financing $A, B, C, \ldots, Z$. In this way we arrive at:

$$
\text{MIN } C_{ABC\ldots Z}(X) = \text{MIN}\left(C_A(x_1), C_B(x_1), C_C(x_1), \ldots, C_Z(x_1)\right)
$$

If financing is required in a greater amount corresponding to $x_1\,(+)\,x_2$ monetary units, it will be necessary to find the minimum cost of the two bands from among each of the sources of financing under consideration. In this way we arrive at.

$$
\begin{aligned}
\text{MIN } C_{ABC\ldots Z}(X) = \text{MIN}(&C_A(x_1)(+)C_A(x_1), C_B(x_1). \\
&(+)C_B(x_1), \ldots, C_Z(x_1)(+)C_Z(x_1))
\end{aligned}
$$

and so on successively for the different monetary bands.

Due to the condition of monotony we have that $C_j(x_n) \leq C_j(x_{n-1})$ with which it is obvious that if a determined monetary band x_n has a cost that is less than x_{n-1} the minimum cost for a determined volume of monetary units X will be arrived at by taking successive bands from one and the same financing source, the first band of which was the minimum.

In this way, if:

$$\text{MIN } C_{A,B,\ldots,Z}(X) = C_k(x_1)$$

on taking the following monetary band from among those offered by the sources of financing under consideration, we only obtain the minimum cost if $C_k(x_2)$ is taken given that:

$$C_k(x_1) \geq C_k(x_2)$$

and therefore, any other monetary band arising from the use of a different source of financing will bring us face to face with a higher cost:

$$C_k(x_1)(+)C_k(x_2) \prec C_k(x_1)(+)C_l(x_1)$$

due to the fact that:

$$C_k(x_1) \prec C_l(x_1)$$

16.5 Possibilities of Arriving at Optimum Combinations

The process we have followed has allowed us to arrive at a combination, which has given rise to minimum costs. Nevertheless, we could very well ask ourselves, at this point, what possibility is there of reaching the intermediary objectives first and later the final objectives. In order to tackle this aspect of the problem we are going to resort to the construction of an inference chain, by means of which we can establish to what extent objectives N, O, P are reached. Below we reproduce the graph in which only the relations between the set of sources $\{A, B, C, D, E, F\}$ have been drawn as well as that of intermediary objectives $\{G; H; I; J; K; L; M\}$.

This inference chain has been shown by means of a graph and is formed by vertices from which one or more arcs leave. The arcs that join the sources with intermediary objectives end up at vertices to which only a single arc arrives. The same does not occur for the vertices that relate to intermediary and final objectives. When more than one arc arrives at a knot, it will be necessary to establish the criteria based on which the exit vertices should be related with the arrival vertex. These criteria can be represented by the operator $(\vee)$ which means that in order for carrying it out it will be necessary that one and/or another of the cases giving rise to the arcs occurs. The other operator, which can be used, is $(\wedge)$ which means that in order to attain the element it will be necessary that one and the other cases giving rise to the arcs occur.

Basing ourselves on the graph that was represented before, it will be seen that the vertices to which more than one arc arrives are those representing the final objectives consisting in the increase in the share quotation (vertex N), increase in sales (vertex O) and the increase in profit levels (vertex P).

With regard to vertex N, arriving at it are the arcs from vertex H (distribution network), I (quality control) and L (production equipment). Therefore

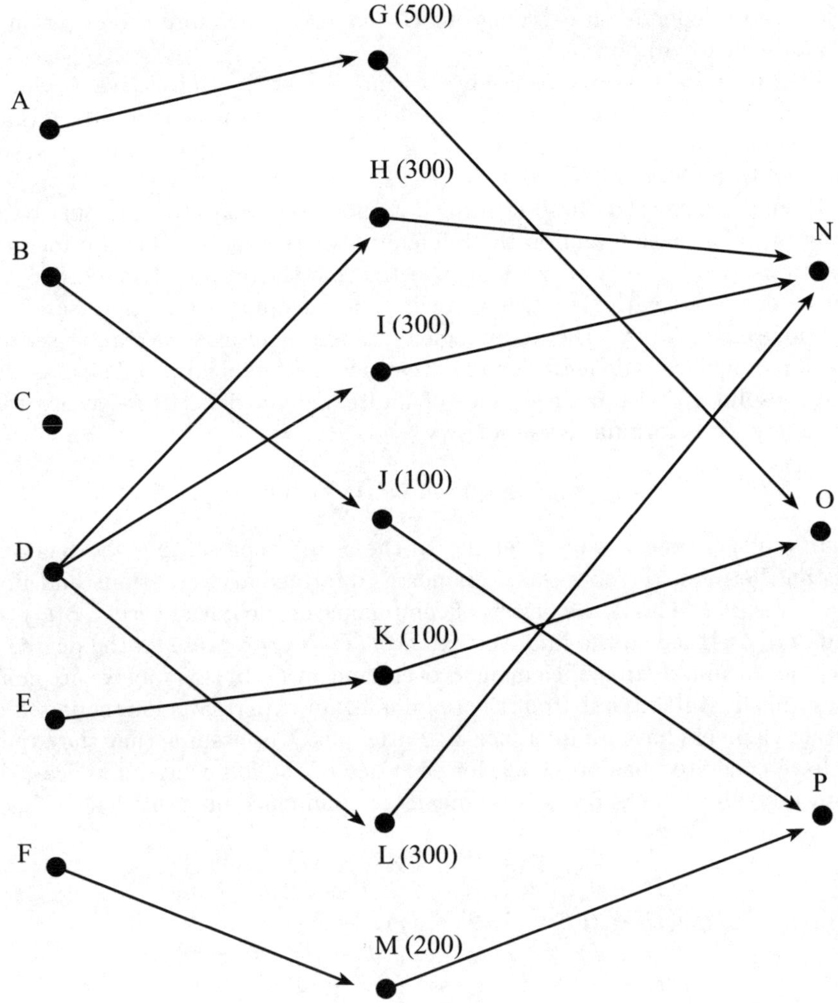

Fig. 16.3.

attaining N depends on the realisation of H and/or I and/or L. In this way the criterion that must be established at the time of knowing to what degree will final objective N be reached will give rise to the use of operator $(\vee)$.

Likewise, the criterion is studied from which we will get to know the valuation warranted by arriving at final objective O (sales increase). For this it is necessary to analyses the elements that originate each one of the arcs arriving at O; on the one hand appears intermediary objective G (expansion of industrial building) and, on the other, intermediary objective K (improvement in the supply system). In this way it is established that attaining final objective O depends on carrying out G and/or K therefore the criterion to be established will be ($\vee$).

Finally it is necessary to study to what extent final objective P will be reached. Arriving at this depends on element J (investment in advertising) and element M (improve liquidity). For this reason and in this case the criterion to be used will be ($\wedge$).

Having considered these criteria it is now necessary to find out, based on an initial situation and on the influence that this has on the intermediary situations, the degree to which it is expected that the proposed final objectives can be complied with. For this we will base our analysis on the use of the Lee[3] inference. Among the many logical inferences in existence and use, this one is particularly adequate for the treatment of financial problems, as well as its resolution within the sphere of multivalent logic. All are aware that this inference is formulated as follows:

$$v(P \to Q) = \overline{v(P)}\,(\vee)\,v(Q)$$

In our case and always referring to the graph representing the relations existing between the sources of financing, intermediary situations and final objectives, $v(P)$ shows the degree of compliance of the initial vertex, $v(Q)$ the valuation assigned to the final vertex and $v(P \to Q)$ represents the degree of inference or implication of element P over element Q. In this way we are going to establish, starting out from the opinion of an expert, the degree in which certain elements have an inference over others. Let us assume that the expert we have consulted has provided, for each one of the following inferences, the following valuations expressed by means of confidence intervals:

$$\begin{aligned} v(A \to G) &= [.8, .9] & \quad v(G \to O) &= [.9, 1] \\ v(B \to J) &= [.7, .8] & v(H \to N) &= [.7, .8] \\ v(D \to H) &= [.7, .9] & v(I \to N) &= [.6, .7] \\ v(D \to I) &= .8 & v(J \to P) &= [.8, .9] \\ v(D \to L) &= [.8, 1] & v(K \to O) &= [.9, 1] \\ v(E \to K) &= 1 & v(L \to N) &= [.7, .8] \\ v(F \to M) &= 1 & v(M \to P) &= .8 \end{aligned}$$

Apart from this data and prior to the application of the proposed model, it is necessary to establish to what degree is it possible to raise the amounts required in the optimisation of the corresponding financial sources. The following valuations are estimated also represented by confidence intervals:

$$v(A) = 1$$
$$v(B) = [.9, 1]$$
$$v(D) = 1$$

[3] Kaufmann A and Gil Aluja J (1992) Técnicas de gestión de empresa. Previsiones, decisiones y estrategias (in spanish). (Ed). Pirámide. Madrid. pp. 227–228.

$$v(E) = [.9, 1]$$
$$v(F) = 1$$

Being aware of this information is going to constitute a basis for obtaining, in the first place, the degree to which each one of the intermediary objectives from the inference chain shown on previous pages can be reached. In this case, then, since we know the valuations of the initial vertices and those corresponding to the inferences of the elements of the first set over the second, we must apply the so-called "forward chaining" the object of which is to arrive at the valuation corresponding to the arrival vertex. In general terms we would have:

$$v(a \to b) = \overline{v(a)}\,(\vee)\,v(b)$$

of which we know the valuations corresponding to:

$v(a \to b)$: valuation corresponding to the inference of an element of the first set over the second

$v(a)$: valuation representing the degree of compliance of the initial vertices.

With all this it can be seen that the valuation that is required to know corresponds to $v(b)$, relative to that of the arrival vertex.

Therefore, by applying the Lee inference to the example we arrive at:

$$v(A \to G) = \overline{v(A)}\,(\vee)\,v(G)$$
$$[.8, .9] = \overline{1}\,(\vee)\,v(G)$$
$$v(G) = [.8, .9]$$

$$v(D \to H) = \overline{v(D)}\,(\vee)\,v(H)$$
$$[.7, .9] = \overline{1}\,(\vee)\,v(H)$$
$$v(H) = [.7, .9]$$

$$v(B \to J) = \overline{v(B)}\,(\vee)\,v(J)$$
$$[.7, .8] = \overline{[.9, .1]}\,(\vee)\,v(J)$$
$$v(J) = [.7, .8]$$

$$v(D \to I) = \overline{v(D)}\,(\vee)\,v(I)$$
$$.8 = \overline{1}\,(\vee)\,v(I)$$
$$v(I) = .8$$

$$v(D \to L) = \overline{v(D)}\,(\vee)\,v(L)$$
$$[.8, 1] = \overline{1}\,(\vee)\,v(L)$$
$$v(L) = [.8, 1]$$

$$v(F \to M) = \overline{v(F)}\,(\vee)\,v(M)$$
$$1 = \overline{1}\,(\vee)\,v(M)$$
$$v(M) = 1$$

$$v(E \to K) = \overline{v(E)}\,(\vee)\,v(K)$$
$$1 = \overline{[.9, 1]}\,(\vee)\,v(K)$$
$$v(K) = 1$$

Once the valuations are known corresponding to the vertices representing the degree of compliance of the intermediary objectives, it will be necessary to continue with the process in order to find to what extend the final objectives are reached. For this we take as the valuation of the initial vertices the elements obtained in the previous phase corresponding to the intermediary objectives, as well as the estimates made relative to the inferences of the intermediary objectives over the final vertices. In this way we arrive at:

$$v(G \to O) = \overline{v(G)} (\vee) v(O)$$
$$[.9, 1] = \overline{[.8, .9]} (\vee) v(O)$$
$$v(O) = [.9, 1]$$

$$v(K \to O) = \overline{v(K)} (\vee) v(O)$$
$$[.9, .1] = \overline{1} (\vee) v(O)$$
$$v(O) = [.9, 1]$$

$$v(H \to N) = \overline{v(H)} (\vee) v(N)$$
$$[.7, .8] = \overline{[.7, .9]} (\vee) v(N)$$
$$v(N) = [.7, .8]$$

$$v(I \to N) = \overline{v(I)} (\vee) v(N)$$
$$[.6, .7] = \overline{.8} (\vee) v(N)$$
$$v(N) = [.6, .7]$$

$$v(L \to N) = \overline{v(L)} (\vee) v(N)$$
$$[.7, .8] = \overline{[.8, 1]} (\vee) v(N)$$
$$v(N) = [.7, .8]$$

$$v(J \to P) = \overline{v(J)} (\vee) v(P)$$
$$[.8, .9] = \overline{[.7, .8]} (\vee) v(P)$$
$$v(P) = [.8, .9]$$

$$v(M \to P) = \overline{v(M)} (\vee) v(P)$$
$$.8 = \overline{1} (\vee) v(P)$$
$$v(P) = .8$$

Finally, in order to get to know the degree of compliance of each one of the three final objectives it is necessary to use the corresponding operators from the previously established criteria. Therefore:

$$v_H(N) (\vee) v_I(N) (\vee) v_L(N) = v(N)$$
$$[.7, .8] (\vee) [.6, .7] (\vee) [.7, .8] = [.7, .8]$$
$$v(N) = \underline{[.7, .8]}$$
$$v_G(O) (\vee) v_K(O) = v(O)$$
$$[.9, 1] (\vee) [.9, 1] = [.9, 1]$$
$$v(O) = \underline{[.9, 1]}$$
$$v_J(P) (\wedge) v_M(P) = v(P)$$
$$[.8, .9] (\wedge) [.8, .8] = [.8, .8]$$
$$v(P) = \underline{.8}$$

We can see that the three objectives approached by the business, are effectively complied with to a relatively high degree. For example, objective N (increase the quotation of the shares) reaches a minimum degree of 0.7 and a maximum of 0.8. Likewise objective O (increase sales) is complied with practically a maximum degree [0.9, 1]. Finally objective P (increase in the profit level) is also reached in a very high degree 0.8.

Having analysed the inference chain, which has led us to these results in its entirety, we arrive at the graph (see Fig. 16.4).

16.6 Final Considerations

As we have seen throughout the previous pages, the use of logical inferences has permitted us to resolve the problem of the most suitable assignment of financial products to the needs brought to light by the activity of the business.

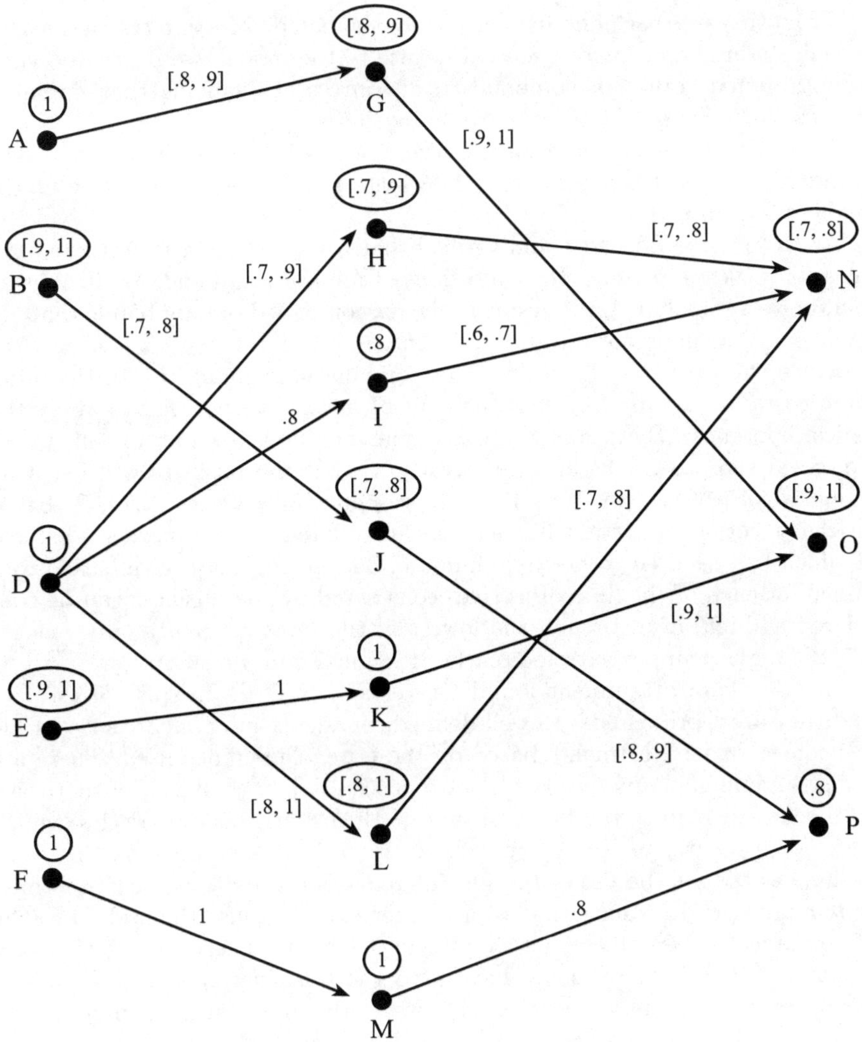

Fig. 16.4.

The sequential nature of the model shows the relations existing between the elements that make up the possible financial means that are susceptible to being used and those that allow the business to reach its objectives.

The characteristics that govern the connections represented in the model have their basis in the minimisation of the costs of the possible financial products that could be included in the financial system of the business and, on the other hand, in the non-increasing monotony of the same as the amounts taken from one and the same financial source increase.

From this approach and having previously included certain restrictions for the effects of giving a more general nature to the model, we have proceeded to obtain that source or combination of sources of financing that minimise the cost for the use of the same by the business.

The use of the Bellman optimisation principle on which dynamic programming is based is the instrument which has allowed us to arrive at the results we obtained.

With this it will be seen and verified that when we are faced with an optimisation process in which the conditions of minimisation and non increasing monotony are included, the result is always comprised of only bands that are members of a single optimisation element.

Once we are aware of the different combinations giving rise to the minimum costs for the use of different financial means, we proceed to apply the logical inferences. By means of the Lee inference we arrived at a result which expressed the possibility in an interval [0,1] that the business will reach its proposed objectives (in the scheme represented by vertices N, O, P) based on the selection of certain financial means offered by the market, the cost of which has resulted to be the minimum. The corresponding valuations obtained for each of the final objectives considered by the business can be considered as high, even though the lowest of the lower extremes only reaches 0.7, the highest upper extreme reaches the maximum presumption.

The levels of attainment found for vertices N, O, P, in spite of the fact that they are represented by a well-defined confidence interval, are susceptible to nuances or readjustments based on the type of inference used. There are a large number of inferences that are susceptible to being applied to the mathematics of uncertainty, each one of them with characteristics, which allow for a greater or lesser applicability in each specific area. In this work we have opted for the use of the Lee inference, since its formulation permits arriving at certain results that to a greater extent adjust themselves to the performance of the elements intervening in the financial activity of businesses, but in the same way we could have used the Lukaciewicz[4] which has been incorporated in many other works relative to the financial environment. The result if we had applied the Lukaciewicz inference to this model would have produced more pessimistic results, that is, much lower valuations. This is due to its formulation. As we have seen the previous pages the Lee inference is defined as:

$$v(P \to Q) = \overline{v(P)}\,(\vee)\,v(Q)$$

and the Lukaciewicz inference as:

$$v(P \to Q) = 1\,(\wedge)\left(\overline{v(P)}\,(+)\,v(Q)\right)$$

[4] Kaufmann A and Gil Aluja J (1992) Técnicas de gestión de empresa. Previsiones, decisiones y estrategias (in spanish). (Ed). Pirámide. Madrid. pp. 227–229.

the similarity can be seen of both formulations in which, while the Lee inference uses a non-linear operator ($\vee$), the Lukaciewicz inference uses a linear operator ($+$). What this does is that in order to reach the same level of implication between two elements, this is done with lower valuations in the case of applying the Lee inference.

17 Differentiation of Financial Products

17.1 The Elements that Differentiate Financial Products

The existence is a characteristic fact of the financial system of our day and age of a wide variety of financial products that are susceptible to being used by those businesses that resort to outside finance in order to cover their eventual requirements of payment means with which to satisfy their financial obligations.

These products are presented, by financial institutions, in such a way that they may induce possible clients to think that they are new forms of credit with characteristics that are different relative to those that are offered by the competition.

However, it has become normal that, after a brief analysis, the conclusion is reached that many of the financial products existing in the market are only different in appearance and that their most important characteristics for the businessman (price, validity, payback conditions, etc.) are practically the same.

As a prior step to the selection of the most adequate financial product for the requirements of the business it will be necessary to clarify the credit panorama, pointing out which are the products that can be considered as truly different and which are those products which only appear to be different. This is a new problem[1] which appears when there are in existence very evolved financial markets in which there are a very large number of institutions attempting to increase their market share both relative to the taking and placing of payment means.

Under these circumstances, the grouping of products of similar characteristics is important for those executives who have to take decisions in the world of finance.

[1] This approach, as well as the example we have developed has been taken from Kaufmann A and Gil Aluja, J (July 1991) Selection of affinities by means of fuzzy relations and Galois lattices, at the *X European Congress on Operative Investigation,* Aachen (Germany).

Anna Maria Gil-Lafuente: *Fuzzy Logic in Financial Analysis*, StudFuzz **175**, 303–317 (2005)
www.springerlink.com

Table 17.1.

C_i Characteristics	P_1	P_2	P_3	P_4	P_5	P_6
C_1 Price of the money	14%	16%	18%	15%	16%	20%
C_2 Payback period	3 years	4 years	4 years	2 years	3 years	6 years
C_3 Possibility of renewal on expiry	0,6	0,5	0,4	0,8	0,7	0,3
C_4 Fractioning payback	Monthly	Quaterly	Biannual	Monthly	Anually	Anual
C_5 Speed in granting credit	0,7	0,6	0,9	0,8	1	0,4

Kaufmann and Gil Aluja developed a scheme for the solution of this problem based on Moore families and Galois lattices[2].

We start out with the knowledge of the existence of a financial market in which there are m products, $P_j, j = 1, 2, 3, \ldots, m$ with certain characteristics, $C_i, j = 1, 2, 3, \ldots, n$ which for business may have a certain degree of acceptance which is valued by $\alpha \in [0; 1]$ in such a way that when the consideration if greater values closed to the unit will be given, and to the extent the consideration is less, figures closer to zero will be assigned. Each product P_j may be described by means of a fuzzy sub-set the referential of which will be given by its characteristics C_i.

	C_1	C_2	C_3		C_4	C_5
$\underset{\sim}{P}_{ij} =$	$\mathscr{M}_{j1}(x)$	$\mathscr{M}_{j2}(x)$	$\mathscr{M}_{j3}(x)$	...	$\mathscr{M}_{jn\ 1}(x)$	$\mathscr{M}_{jn}(x)$

We have taken an example from the work of said professors, in which is assumed that in a business a grouping must be made of six financial products (medium and long-term credits), P_1, P_2, P_3, P_4, P_5, P_6, the characteristics of which are shown in Table 17.1

It can be seen that characteristics C_3 and C_5 have already been valued in $\alpha \in [0; 1]$ in accordance with the estimate given by the financial experts of the business. The remaining data has warranted, also in the opinion of the same experts a valuation in $[0; 1]$ which allows us to present the following matrix:

	P_1	P_2	P_3	P_4	P_5	P_6
C_1	1	0,8	0,7	0,9	0,8	0,5
C_2	0,5	0,7	0,7	0,3	0,5	1
C_3	0,6	0,5	0,4	0,8	0,7	0,3
C_4	0,3	0,5	0,7	0,3	1	1
C_5	0,7	0,6	0,9	0,8	1	0,4

[2] Evariste Galois, French mathematician (1811–1832) carried out important research on the role of groups in resolving algebraic equations. He died in a duel at the age of 21.

With the object of simplifying the notations we have done $P_1 = A$, $P_2 = B$, $P_3 = C$, $P_4 = D$, $P_5 = E$, $P_6 = F$, $C_1 = a$, $C_2 = b$, $C_3 = c$, $C_4 = d$, $C_5 = e$.

We now proceed to study these relations by means of α-cuts. We arrive at:

$\alpha = 1$	A	B	C	D	E	F
a	1					
b						1
c						
d					1	1
e					1	

$\alpha \geq 0{,}9$	A	B	C	D	E	F
a	1			1		
b						1
c						
d					1	1
e			1		1	

$\alpha \geq 0{,}8$	A	B	C	D	E	F
a	1	1		1	1	
b						1
c				1		
d					1	1
e			1	1	1	

$\alpha \geq 0{,}7$	A	B	C	D	E	F
a	1	1	1	1	1	
b		1	1			1
c				1	1	
d			1		1	1
e	1		1	1	1	

$\alpha \geq 0{,}6$	A	B	C	D	E	F
a	1	1	1	1	1	
b		1	1			1
c	1			1	1	
d			1		1	1
e	1	1	1	1	1	

$\alpha \geq 0{,}5$	A	B	C	D	E	F
a	1	1	1	1	1	1
b	1	1	1		1	1
c	1	1		1	1	
d		1	1		1	1
e	1	1	1	1	1	

$\alpha \geq 0{,}4$	A	B	C	D	E	F
a	1	1	1	1	1	1
b	1	1	1		1	1
c	1	1	1	1	1	
d		1	1		1	1
e	1	1	1	1	1	1

$\alpha \geq 0{,}3$	A	B	C	D	E	F
a	1	1	1	1	1	1
b	1	1	1	1	1	1
c	1	1	1	1	1	1
d	1	1	1	1	1	1
e	1	1	1	1	1	1

In order to establish the "relations of affinity" we resort to the Moore family method. The concept of the relation of affinity (rectangular matrix), from a mathematical point of view, is based on the existence of a sub-set of elements or objects that have in common several properties or characteristics at different levels above and below a limit. This concept of affinity constitutes the support of assemblages in the main classifications, human groups, association of products, etc. In this way it constitutes a generalisation of the relation of similarity (square matrix).

The set of P_j are denominated $E^{(1)}$ and the set of C_i are denominate $E^{(2)}$, therefore:

$$E^{(1)} = \{P_1, P_2, P_3, P_4, P_5, P_6\} = \{A, B, C, D, E, F\}$$
$$E^{(2)} = \{C_1, C_2, C_3, C_4, C_5\} = \{a, b, c, d, e\}$$

We now immediately constitute the largest possible set with the elements of $E^{(2)}$, (the so-called power set $E^{(2)}$), which will be:

$$P(E^{(2)}) = \begin{Bmatrix} \emptyset, a, b, c, d, e, ac, ad, ae, bc, bd, be, cd, ce, de, abc, abd, abe \\ acd, ace, ade, bcd, bce, bde, cde, abcd, abce, abde, bcde, E^{(2)} \end{Bmatrix}$$

For each element of $P(E^{(2)})$ we then make a correspondence of the element or elements of $P(E^{(1)})$ for the different levels of α taking into account the previous tables. With this we draw up the table shown in Table 17.2 below

17.2 Introduction of the Galois Lattice to the Problem of Groupings

Very briefly we will recall the properties of Galois lattices. For the two finite sets $E^{(1)}$ and $E^{(2)}$ long with their respective power sets $P(E^{(1)})$ and $P(E^{(2)})$.

After this we take the relations in the following order:

$$\forall X, X' \in P(E^{(1)}), \quad \forall Y, Y' \in P(E^{(2)}):$$

$$(X, Y) \preceq (X', Y') \,\square\, (X \subset X', Y \supset Y')$$

therefore, the upper limit $X \nabla Y$ of (X, Y) and the lower limit $X' \Delta Y'$ of (X', Y') are the consequence of the previous relation of order.

At the same time the relation of opposing order is introduced:

$$\forall X, X' \in P(E^{(1)}), \quad \forall Y, Y' \in P(E^{(2)}):$$

$$(X, Y) \succeq (X', Y') \,\square\, (X \supset X', Y \subset Y')$$

where the upper limit $X \nabla Y$ of (X, Y) and the lower limit $X' \Delta Y'$ of (X', Y') are the consequence of the previous relation.

We now introduce the pairs $(\emptyset, E)$ and $(E, \emptyset)$ as the lower limit (respectively upper limit) in order to complete the relation of order.

Table 17.2.

	$\alpha = 1$	$\alpha \geq 0{,}9$	$\alpha \geq 0{,}8$	$\alpha \geq 0{,}7$	$\alpha \geq 0{,}6$	$\alpha \geq 0{,}5$	$\alpha \geq 0{,}4$	$\alpha \geq 0{,}3$
a	A	AD	ABCE	ABCDE	ABCDE	ABCDEF	ABCDEF	ABCDEF
b	F	F	F	BCF	BCF	ABCEF	ABCEF	ABCDEF
c	Ø	Ø	D	DE	ADE	ABDE	ABCDE	ABCDEF
d	EF	EF	EF	CEF	CEF	BCEF	BCEF	ABCDEF
e	E	CE	CDE	ACDE	ABCDE	ABCDE	ABCDEF	ABCDEF
ab	Ø	Ø	Ø	BC	BC	ABCEF	ABCEF	ABCDEF
ac	Ø	Ø	D	DE	ADE	ABDE	ABCDE	ABCDEF
ad	Ø	Ø	E	CE	CE	BCEF	BCEF	ABCDEF
ae	Ø	Ø	DE	ACDE	ABCDE	ABCDE	ABCDEF	ABCDEF
bc	Ø	Ø	Ø	Ø	Ø	ABE	ABCE	ABCDEF
bd	F	F	F	CF	CF	BCEF	BCEF	ABCDEF
be	Ø	Ø	Ø	C	BC	ABCE	ABCEF	ABCDEF
cd	Ø	Ø	Ø	E	E	BE	BCE	ABCDEF
ce	Ø	Ø	D	DE	ADE	ABDE	ABCDE	ABCDEF
de	E	E	E	CE	CE	BCE	BCEF	ABCDEF
abc	Ø	Ø	Ø	Ø	Ø	ABE	ABCE	ABCDEF
abd	Ø	Ø	Ø	C	C	BCEF	BCEF	ABCDEF
abe	Ø	Ø	Ø	C	BC	ABCE	ABCEF	ABCDEF
acd	Ø	Ø	Ø	E	E	BE	BCE	ABCDEF
ace	Ø	Ø	Ø	DE	ADE	ABDE	ABCDE	ABCDEF
ade	Ø	Ø	DE	CE	CE	BCE	BCEF	ABCDEF
bcd	Ø	Ø	Ø	Ø	Ø	BE	BCE	ABCDEF
bce	Ø	Ø	Ø	Ø	Ø	ABE	ABCE	ABCDEF
bde	Ø	Ø	Ø	C	C	BCE	BCEF	ABCDEF
cde	Ø	Ø	Ø	E	E	BE	BCE	ABCDEF
abcd	Ø	Ø	Ø	Ø	Ø	BE	BCE	ABCDEF
abce	Ø	Ø	Ø	Ø	Ø	ABE	ABCE	ABCDEF
abde	Ø	Ø	Ø	C	C	BCE	BCEF	ABCDEF
acde	Ø	Ø	Ø	E	E	BE	BCE	ABCDEF
bcde	Ø	Ø	Ø	Ø	Ø	BE	BCE	ABCDEF
abcde	Ø	Ø	Ø	Ø	Ø	BE	BCE	ABCDEF

If the following properties are verified:

$$(U,V) = (X,Y)\ \nabla\ (X',Y') \Rightarrow (U \supset X \cup X' \text{ and } V \subset Y \cap Y')$$

and also:

$$(Z,T) = (X,Y)\ \Delta\ (X',Y') \Rightarrow (Z \subset X \cup X' \text{ and } T \supset Y \cap Y')$$

the "Galois lattice" is defined.

If we look at the set of maximum sub-relations in the relation $R : E \cdot E'$ (also called "primary matrices" or coverage of R) this set has the configuration of a Galois lattice by construction. The maximum sub-relation will be called affinity.

Now let us assume that the executives consider level $\alpha \geq 0{,}7$ as sufficient for finding the relations of affinity. At this level we take into account the non-void elements of $P(E^{(1)})$, and the sub-sets of $P(E^{(2)})$ that are not included in any others. We arrive at:

C	abde
E	acde
BC	ab
CE	ade
CF	bd
DE	ace
BCF	b
CEF	d
ACDE	ae
ABCDE	a

Therefore, the affinities will be the following. $(abde, C)$, $(acde, E)$, (ab, BC), (ade, CE), (bd, CF), (ace, DE), (b, BCF), (d, CEF), $(ac, ACDE)$ and $(a, ABCDE)$. This allows us to present the Galois lattice as shown in Fig. 17.1

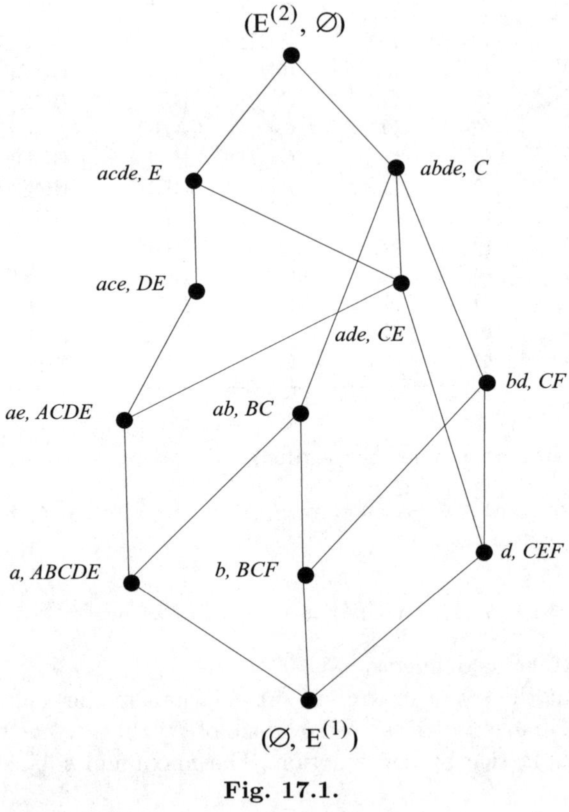

Fig. 17.1.

This reticular structure shows very visually the affinities existing between the different products relative to the characteristics of the same. According to the importance assigned by the executives of the business to each characteristic they will be able to consider some products as having affinity with others. Therefore, if the characteristics of a = price for the money and e = speed in granting, we find that the products with affinity are $P_1 = A$, $P_3 = C$, $P_4 = D$ and $P_5 = E$. It becomes obvious that, to the extent that more characteristics are required, at one and the same level α, the number of similar financial products diminishes and in the reverse, the requirement for a lower number of characteristics means more products with similarities.

A example of the interest that the use of Galois lattices acquire can be seen when we observe the fact that the affinities are regressive in the lattice when referring to $E^{(1)}$ and progressive relative to $E^{(2)}$, when coming from behind to the front. It is mainly when a high number of relations are being considered that the selection acquired importance from an affinity point of view and when this selection is simpler.

Obviously different simulations can be done in this field in order to study diverse changes in the estimates or variations in the referential sets $E^{(1)}$ and/or $E^{(2)}$.

As stated by Kaufmann and Gil Aluja[3] this method can also be used in order to find the maximum sub-relation of resemblance in relations of similarity. In this case, Galois lattices are trivial and the principal question is then centred on the search by means of the Moore family.

When there is not sufficient data available, the human mind submits itself to several processes in order to find similarities. A Galois lattice is a fundamental scheme, which associates with ease to the fuzzy logic of our normal way of thinking.

17.3 Grouping by Means of Maximum Sub-relations of Similarity

In certain cases, the problem of grouping financial products which are offered in the market according to their characteristics that are considered in common to a certain degree (to be determined) can be resolved by means of a different path from the one we have just described.

In the scheme we are going to develop we will be using well known elements in the theory of fuzzy sub-sets such as the Hamming distance, relations of dissimilarity and maximum sub-relations of similarity.

[3] Kaufmann A and Gil Aluja J (July 1991), op. cit.

For comparative effects and greater simplicity we are going to start out from the example used in the previous section, for which we reproduce the following fuzzy matrix:

	P_1	P_2	P_3	P_4	P_5	P_6
C_1	1	0,8	0,7	0,9	0,8	0,5
C_2	0,5	0,7	0,7	0,3	0,5	1
C_3	0,6	0,5	0,4	0,8	0,7	0,3
C_4	0,3	0,5	0,7	0,3	1	1
C_5	0,7	0,6	0,9	0,8	1	0,4

The relative Hamming distance between each one of the products and the others will be:

$$\delta(P_1, P_2) = 1/5(|1-0{,}8| + |0{,}5-0{,}7| + |0{,}6-0{,}5| + |0{,}3-0{,}5| + |0{,}7-0{,}6|) = 0{,}16$$
$$\delta(P_1, P_3) = 1/5(|1-0{,}7| + |0{,}5-0{,}7| + |0{,}6-0{,}4| + |0{,}3-0{,}7| + |0{,}7-0{,}9|) = 0{,}26$$
$$\delta(P_1, P_4) = 1/5(0{,}1 + 0{,}2 + 0{,}2 + 0 + 0{,}1) = 0{,}12$$
$$\delta(P_1, P_5) = 1/5(0{,}2 + 0 + 0{,}1 + 0{,}7 + 0{,}3) = 0{,}26$$
$$\delta(P_1, P_6) = 0{,}46 \quad \delta(P_2, P_3) = 0{,}14 \quad \delta(P_2, P_4) = 0{,}24 \quad \delta(P_2, P_5) = 0{,}26$$
$$\delta(P_2, P_6) = 0{,}30 \quad \delta(P_3, P_4) = 0{,}30 \quad \delta(P_3, P_5) = 0{,}20 \quad \delta(P_3, P_6) = 0{,}28$$
$$\delta(P_4, P_5) = 0{,}26 \quad \delta(P_4, P_6) = 0{,}54 \quad \delta(P_5, P_6) = 0{,}36$$

If the results we have arrived at are conveniently grouped we obtain matrix $\underset{\sim}{R}$, which can be considered as a matrix of "dissimilarity". This matrix is anti-reflexive as all the values of the main diagonal are all zero and also symmetrical, since the distance between P_i and P_j is the same as the distance existing between P_{i*} and P_j.

$\underset{\sim}{R}$	P_1	P_2	P_3	P_4	P_5	P_6
P_1	0	0,16	0,26	0,12	0,26	0,43
P_2	0,16	0	0,14	0,24	0,26	0,30
P_3	0,26	0,14	0	0,30	0,20	0,28
P_4	0,12	0,24	0,30	0	0,26	0,54
P_5	0,26	0,26	0,20	0,26	0	0,36
P_6	0,43	0,30	0,28	0,54	0,36	0

With the object of obtaining the relations of similarity it will be sufficient to arrive at the complementary matrix by calculating for each element the complement to the unit:

$$[\overline{\underset{\sim}{\mathbf{R}}}] = [1] - [\underset{\sim}{\mathbf{R}}]$$

In this way the following "matrix of similarity" will be found and which will also be symmetrical but reflexive:

$\overline{\underset{\sim}{R}}$	P_1	P_2	P_3	P_4	P_5	P_6
P_1	1	0,84	0,74	0,88	0,74	0,57
P_2		1	0,86	0,76	0,74	0,70
P_3			1	0,70	0,80	0,72
P_4				1	0,74	0,46
P_5					1	0,64
P_6						1

With the object of making the corresponding analysis we will find the ordinary matrices for all levels $\alpha \in [0; 1]$. We will find successively.

$\overline{\underset{\sim}{R}}_1$	P_1	P_2	P_3	P_4	P_5	P_6
P_1	1					
P_2		1				
P_3			1			
P_4				1		
P_5					1	
P_6						1

$\overline{\underset{\sim}{R}}_{0,88}$	P_1	P_2	P_3	P_4	P_5	P_6
P_1	1			1		
P_2		1				
P_3			1			
P_4	1			1		
P_5					1	
P_6						1

$\overline{\underset{\sim}{R}}_{0,80}$	P_1	P_2	P_3	P_4	P_5	P_6
P_1	1	1		1		
P_2	1	1	1			
P_3		1	1		1	
P_4	1			1		
P_5			1		1	
P_6						1

$\overline{\underset{\sim}{R}}_{0,76}$	P_1	P_2	P_3	P_4	P_5	P_6
P_1	1	1		1		
P_2	1	1	1	1		
P_3		1	1		1	
P_4	1	1		1		
P_5			1		1	
P_6						1

$\underset{\sim}{\overline{R}}_{0,86}$	P_1	P_2	P_3	P_4	P_5	P_6
P_1	1			1		
P_2		1	1			
P_3		1	1			
P_4	1			1		
P_5					1	
P_6						1

$\underset{\sim}{\overline{R}}_{0,84}$	P_1	P_2	P_3	P_4	P_5	P_6
P_1	1	1		1		
P_2	1	1	1			
P_3		1	1			
P_4	1			1		
P_5					1	
P_6						1

$\underset{\sim}{\overline{R}}_{0,74}$	P_1	P_2	P_3	P_4	P_5	P_6
P_1	1	1	1	1	1	
P_2	1	1	1	1	1	
P_3	1	1	1		1	
P_4	1	1		1	1	
P_5	1	1	1	1	1	
P_6						1

$\underset{\sim}{\overline{R}}_{0,72}$	P_1	P_2	P_3	P_4	P_5	P_6
P_1	1	1	1	1	1	
P_2	1	1	1	1	1	
P_3	1	1	1		1	1
P_4	1	1		1	1	
P_5	1	1	1	1	1	
P_6			1			1

We are now going to show how a determined number of experts from a business consider that it can be accepted that financial products can form a homogeneous group when their distance is equal to or less than $\alpha = 0{,}24$, while others are inclined to accept the limit of $\alpha = 0{,}26$ and a last group $\alpha = 0{,}28$ (resemblances $\alpha = 0{,}76$, $\alpha = 0{,}74$, $\alpha = 0{,}72$). In this event of discrepancy it is convenient to study each one of the alternatives, to see how the grouping takes place in each one of the cases. Obviously in the event of the criteria coinciding, we are faced with a particular case and the analysis will be limited to a single level of α.

We have shown the graphs relative to each of the three levels in Fig. 17.2.

This intuitive way of arriving at the maximum sub-relations of similarity is not operative, since it can lead to confusion and above all its use is made practically impossible when the number of financial products increases.

17.4 The Pichat Algorithm

Several algorithms have been drawn up in order to provide a mechanism that is capable of arriving at sub-matrices or transitive graphs that represent maximum sub-relations of similarity. In this case we have selected, because

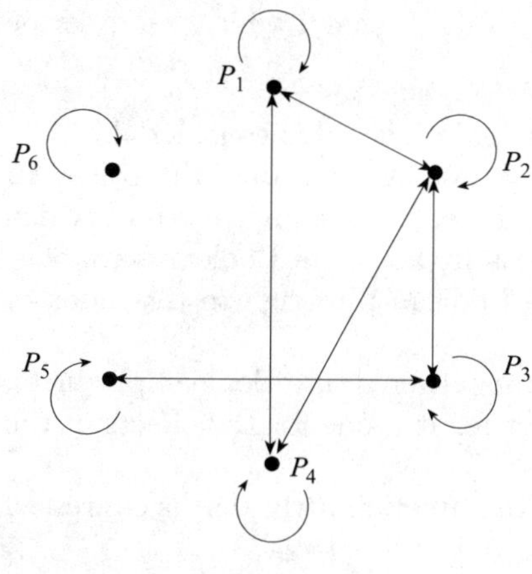

$\alpha \geq 0{,}76$

It can clearly be seen in this case that (P_1, P_2, P_3), (P_2, P_3), (P_3, P_5) and P_6 form groups.

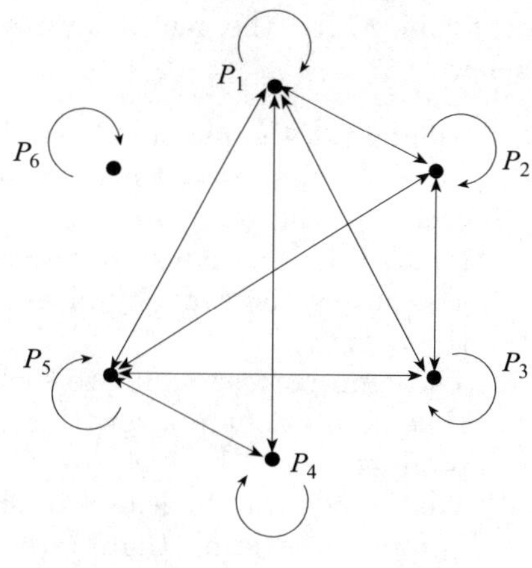

$\alpha \geq 0{,}74$

In this group it is not quite so simple to see that (P_1, P_2, P_4, P_5), (P_1, P_2, P_3, P_5) and P_6 form groups.

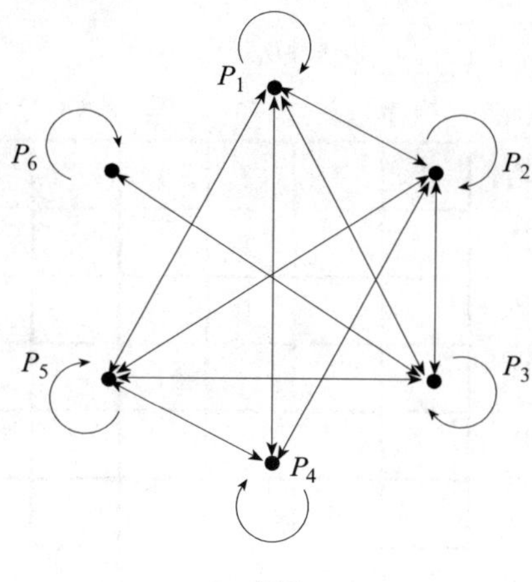

$\alpha = 0{,}72$

Neither in this case is it easy to determine that (P_3, P_6), (P_1, P_2, P_4, P_5), (P_1, P_2, P_3, P_5) form groups.

Fig. 17.2.

of its simplicity, the algorithm drawn up by E. Pichat[4], which we describe below:

(a) The half of the matrix above the principal diagonal is considered.
(b) Take the first row, then the second and so on one after the other. In each row the zeros are considered in the corresponding boxes of which (void) take the indices of the column as if they were Boolean variables, associating the same by means of the Boolean plus sign to the index of the row.
(c) The sums corresponding to each row are gathered into Boolean products. If no zero exists in a row then consider for it a one for the effects of the product.
(d) We then do the Boolean product and the function arrived at is expressed in minimum terms, that is, that $x + x = x$, $x + xy = x$.
(e) Having arrived at the result, we then find the complement of each one of the terms of the same. The complementary terms give rise to the maximum sub-relations that posses the property of transitivity.

In the case we are studying, the ordinary matrix corresponding to level $\alpha \geq 0{,}76$ is as follows:

Level $\alpha \geq 0{,}76$

$\overline{\underset{\sim}{R}}_{0,76}$	P_1	P_2	P_3	P_4	P_5	P_6
P_1	1	1		1		
P_2		1	1	1		
P_3			1		1	
P_4				1		
P_5					1	
P_6						1

In order to abbreviate the nomenclature we substitute P_1 by 1, P_2 by 2, ... Therefore for each one of the rows we arrive at:

Row P_1: $1 \dotplus 356$
Row P_2: $2 \dotplus 56$
Row P_3: $3 \dotplus 46$
Row P_4: $4 \dotplus 56$
Row P_5: $5 \dotplus 6$
Row P_6: 1

[4] Pichat E (1970) Contribution a l'algorithmique non numerique dans les ensembles ordonnés, Doctoral Science Thesis, Science Faculty of Grenoble, France.

We then do the corresponding product, which will be:

$$\begin{aligned}
S &= (1 \dot{+} 356)\ (2 \dot{+} 56)\ (3 \dot{+} 46)\ (4 \dot{+} 56)\ (5 \dot{+} 6)\\
&= (12 \dot{+} 156 \dot{+} \cancel{2356} \dot{+} 356)\ (3 \dot{+} 46)\ (4 \dot{+} 56)\ (5 \dot{+} 6)\\
&= (123 \dot{+} 1246 \dot{+} \cancel{1356} \dot{+} 1456 \dot{+} 356 \dot{+} \cancel{3456})\ (4 \dot{+} 56)\ (5 \dot{+} 6)\\
&= (1234 \dot{+} \cancel{12356} \dot{+} 1246 \dot{+} \cancel{12456} \dot{+} \cancel{1456} \dot{+} 1456 \dot{+} \cancel{3456} \dot{+} 356)\ (5 \dot{+} 6)\\
&= (12345 \dot{+} \cancel{12346} \dot{+} \cancel{12456} \dot{+} 1246 \dot{+} 1456 \dot{+} \cancel{1456} \dot{+} \cancel{356} \dot{+} 356)\\
&= (12345 \dot{+} 1246 \dot{+} 1456 \dot{+} 356)
\end{aligned}$$

We obtain the complements of each of the terms:

$$S' = 6 \dot{+} 35 \dot{+} 23 \dot{+} 124$$

Finally obtaining the following maximum sub-relations of similarity:

$$(P_6),\ (P_1, P_2, P_4, P_5) \text{ and } (P_1, P_2, P_3, P_5)$$

Therefore, when a minimum level of 0,76 is accepted as good it can be stated that the executives should choose indistinctly between financial products P_6, either P_3 or P_5, either P_2 or P_3, and either P_1 or P_2 or P_4. The result then is that the choice falls to 4 options instead of 6.

When the requirement of conformity is lowered to lower level $\alpha \geq 0{,}74$ we arrive at the following ordinary matrix:

Level $\alpha \geq 0{,}74$

$\underset{\sim}{R}_{0,76}$	P_1	P_2	P_3	P_4	P_5	P_6
P_1	1	1		1		
P_2		1	1	1		
P_3			1		1	
P_4				1		
P_5					1	
P_6						1

For each row we will find:

Row P_1: $1 \dot{+} 6$
Row P_2: $2 \dot{+} 6$
Row P_3: $3 \dot{+} 46$
Row P_4: $4 \dot{+} 6$
Row P_5: $5 \dot{+} 6$
Row P_6: 1

We then do the product:

$$
\begin{aligned}
S &= (1 \dotplus 6) \dotplus (2 \dotplus 6)\ (3 \dotplus 46)\ (4 \dotplus 6)\ (5 \dotplus 6) \\
&= (12 \dotplus \cancel{16} \dotplus \cancel{26} \dotplus 6)\ (3 \dotplus 46)\ (4 \dotplus 6)\ (5 \dotplus 6) \\
&= (123 \dotplus \cancel{1246} \dotplus 36 \dotplus 46)\ (4 \dotplus 6)\ (5 \dotplus 6) \\
&= (1234 \dotplus \cancel{1236} \dotplus \cancel{346} \dotplus 36 \dotplus 46 \dotplus \cancel{46})\ (5 \dotplus 6) \\
&= 12345 \dotplus \cancel{12346} \dotplus \cancel{356} \dotplus 36 \dotplus \cancel{456} \dotplus 46
\end{aligned}
$$

The complement will be:

$$S' = 6 \dotplus 1245 \dotplus 1235$$

We then arrive at the following maximum sub-relations of similarity:

$$(P_6),\ (P_1, P_2, P_4, P_5) \text{ and } (P_1, P_2, P_3, P_5)$$

What has taken place here is a re-grouping forming three groups P_6 continues as the single financial product, while now grouped together are on the one hand P_1, P_2, P_4 and P_5 and on the other P_1, P_2, P_3 and P_5. The differences existing between P_3 and P_4 make it impossible to consider then jointly, while they are suitable for forming a group separately with P_1, P_2 and P_5.

In the last supposed we analyse, the requirement of conformity will reduce to $\alpha \geq 0{,}72$. In that case we have as the ordinary matrix:

Level $\alpha \geq 0{,}72$

$\bar{\underset{\sim}{R}}_{0,72}$	P_1	P_2	P_3	P_4	P_5	P_6
P_1	1	1	1	1	1	
P_2		1	1	1	1	
P_3			1		1	1
P_4				1	1	
P_5					1	
P_6						1

For each row we will arrive at the following:

Row P_1: $1 \dotplus 6$
Row P_2: $2 \dotplus 6$
Row P_3: $3 \dotplus 4$
Row P_4: $4 \dotplus 6$
Row P_5: $5 \dotplus 6$
Row P_6: 1

The product will be:

$$
\begin{aligned}
S &= (1 \dot{+} 6) \dot{+} (2 \dot{+} 6)\ (3 \dot{+} 4)\ (4 \dot{+} 6)\ (5 \dot{+} 6) \\
&= (12 \dot{+} \not{1}6 \dot{+} \not{2}6 \dot{+} 6)\ (3 \dot{+} 4)\ (4 \dot{+} 6)\ (5 \dot{+} 6) \\
&= (124 \dot{+} 123 \dot{+} 36 \dot{+} 46)\ (4 \dot{+} 6)\ (5 \dot{+} 6) \\
&= (124 \dot{+} 12\not{3}4 \dot{+} 3\not{4}6 \dot{+} \not{4}6 \dot{+} 12\not{4}6 \dot{+} 12\not{3}6 \dot{+} 36 \dot{+} 46)\ (5 \dot{+} 6) \\
&= 1245 \dot{+} 12\not{4}6 \dot{+} 3\not{5}6 \dot{+} 36 \dot{+} 4\not{5}6 \dot{+} 46
\end{aligned}
$$

Therefore the complement:

$$S' = 36 \dot{+} 1245 \dot{+} 1235$$

Finally arriving at the following maximum sub-relations of similarity:

$$(P_3, P_6),\ (P_1, P_2, P_4, P_5) \text{ and } (P_1, P_2, P_3, P_5)$$

It will be seen that it is necessary to go down to level $\alpha = 0{,}72$ in order for P_6 to be considered as forming a group with another, in this way loosing its singularity. At this level the same groups continue to exist as for level $\alpha \geq 0{,}74$ for (P_1, P_2, P_4, P_5) and (P_1, P_2, P_3, P_5).

If we continue reducing the levels, a more extensive grouping will take place, right down to the lowest level $\alpha = 0{,}46$, in which the very low requirement for homogeneity means that all the products can be considered as having no significant differences at this level.

In this chapter we have described two paths for resolving the problem when attempting to determine if a financial product can be considered as similar to another or others, or we are truly dealing with a singular product. Since the technical bases of one and the other are different, the results cannot coincide.

When "Moore families" and "Galois lattices" are taken as the basis a wealth of information is obtained lacking in the technique based on the "Hamming distance". In fact, the Galois lattice supplies for each level α all the possible groupings of financial products taking into account one characteristic when considering two of them, etc., in such a way that it is possible to make a grouping in view of those that may be considered as most important. This does not occur when using the Hamming distance, where all the qualities are considered without any possibility of preference between them. The ease of use of relations of similarity (square matrix) is weakened with the introduction off the more general concept of affinity (rectangular matrix).

18 Selection of Financial Products

18.1 Approach to the Problem

The raising of financial means by businesses brings up a problem of decision as a consequence of the variety of financial products that the banks and other credit institutions place at the disposal of their eventual clients.

With increasing frequency it can be seen that new products appear on the market under many different forms that, either real or apparent, have different characteristics. It should not be forgotten that the strong competition characterising the financial world obliges those offering payment means to a great effort of diversification and differentiation of products that permits them, on the one hand, to cover the widest range of possible users and, on the other, provoke a flaw by means of the presentation of different products with the object of get around the inexorable laws of the perfect market.

When the need arises for resorting to outside financing, executives in business find themselves faced with a certain number, obviously finite, of options offered by the market, from among which a selection must be made of the one that is best suited to the specific requirements of the business.

Evidently that for each business, and even for each specific situation, there will be a different valuation of each one of the characteristics of the financial products. Therefore, in certain cases, the speed of obtaining the financial means will be very important, on other occasions what is more important is the repayment period. In short, the businessman will estimate for each circumstance an order of precedence of the characteristic that go to make up the products.

In this context two fundamental elements appear that make up the problem:

(1) Differentiation in the characteristics of each one of the financial products on offer.
(2) Different estimate, by the acquirer, of each of the characteristics relative to the rest, which provides an order of preference.

Evidently, the degree of preference for each one of the characteristics relative to the others may sometimes be determined by means of measurements, that is, with an objective nature, but on other occasions it will be necessary to

Anna Maria Gil-Lafuente: *Fuzzy Logic in Financial Analysis*, StudFuzz **175**, 319–340 (2005)
www.springerlink.com

resort to subjective numerical situations, that is by means of valuations. The same thing occurs when a comparison must be made, for each characteristic, of the degree of preference between one product and the rest.

The possible participation of objective data and subjective estimates makes it advisable to use management techniques that are valid for the field of uncertainty, taking into account the fact that the mathematics of certainty can be considered as a particular case of the mathematics of uncertainty, the schemes of which, of a "soft" nature, can also be applied to the case of crisp data of a "hard" nature.

On the other hand, the existence of relations between products, as well as the relations in the estimates of the different characteristics bring to mind the convenience of presenting this problem by means of subjective matrices, taking advantage of all the possibilities offered by matrix calculations.

With all this an attempt is made to arrive at certain results that express the order of preference between different financial products to which a business may opt. The subjective nature of the estimated values should lead to certain conclusions that can be expressed by means of fuzzy sets.

18.2 A First Approximation: The Coefficient of Adequacy

With the object of introducing a range of techniques that allow for the treatment of the selection of financial products when faced with different situations, we are going to develop some schemes, commencing by one that is based on the so-called "coefficient of adequacy" the simplicity of which makes it most attractive for use in reality.

We will show this by means of an example, taking into account that its generalisation poses no difficulties.

Let us assume that in the financial market there are three products P_1, P_2 and P_3 with different characteristics relative to:

C_1 = price of the money;
C_2 = repayment period;
C_3 = possibilities for renewal;
C_4 = fractioning of repayment;
C_5 = speed in granting.

For each characteristic a property is considered. For C_1 "inexpensive money"; for C_2 "good repayment period"; for C_3 "possibility of renewal"; for C_4 "suitable for fractioning repayment"; for C_5 "speed in granting".

For each one of these characteristics the following information is obtained.

For C_1:

- the price for P_1 is 20%;
- the price for P_2 is 22%;
- the price for P_3 is 18%.

The financial director establishes as the descriptor for the concept "inexpensive money" the following normal fuzzy sub-set:

	P_1	P_2	P_3
$\underset{\sim}{D}_1 = \underset{\sim}{f}(C_1) =$	0,9000	0,8181	1

For C_2:

- payback period for P_1 is 5 years;
- payback period for P_2 is 6 years;
- payback period for P_3 is 4 years.

The descriptor of the concept "good payback period" for the business is:

	P_1	P_2	P_3
$\underset{\sim}{D}_2 = \underset{\sim}{f}(C_2) =$	$0,8\widehat{3}$	1	$0,\widehat{6}$

For C_3:

- The "possibilities for renewal" of P_1 are half those of P_2 and 1/3 those of P_3.

The following normal fuzzy sub-set is estimated as the descriptor of the concept of "possibilities for renewal":

	P_1	P_2	P_3
$\underset{\sim}{D}_3 = \underset{\sim}{f}(C_3) =$	$0,\widehat{3}$	$0,\widehat{6}$	1

For C_4:

- repayment of P_1 is quarterly;
- repayment of P_2 is monthly;
- repayment of P_3 is quarterly.

The business considers as the descriptor for "fractioning of repayment" the following normal fuzzy sub-set:

	P_1	P_2	P_3
$\underset{\sim}{D}_4 = \underset{\sim}{f}(C_4) =$	1	$0,\widehat{3}$	1

For C_5:

– renewal of P_1 will be three times faster and more fluid than P_2 and five times more than P_3:

The descriptor of the concept of "speed in granting" is shown in the following normal fuzzy sub-set:

	P_1	P_2	P_3
$\underset{\sim}{D}_5 = \underset{\sim}{f}(C_5) =$	1	$0,\widehat{3}$	$0,2$

With this information we can arrive at a matrix formed by the descriptors placed as rows of the same. In this way the columns will represent the characteristics of each on of the products P_j, $j = 1,2,3$.

The following is the matrix of the descriptors:

$[\underset{\sim}{D}] =$	P_1	P_2	P_3
C_1	0,9000	0,8181	1
C_2	$0,8\widehat{3}$	1	$0,\widehat{6}$
C_3	$0,\widehat{3}$	$0,\widehat{6}$	1
C_4	1	$0,\widehat{3}$	1
C_5	1	$0,\widehat{3}$	$0,2$

that allows us to find a fuzzy sub-set for each financial product. The result is:

	C_1	C_2	C_3	C_4	C_5
$\underset{\sim}{P}_1 =$	0,9000	$0,8\widehat{3}$	$0,\widehat{3}$	1	1

	C_1	C_2	C_3	C_4	C_5
$\underset{\sim}{P}_2 =$	0,8181	1	$0,\widehat{6}$	$0,\widehat{3}$	$0,\widehat{3}$

	C_1	C_2	C_3	C_4	C_5
$\underset{\sim}{P}_3 =$	1	$0,\widehat{6}$	1	1	$0,2$

which brings to light the degree in which each product posses each one of the characteristics C_i, $i = 1,2,3,4,5$.

If we obtain the intersection of the descriptors $\underset{\sim}{\mathbf{D}}_1 \cap \underset{\sim}{\mathbf{D}}_2 \cap \underset{\sim}{\mathbf{D}}_3 \cap \underset{\sim}{\mathbf{D}}_4 \cap \underset{\sim}{\mathbf{D}}_5$ the result will be:

	P_1	P_2	P_3
$\underset{\sim}{D}_1 \cap \underset{\sim}{D}_2 \cap \underset{\sim}{D}_3 \cap \underset{\sim}{D}_4 \cap \underset{\sim}{D}_5 =$	$0,\widehat{3}$	$0,\widehat{3}$	0,2

which indicates that financial products P_1 and P_2, posses, at the very least in a degree of $0,\widehat{3}$ all the required characteristics, while P_3 posses them, at least, in a degree of 0,2.

It is quite evident that the information we have received is very poor for taking the decision to select one or other of the three financial products. For this reason we are going to develop a new procedure that requires the incorporation of the concept of a "sub-set of thresholds".

The end we are seeking with the introduction of the sub-set of thresholds is to determine, in a certain manner, up to what point (degree) is each one of the characteristics important for the financial product to be considered suitable for the needs of a business. We will establish, then, a sub-set of thresholds by assigning to each characteristic a level (degree) that it is considered its product should posses as a minimum in order for it to be considered satisfactory. In order to make this concept operative the following can be agreed on:

(a) Establish a penalisation when a characteristic is possessed in a lesser degree than required and, on the other hand, give no award when this is exceeded.
(b) Fully accept the existence of a characteristic when a product posses it in the required degree and totally reject its existence when it does not reach the required degree.

When hypothesis (a) is accepted it is possible to use the "coefficient of adequacy"[1]. The consideration of hypothesis (b) permits the use of the clan theory.

With the object of making this clearer we propose continuing with the previous example, now defining the fuzzy sub-set of thresholds for the business that we designate by $\underset{\sim}{\mathbf{P}}^*$:

	C_1	C_2	C_3	C_4	C_5
$\underset{\sim}{P}^* =$	0,9000	$0,8\widehat{3}$	$0,\widehat{6}$	$0,\widehat{6}$	$0,\widehat{3}$

[1] Kaufmann A and Gil Aluja J (1986) Introducción a la teoría de subconjuntos borrosos a la gestión de las empresas. (Ed) Milladoiro, Santiago de Compostela, pp. 142–143.

This sub-set indicates that the executives of the business consider that a financial product is totally acceptable if its price is good in a degree of 0,90; if the repayment period is good in a degree of $0{,}8\widehat{3}$; if its possibilities for renewal are $0{,}\widehat{6}$; if suitable repayment is $0{,}\widehat{6}$ and if the speed of granting is $0{,}\widehat{3}$.

We now continue with the hypothesis of penalisation in those products in which their characteristics do not reach the required degree. This penalisation is not total (all or nothing), but is progressive in line with the deficit. In this case resort can be made to the "coefficient of adequacy", which we designate by means of $K(\underset{\sim}{\mathbf{P}}_j; \underset{\sim}{\mathbf{P}}^*)$ and is constructed as follows:

When $\mu_{P_j}(C_i) \geq \mu_{P^*}(C_i)$ we do $k_i(\underset{\sim}{\mathbf{P}}_j \to \underset{\sim}{\mathbf{P}}^*) = 1$

When $\mu_{P_j}(C_i) < \mu_{P^*}(C_i)$ we do $k_i(\underset{\sim}{\mathbf{P}}_j \to \underset{\sim}{\mathbf{P}}^*) = 1 - \mu_{P^*}(C_i) + \mu_{P_j}(C_i)$

$K(\underset{\sim}{\mathbf{P}}_j; \underset{\sim}{\mathbf{P}}^*)$ is obtained by adding the $k_i(\underset{\sim}{\mathbf{P}}_j \to \underset{\sim}{\mathbf{P}}^*)$ and dividing the result by the number of characteristics (in this example five).

After doing the corresponding calculations we arrive at:

$$K(\underset{\sim}{\mathbf{P}}_1; \underset{\sim}{\mathbf{P}}^*) = \frac{1}{5}(1 + 1 + 0{,}\widehat{6} + 1) = \frac{4{,}\widehat{6}}{5} = 0{,}9\widehat{3}$$

$$K(\underset{\sim}{\mathbf{P}}_2; \underset{\sim}{\mathbf{P}}^*) = \frac{1}{5}(0{,}9181 + 1 + 1 + 0{,}\widehat{6} + 1) = \frac{4{,}5847}{5} = 0{,}9169$$

$$K(\underset{\sim}{\mathbf{P}}_3; \underset{\sim}{\mathbf{P}}^*) = \frac{1}{5}(1 + 0{,}8\widehat{3} + 1 + 1 + 0{,}8\widehat{6}) = \frac{4{,}7}{5} = 0{,}94$$

Even though taking into account the proximity between the three financial products, the conclusion is reached that, under the admitted circumstances, the most suitable for the business will be financial product P_3.

18.3 The Result by Means of the Clan Theory

Information is one of the fundamental elements for taking decisions in a modern economic system. The financial environment is not, evidently, an exception. For this we are going to develop a scheme which, under certain conditions, permits for treating data in a very wide way, giving rise to extraordinarily useful information in order to be able to decide on the suitability of taking a determined financial product.

For this we will accept the hypothesis according to which when the degree of the characteristic of a financial product does not reach the required level it is considered that this characteristic is not posses by it. For showing this in a better way we will divide the process into the following sections:

1. A comparison is made of each $\mu_{\underset{\sim}{\mathbf{D}}_j}$, $j = 1, 2, 3$, of each descriptor $\underset{\sim}{\mathbf{D}}_i$, $i = 1, 2, 3, 4, 5$ with the corresponding $\mu_{\underset{\sim}{\mathbf{P}}^*}$.

$$\text{When } \mu_{\underset{\sim}{\mathbf{D}}_j} < \mu_{\underset{\sim}{\mathbf{P}}^*} \text{ assign a } 0$$
$$\text{When } \mu_{\underset{\sim}{\mathbf{D}}_j} \geq \mu_{\underset{\sim}{\mathbf{P}}^*} \text{ assign a } 1 .$$

In this way we have:

	P_1	P_2	P_3	
$D_1^{(0,9)} = f^{(0,9)}(C_1) =$	0,9000	0,8181	1	$= \{P_1, P_3\}$

	P_1	P_2	P_3	
$D_2^{(0,8\widehat{3})} = f^{(0,8\widehat{3})}(C_2) =$	$0,8\widehat{3}$	1	$0,\widehat{6}$	$= \{P_1, P_2\}$

	P_1	P_2	P_3	
$D_3^{(0,\widehat{6})} = f^{(0,\widehat{6})}(C_3) =$	$0,\widehat{3}$	$0,\widehat{6}$	1	$= \{P_2, P_3\}$

	P_1	P_2	P_3	
$D_4^{(0,\widehat{6})} = f^{(0,\widehat{6})}(C_4) =$	1	$0,\widehat{3}$	1	$= \{P_1, P_3\}$

	P_1	P_2	P_3	
$D_5^{(0,\widehat{3})} = f^{(0,\widehat{3})}(C_5) =$	1	$0,\widehat{3}$	0,2	$= \{P_1, P_2\}$

which gives rise to the "family":

$$F = \{P_1, P_3\}, \{P_1, P_2\}, \{P_2, P_3\}, \{P_1, P_3\}, \{P_1, P_2\}$$

2. A matrix is composed by including the descriptors obtained at the required level, with which we arrive at:

	P_1	P_2	P_3
C_1	1	0	1
C_2	1	1	0
C_3	0	1	1
C_4	1	0	1
C_5	1	1	0

We could reach this same result by taking the matrix of descriptors $[\underset{\sim}{\mathbf{D}}]$ and the sub-set of thresholds $\underset{\sim}{\mathbf{P}}^*$:

$[\underset{\sim}{D}] =$

	P_1	P_2	P_3
C_1	1	0	1
C_2	1	1	0
C_3	0	1	1
C_4	1	0	1
C_5	1	1	0

$\underset{\sim}{P}^* =$

C_1	0,9000
C_2	$0,8\widehat{3}$
C_3	$0,\widehat{6}$
C_4	$0,\widehat{6}$
C_5	$0,\widehat{3}$

and the elements of each row are compared with the corresponding fuzzy subset assigning a 1 when the values of the matrix are equal to or higher and a zero when they are lower. Thus in the first row, as $0{,}900 = 0{,}900$, (C_1, P_1) will be assigned a 1; as $0{,}8181 < 0{,}900$, (C_1, P_2) a 0; as $1 > 0{,}900$, (C_1, P_3) will be assigned a 1; and so on successively arriving at:

	P_1	P_2	P_3
C_1	1	0	1
C_2	1	1	0
C_3	0	1	1
C_4	1	0	1
C_5	1	1	0

3. From the "family" obtained a determination is made of those products that have and those that do not have the five characteristics in the required degree. The result is:

$$
\begin{aligned}
f^{(0,9)}(C_1) &= \{P_1, P_3\} & f^{(0,9)}(\bar{C}_1) &= \{P_2\} \\
f^{(0,8\widehat{3})}(C_2) &= \{P_1, P_2\} & f^{(0,8\widehat{3})}(\bar{C}_2) &= \{P_3\} \\
f^{(0,\widehat{6})}(C_3) &= \{P_2, P_3\} & f^{(0,\widehat{6})}(\bar{C}_3) &= \{P_1\} \\
f^{(0,\widehat{6})}(C_4) &= \{P_1, P_3\} & f^{(0,\widehat{6})}(\bar{C}_4) &= \{P_2\} \\
f^{(0,\widehat{3})}(C_5) &= \{P_1, P_2\} & f^{(0,\widehat{3})}(\bar{C}_5) &= \{P_3\}
\end{aligned}
$$

4. We now move on to find the mini-terms or "atoms" by means of the intersection of the common sub-sets found in the previous section[2]. We arrive at:

[2] With the object of avoiding a too complex nomenclature we will not indicate the super-index in functions.

$$f(C_1)\cap f(C_2)\cap f(C_3)\cap f(C_4)\cap f(C_5)=\emptyset$$
$$f(C_1)\cap f(C_2)\cap f(C_3)\cap f(C_4)\cap f(\bar{C}_5)=\emptyset$$
$$f(C_1)\cap f(C_2)\cap f(C_3)\cap f(\bar{C}_4)\cap f(C_5)=\emptyset$$
$$f(C_1)\cap f(C_2)\cap f(\bar{C}_3)\cap f(C_4)\cap f(C_5)=\{P_1\}$$
$$f(C_1)\cap f(\bar{C}_2)\cap f(C_3)\cap f(C_4)\cap f(C_5)=\emptyset$$
$$f(\bar{C}_1)\cap f(C_2)\cap f(C_3)\cap f(C_4)\cap f(C_5)=\emptyset$$
$$f(C_1)\cap f(C_2)\cap f(C_3)\cap f(\bar{C}_4)\cap f(\bar{C}_5)=\emptyset$$
$$f(C_1)\cap f(C_2)\cap f(\bar{C}_3)\cap f(C_4)\cap f(\bar{C}_5)=\emptyset$$
$$f(C_1)\cap f(\bar{C}_2)\cap f(C_3)\cap f(C_4)\cap f(\bar{C}_5)=\{P_3\}$$
$$f(\bar{C}_1)\cap f(C_2)\cap f(C_3)\cap f(C_4)\cap f(\bar{C}_5)=\emptyset$$
$$f(C_1)\cap f(C_2)\cap f(\bar{C}_3)\cap f(\bar{C}_4)\cap f(C_5)=\emptyset$$
$$f(C_1)\cap f(\bar{C}_2)\cap f(C_3)\cap f(\bar{C}_4)\cap f(C_5)=\emptyset$$
$$f(\bar{C}_1)\cap f(C_2)\cap f(C_3)\cap f(\bar{C}_4)\cap f(C_5)=\{P_2\}$$
$$f(C_1)\cap f(\bar{C}_2)\cap f(\bar{C}_3)\cap f(C_4)\cap f(C_5)=\emptyset$$
$$f(\bar{C}_1)\cap f(C_2)\cap f(\bar{C}_3)\cap f(C_4)\cap f(C_5)=\emptyset$$
$$f(\bar{C}_1)\cap f(\bar{C}_2)\cap f(C_3)\cap f(C_4)\cap f(C_5)=\emptyset$$

We interrupt the process at this point, since there are not three $f(\bar{C}_1)$ that have a same P_j and, therefore, the result of the intersection is, from here on, the void set.
The mini-terms or atoms that are not void then are $\{P_1\}$, $\{P_2\}$, $\{P_3\}$.
These terms or atoms that are not void can also be obtained from matrix:

	P_1	P_2	P_3
C_1	1	0	1
C_2	1	1	0
C_3	0	1	1
C_4	1	0	1
C_5	1	1	0

By successively changing the rows in order to include the $\bar{C}_1$. In this way we arrive at:

	P_1	P_2	P_3
C_1	1	0	1
C_2	1	1	0
C_3	0	1	1
$\bar{C}_4$	0	1	0
C_5	1	1	0

	P_1	P_2	P_3
C_1	1	0	1
C_2	1	1	0
$\overline{C_3}$	1	0	0
C_4	1	0	1
C_5	1	1	0

	P_1	P_2	P_3
C_1	1	0	1
$\overline{C_2}$	0	0	1
C_3	0	1	1
C_4	1	0	1
$\overline{C_5}$	0	0	1

	P_1	P_2	P_3
$\overline{C_1}$	0	1	0
C_2	1	1	0
C_3	0	1	1
$\overline{C_4}$	0	1	0
C_5	1	1	0

	P_1	P_2	P_3
C_1	1	0	1
C_2	1	1	0
C_3	0	1	1
C_4	1	0	1
$\overline{C_5}$	0	0	1

The columns that have a 1 in all their elements give rise to the mini-terms, which in this case are $\{P_1\}$, $\{P_3\}$, $\{P_2\}$ also arrived at by the previous procedure.

5. Clan $K(F)$ is obtained produced by the "family" by taking the atoms and all their possible unions to which f will be added:

$$K(F) = \{\emptyset, \{P_1\}, \{P_2\}, \{P_3\}, \{P_1, P_2\}, \{P_1, P_3\}, \{P_2, P_3\}, \{P_1, P_2, P_3\}\}$$

It can be seen that this clan is a Boole sub-lattice (see Fig. 18.1).

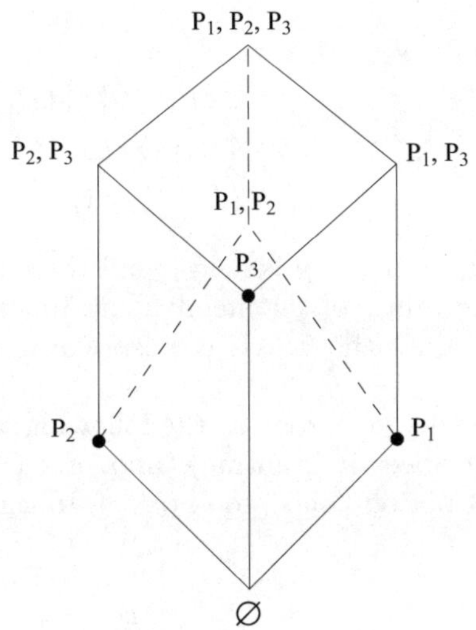

Fig. 18.1.

The non-void atoms have been represented by Σ.

By doing this process a wide range of information is obtained all of which can be most useful for taking decisions relative to the most suitable financial product for the interests of the business.

Thus, it will be seen that product P_1 has all the required characteristics except for the possibilities of renewal. Product P_2 does not have a good price nor adequate fractioning of repayments. On the other hand, product P_3 is not suitable for the business because of the repayment period set and also relative to the time necessary for granting the credit. All of this can easily be deduced due to the zeros that appear in the tables on the previous page.

Even in a case as simple as the one we have shown, the decision does not have to be the only one and the financial product chosen will depend on the importance that the executives of the business assign to each of the characteristics.

This is a new element that undoubtedly takes part in the selection of a financial product and which, due to its interest, should be explicitly taken into account on drawing up a model for the selection of financial products. But this will be the object for treatment in later sections.

Another type of information, perhaps less useful for this particular problem, but not exempt of interest, refers to the determination of the product or products that have some characteristics and not others, for which certain "keys" are established. Thus, for example, if we are looking for key: (low price "and" possibilities for renewal "and" suitable fractioning of payback) "and/or" (good repayment period "and" speed in granting), we arrive at:

$$\begin{aligned}(D_1 \cap D_3 \cap D_4) \cup (D_2 \cap D_5) &= (\{P_1, P_3\} \cap \{P_2, P_3\} \cap \{P_1, P_3\} \\ &\quad \cup \{P_1, P_2\} \cap \{P_1, P_2\} \\ &= \{P_3\} \cup \{P_1, P_2\} \\ &= \{P_1, P_2, P_3\}\end{aligned}$$

The result here is that this key is held by all three products. This can be verified by seeing if the values of the membership function of $\underset{\sim}{\mathbf{P}}_1$, $\underset{\sim}{\mathbf{P}}_2$ and $\underset{\sim}{\mathbf{P}}_3$ are equal to or higher than the values corresponding to the threshold subset $\underset{\sim}{\mathbf{P}}^*$.

One could also consider keys such as the following: (low price "and" suitable payback "and " no speed in granting) "and/or" (possibilities for renewal "and" good repayment period "and" good price). In this case we would have:

$$\begin{aligned}(D_1 \cap D_4 \cap \bar{D}_5) \cup (D_3 \cap \bar{D}_2 \cap D_1) &= (\{P_1, P_3\} \cap \{P_1, P_3\} \cap \{P_3\} \\ &\quad \cup \{P_2, P_3\} \cap \{P_3\} \cap \{P_1, P_3\} \\ &= \{P_3\} \cup \{P_3\} \\ &= \{P_3\}\end{aligned}$$

In this way, as is easily seen, on financial product P_3 complies with this particular key.

A large number of keys can be composed that can provide useful information for the financial institutions themselves who set up and sell the products as well as for businesses which are the eventual end users.

18.4 Approach to a Model of Subjective Preferences

The schemes we have developed up to this point do not take explicitly into account an element that we feel is fundamental for the decision of choosing a financial product. This is the relative importance that a determined business grants to each characteristic of the product relative to the rest. With the object of avoiding this inconvenience we are going to draw up a model that posses the possibility of operation for the treatment of the situation that current reality presents frequently. For this we are going to operate in the discontinuous functional field.

We start out from the existence of a finite and re-countable number of financial products $P_1, P_2, \ldots, P_n$ which each posses certain determined characteristics $C_1, C_2, \ldots, C_m$ in such a way that for each characteristic it is possible to establish a quantified (objective or subjective) relation of preferences. Therefore for C_j we have that: P_1 is preferred μ_1/μ_2 times over P_2, μ_1/μ_3 times over $P_3, \ldots, \mu_1/\mu_n$ times over $P_n, \ldots, P_n$ is preferred μ_n/μ_1 times over P_1, μ_n/μ_2 times over $P_2, \ldots, \mu_n/\mu_{n-1}$ times over P_{n-1}.

With this we will be able to construct the following matrix, which will be reflexive and reciprocal by construction:

$$[C_{ij}] = \begin{bmatrix} 1 & \dfrac{\mu_1}{\mu_2} & \dfrac{\mu_1}{\mu_3} & \cdots & \dfrac{\mu_1}{\mu_n} \\ \dfrac{\mu_2}{\mu_1} & 1 & \dfrac{\mu_2}{\mu_3} & \cdots & \dfrac{\mu_2}{\mu_n} \\ \cdots & \cdots & \cdots & \cdots & \cdots \\ \dfrac{\mu_n}{\mu_1} & \dfrac{\mu_n}{\mu_2} & \dfrac{\mu_n}{\mu_3} & \cdots & 1 \end{bmatrix}$$

This matrix is also coherent or consistent since the following is complied with:

$$\forall i, j, k \in \{1, 2, \ldots, n\}, \frac{\mu_i}{\mu_j} \cdot \frac{\mu_j}{\mu_k} = \frac{\mu_i}{\mu_k}$$

Now then, the condition of consistency is not always complied with in the treatment of financial phenomena. For this reason we are going to consider certain properties[3] of positive matrices, that is, those in which all the elements that are members of R_0^+ (are positive):

(a) A positive square matrix posses a dominant value of its own l real positive which is unique for which what is complied is that $\lambda \geq n$, where n is the order of the square matrix.
(b) The vector that corresponds to the dominant own value is found also formed by positive terms and when normalised, is unique.

When λ is a number close to n it is said that the matrix is nearly coherent; on the contrary it will be necessary to make an adjustment between the elements of the matrix if wanting to use this scheme correctly. It is considered that $\lambda - n$ or $\frac{\lambda - n}{n}$ is an index of coherence.

As is very well known, when a reciprocal matrix is also coherent it complies with:

$$[C_{ij}] \cdot [v_i]^T = n \cdot [v_i]^T$$

where $[v_i]^T$ is the transpose of row i.

When the reciprocal matrix is not coherent, we write:

[3] Kaufmann A and Gil Aluja J (1987) Técnicas operativas de gestión para el tratamiento de la incertidumbre. (Ed) Hispano Europea, Barcelona, p. 225.

$$[C_{ij}] \cdot [v'_i]^T = \lambda \cdot [v'_i]^T$$

We accept $[v'_i]$ as the result when the index of coherence $\frac{\lambda - n}{n}$ is sufficiently small.

For each characteristic C_j, $j = 1, 2, \ldots, m$ the corresponding reflexive and reciprocal matrix $[C_{ij}]$ is obtained. Once the m matrices are constructed the dominant own values λ_j and their corresponding vectors $\left[\begin{smallmatrix} X_{ij} \\ \cdots \\ X_{nj} \end{smallmatrix}\right]$ must be found for each one, verifying if they posses sufficient consistency by means of the "index of coherence". The elements of each corresponding own vector will give rise to a fuzzy sub-set:

	P_1	P_2	P_3	…	P_n
$\underset{\sim}{X}_j =$	x_{1j}	x_{2j}	x_{3j}	…	x_{4j}

which once normalised in sum equal to one will be:

	P_1	P_2	P_3	…	P_n
$\underset{\sim}{D}_j =$	p_{1j}	p_{2j}	p_{3j}	…	p_{4j}

The m own vectors are regrouped forming a matrix the form of which will be:

$[p_{ij}] =$

	C_1	C_2	C_3	C_4		C_m
P_1	p_{11}	p_{12}	p_{13}	p_{14}	…	p_{1m}
P_2	p_{21}	p_{22}	p_{23}	p_{24}	…	p_{2m}
P_3	p_{31}	p_{32}	p_{33}	p_{34}	…	p_{3m}
P_4	p_{41}	p_{42}	p_{43}	p_{44}	…	p_{4m}
	⋮	⋮	⋮	⋮	⋮	⋮
P_n	p_{n1}	p_{n2}	p_{n3}	p_{n4}	…	p_{nm}

Each column of this matrix brings to light the relative degree in which a characteristic is possessed by all the financial products. As we have already pointed out, this can be represented by a normalised fuzzy sub-set $\underset{\sim}{D}_j$. From this perspective there exist m fuzzy sub-sets. On the other hand each row expressed, for one product, the degree in which it possess each one of the characteristics, which is also represented by a fuzzy sub-set $\underset{\sim}{Q}_i$ such as:

	C_1	C_2	C_3	C_4	…	C_m
$\underset{\sim}{Q}_i =$	p_{i1}	p_{i2}	p_{i3}	p_{i4}	…	p_{im}

On the other hand, each business has a different appreciation of the importance that each characteristic has. Evidently, this estimate can vary from one moment to another and its quantification has a basically subjective sense, therefore will be expressed by means of valuations. The establishment of these valuations can be done by means of a comparison between the relative importance of a characteristic in relation to the rest. Therefore, for example, it can be said that a characteristic is two times as important as another, or has half the importance of a third.

In this way we can construct a new matrix that obviously will be square, reflexive and anti-symmetrical. Since there are n products, its order will be $m \times m$:

$$[a_{ij}] = \begin{array}{c|cccccc|} & C_1 & C_2 & C_3 & C_4 & & C_m \\ C_1 & 1 & a_{12} & a_{13} & a_{14} & \dots & a_{1m} \\ C_2 & a_{21} & 1 & a_{23} & a_{24} & \dots & a_{2m} \\ C_3 & a_{31} & a_{32} & 1 & a_{34} & \dots & a_{3m} \\ C_4 & a_{41} & a_{42} & a_{43} & 1 & \dots & a_{4m} \\ & \vdots & \vdots & \vdots & \vdots & \vdots & \vdots \\ C_m & a_{m1} & a_{m2} & a_{m3} & a_{m4} & \dots & 1 \end{array}$$

Due to the condition of asymmetry the following will be complied with:

$$a_{ij} = \frac{1}{a_{ij}}$$

Once this matrix has been determined, we proceed to obtain the corresponding dominant value and vector. This vector will bring to light the preferences of the business relative to the characteristics:

$$y_j = \begin{bmatrix} y_1 \\ y_2 \\ y_3 \\ \dots \\ y_m \end{bmatrix}$$

In order for this vector to be susceptible to being used as a weighting element, we are going to convert it into another that possesses the property that the sum of its elements be equal to the unit. For this we do:

$$b_j = \frac{y_j}{\sum_{j=1}^{m} y_j}, \quad j = 1, 2, \dots, m$$

With which we arrive at:

$$b_j = \begin{bmatrix} b_1 \\ b_2 \\ b_3 \\ \dots \\ b_m \end{bmatrix}$$

We are now in a position finally to arrive at the sought after result, by taking matrix $[p_{ij}]$ and multiplying it to the right by vector $[b_j]$. The result will be another vector, which will express the relative importance of each financial product for the business, taking into account its preferences for each one of the characteristics:

$$\begin{bmatrix} p_{11} & p_{12} & p_{13} & \cdots & p_{1m} \\ p_{21} & p_{22} & p_{23} & \cdots & p_{2m} \\ p_{31} & p_{32} & p_{33} & \cdots & p_{3m} \\ \dots & \dots & \dots & \dots & \dots \\ p_{n1} & p_{n2} & p_{n3} & \cdots & p_{nm} \end{bmatrix} \cdot \begin{bmatrix} b_1 \\ b_2 \\ b_3 \\ \dots \\ b_m \end{bmatrix} = \begin{bmatrix} d_1 \\ d_2 \\ d_3 \\ \dots \\ d_m \end{bmatrix}$$

The result can also be expressed by means of a normal fuzzy sub-set, by doing:

	P_1	P_2	P_3	P_4	...	P_m
$\underset{\sim}{H}$ =	h_1	h_2	h_3	h_4	...	h_m

At least one $h_j = 1$ will exist.

This model on the contrary to all those that use as the only basis for selection, the price of the money, has as its greatest advantage the possibility of incorporating a wide range of elements that, in the reality of businesses, at times play a decisive role at the time of taking the decision to select a financial product from among those offered on the market. These elements normally do not have the same weight at the time of making a valuation. As has been seen, this circumstance is included in this scheme, in this way gaining sufficient flexibility and adaptability in order to be used in business reality with a high degree of generality.

18.5 Application of the Proposed Model

With the object of illustrating the model a case has been considered which we have linked to the one shown in this same chapter in which a business, in order to cover certain financial requirements, resorts to three credit institutions which propose as the most adequate, one financial product each. Therefore there is a choice between three products P_1, P_2, P_3.

The characteristics of these products makes them different, but in certain aspects some are more attractive, but in others these are less favourable. Obviously, in the eyes of the businessman not all the characteristics have the same weight at the time of deciding to accept one or another. The five characteristics mentioned previously were considered as important: price of the money, payback period, possibilities for renewal, fractioning repayments, speed of granting.

1. With regard to the price of the money the following data is considered: for P_1 20%, for P_2 22% and for P_3 18%. This then is objective data and it is logical to think that the preference would be for the lowest price in a proportional manner. Therefore P_1 to P_2 would be preferred 22/20 and P_1 to P_3 18/20. P_2 to P_3 would be preferable 18/22. In this way the following matrix can be constructed:

	P_1	P_2	P_3
P_1	1	11/10	9/10
P_2	10/11	1	9/11
P_3	10/9	11/9	1

Once this matrix has been constructed the corresponding dominant own value and vector must be obtained. Among the various procedures existing we are going to use the following:

$$\begin{bmatrix} 1 & 1{,}1 & 0{,}9 \\ 0{,}9090 & 1 & 0{,}8181 \\ 1{,}1111 & 1{,}2222 & 1 \end{bmatrix} \cdot \begin{bmatrix} 1 \\ 1 \\ 1 \end{bmatrix} = \begin{bmatrix} 3 \\ 2{,}7271 \\ 3{,}3333 \end{bmatrix} = 3{,}3333 \cdot \begin{bmatrix} 0{,}9 \\ 0{,}8181 \\ 1 \end{bmatrix}$$

$$\begin{bmatrix} 1 & 1{,}1 & 0{,}9 \\ 0{,}9090 & 1 & 0{,}8181 \\ 1{,}1111 & 1{,}2222 & 1 \end{bmatrix} \cdot \begin{bmatrix} 0{,}9 \\ 0{,}8181 \\ 1 \end{bmatrix} = \begin{bmatrix} 2{,}6999 \\ 2{,}4543 \\ 2{,}9998 \end{bmatrix} = 2{,}9998 \cdot \begin{bmatrix} 0{,}9 \\ 0{,}8181 \\ 1 \end{bmatrix}$$

From here it can be deduced that:

$$\lambda_1 = 2{,}9998 \qquad [X_{i1}] = \begin{bmatrix} 0{,}9 \\ 0{,}8181 \\ 1 \end{bmatrix}$$

For normalisation of the sum equal to 1 we do:

$$\frac{X_{11}}{\sum_{i=1}^{3} X_{i1}} = \frac{0{,}9}{2{,}7181} = 0{,}3311$$

$$\frac{X_{21}}{\sum_{i=1}^{3} X_{i1}} = \frac{0{,}8181}{2{,}7181} = 0{,}3009$$

$$\frac{X_{31}}{\sum_{i=1}^{3} X_{i1}} = \frac{1}{2{,}7181} = 0{,}3679$$

In this way arriving at:

$$[p_{i1}] = \begin{array}{c|c|} & \\ P_1 & 0{,}3311 \\ P_2 & 0{,}3009 \\ P_3 & 0{,}3679 \end{array}$$

2. The payback periods established for each product are as follows: 5 years for P_1; 6 years for P_2 and 4 years for P_3. The business establishes a clear preference for the products with a longer payback period and this preference, on being able to be presented by a proportionality we arrive at the following matrix:

	P1	P2	P3
P1	1	5/6	5/4
P2	6/5	1	6/4
P3	4/5	4/6	1

We follow the same process for arriving at the corresponding dominant own value and vector. Therefore:

$$\begin{bmatrix} 1 & 0{,}8333 & 1{,}25 \\ 1{,}2 & 1 & 1{,}5 \\ 0{,}8 & 0{,}6666 & 1 \end{bmatrix} \cdot \begin{bmatrix} 1 \\ 1 \\ 1 \end{bmatrix} = \begin{bmatrix} 3.0833 \\ 3{,}7 \\ 2{,}4666 \end{bmatrix} = 3{,}7 \cdot \begin{bmatrix} 0{,}8333 \\ 1 \\ 0{,}6666 \end{bmatrix}$$

$$\begin{bmatrix} 1 & 0{,}8333 & 1{,}25 \\ 1{,}2 & 1 & 1{,}5 \\ 0{,}8 & 0{,}6666 & 1 \end{bmatrix} \cdot \begin{bmatrix} 0{,}8333 \\ 1 \\ 0{,}6666 \end{bmatrix} = \begin{bmatrix} 2{,}4998 \\ 2{,}9998 \\ 1{,}9998 \end{bmatrix} = 2{,}9998 \cdot \begin{bmatrix} 0{,}8333 \\ 1 \\ 0{,}6666 \end{bmatrix}$$

Therefore:

$$\lambda_2 = 2{,}9998 \qquad [X_{i2}] = \begin{bmatrix} 0{,}8333 \\ 1 \\ 0{,}6666 \end{bmatrix}$$

For normalisation of the sum equal to one:

$$\frac{X_{12}}{\sum_{i=1}^{3} X_{i2}} = \frac{0{,}8333}{2{,}4999} = 0{,}3333$$

$$\frac{X_{22}}{\sum_{i=1}^{3} X_{i2}} = \frac{1}{2{,}4999} = 0{,}4$$

$$\frac{X_{32}}{\sum_{i=1}^{3} X_{i2}} = \frac{0{,}6666}{2{,}4999} = 2{,}2666$$

Arriving at:

$$[P_{i2}] = \begin{bmatrix} 0{,}3333 \\ 0{,}4 \\ 0{,}2666 \end{bmatrix}$$

3. The possibilities of obtaining renewal of the credit at its expiry are different for each of the financial products. There are double the possibilities of renewal of P_2 than for P_1 and three times the possibilities of renewal of P_3 than for P_1. The possibilities of renewing P_2 appear to be less in 4/5 relative to P_3.[4] We construct the following matrix:

	P1	P2	P3
P1	1	1/2	1/3
P2	2	1	4/5
P3	3	5/4	1

and arrive at the corresponding dominant own value and vector:

$$\lambda_3 = 3{,}0036 \qquad [X_{i3}] = \begin{bmatrix} 0{,}3542 \\ 0{,}7528 \\ 1 \end{bmatrix}$$

Then we do the normalisation in sum equal to 1 arriving at:

$$[P_{i3}] = \begin{bmatrix} 0{,}1681 \\ 0{,}3572 \\ 0{,}4746 \end{bmatrix}$$

4. Repayment of the loan must be done quarterly for financial products P_1 and P_3 and monthly for P_2. A proportional preference is established for the quarterly expiry over the monthly one. The matrix will be:

	P1	P2	P3
P1	1	3	1
P2	1/3	1	1/3
P3	1	3	1

[4] It will be seen that in this characteristic, of a clear subjective meaning, a variation has been introduced which consists of breaking up the total consistency in order to give more generality to the example. In fact it will be seen that:

$$\frac{1}{2} \cdot \frac{4}{5} \neq \frac{1}{3}$$

the corresponding dominant own value and vector:

$$\lambda_4 = 2{,}9999 \qquad [X_{i4}] = \begin{bmatrix} 1 \\ 0{,}3333 \\ 1 \end{bmatrix}$$

We then do the normalisation in sum equal to one:

$$[P_{i4}] = \begin{bmatrix} 0{,}4285 \\ 0{,}1428 \\ 0{,}4285 \end{bmatrix}$$

5. Experience leads us to think that financial product P_1 will be obtained very quickly, P_2 with relative speed and this factor for P_3 will be slow. Based on these feelings it is estimated that P_1 is preferred 3 times over P_2 and 5 times over P_3, while P_2 is preferred 2 times over P_3.[5] We obtain the following matrix:

	P_1	P_2	P_3
P_1	1	3	5
P_2	1/3	1	2
P_3	1/5	1/2	1

The corresponding dominant own value and vector are:

$$\lambda_5 = 3{,}0028 \qquad [X_{i5}] = \begin{bmatrix} 1 \\ 0{,}3542 \\ 0{,}1881 \end{bmatrix}$$

and when we do the normalisation in sum equal to one we arrive at:

$$[P_{i5}] = \begin{bmatrix} 0{,}6483 \\ 0{,}2296 \\ 0{,}1219 \end{bmatrix}$$

Once we have obtained these five vectors $[p_{ij}]$, $j = 1{,}2{,}3{,}4{,}5$, we group them and form the following matrix:

$[p_{ij}] =$

	C_1	C_2	C_3	C_4	C_5
P_1	0,3311	0,3333	0,1681	0,4285	0,6483
P_2	0,3009	0,4000	0,3572	0,1428	0,2296
P_3	0,3679	0,2666	0,4746	0,4285	0,1219

[5] Also for this characteristic we have avoided total consistency.

We now move on to analyse the other side of the problem. This is to bring to light how a business considers each one of these characteristics relative to the others. In this case the following comparisons have been established:

- The first characteristic is worth 2 times the second, 6 times the third, 8 times the fourth and 4 times the fifth.
- The second characteristic is worth 4 times the third, 6 times the fourth and 2 times the fifth.
- The third characteristic is worth 3 times the fourth and 1/2 the fifth.
- The fourth characteristic is worth 1/3 the fifth.

With the following square, reflexive and reciprocal matrix can be arrived at:

	C_1	C_2	C_3	C_4	C_5
C_1	1	2	6	8	4
C_2	1/2	1	4	6	2
C_3	1/6	1/4	1	3	1/2
C_4	1/8	1/6	1/3	1	1/3
C_5	1/4	1/2	2	3	1

In order to obtain the corresponding dominant own value and vector the same process can be used as followed before. In this way we find:

$$\lambda_c = 5{,}0842$$

$$[y_j] = \begin{bmatrix} 1 \\ 0{,}5709 \\ 0{,}1779 \\ 0{,}0916 \\ 0{,}2854 \end{bmatrix}$$

and with the normalisation in sum equal to one:

$$[b_j] = \begin{bmatrix} 0{,}4704 \\ 0{,}2685 \\ 0{,}0836 \\ 0{,}0430 \\ 0{,}1342 \end{bmatrix}$$

Finally, if we take matrix $[p_{ij}]$ and multiply to the right by vector $[b_j]$, which in short constitutes a weighting, we arrive at:

$$[d_j] = \begin{bmatrix} 0{,}3311 & 0{,}3333 & 0{,}1681 & 0{,}4285 & 0{,}6483 \\ 0{,}3009 & 0{,}4000 & 0{,}3572 & 0{,}1428 & 0{,}2296 \\ 0{,}3679 & 0{,}2666 & 0{,}4746 & 0{,}4285 & 0{,}1219 \end{bmatrix} \cdot \begin{bmatrix} 0{,}4704 \\ 0{,}2685 \\ 0{,}0836 \\ 0{,}0430 \\ 0{,}1342 \end{bmatrix} = \begin{bmatrix} 0{,}3647 \\ 0{,}3157 \\ 0{,}3191 \end{bmatrix}$$

Taking into account that we have only considered four decimal points and the last one has not been rounded up, the sum of the elements of the last matrix does not give the unit as the result, which would have occurred if the rounding up were to have been done.

The result we have arrived at can also be expressed by means of a normal fuzzy sub-set, as follows:

	P_1	P_2	P_3
P =	1,0000	0,8656	0,8749

It will be seen in this fuzzy sub-set that financial product P_1 is preferable to products P_2 and P_3, although not too much. There is very little difference between P_2 and P_3.

This example could be taken as typical since it shows what happens often in financial reality, when the businessman is faced with the need to choose between apparently different products but which, when all is said and done, are very similar. This situation should not come as a surprise to us if it is thought that financial institutions attempt to compensate certain disadvantages of a product relative to other of the competition, by means of incentives to certain aspects that make it more attractive and allow in this way for its placing in the market under conditions of competitiveness.

19 Neural Structures for the Selection of Financial Resources

19.1 Basic Principles of Neural Structures

One of the basic problems that is inherent to the financial activity of a business is determined by the selection of financial resources that may have different origins, and the end use of which consists in being able to attend to the normal operation of its activity.

The objective, for the business, of obtaining those financial means at the least possible cost, is a matter that has been looked at in different works. But although on certain occasions the resolution of the problem was based on traditional elements, and more recently resorting to sequential schemes[1] surrounded by the principles of dynamic programming, we feel it is now necessary to take a further step and introduce certain elements belonging to the field of neuro-mimesis with the object of providing a solution to the problem of selecting financial means.

The starting out point, which will allow us to apply neural structures to the problem, is based on the parallelism existing between the configuration of a graph and a neuron network[2].

In this way by assimilating both concepts and applying the principles and properties of the theory of graphs to neural structures, what is established is a scheme that allows for choosing the financial means that are most adequate to the requirements of the business.

As a prior step to the establishment of the model which will permit us to reach the sought after result, we feel it is suitable to make a brief reference to the most elemental concepts of neuro-mimesis.

1 See, for example, Gil Lafuente A M (1996) Estrategías secuenciales para la captación de medios financieros (in Spanish), Proceedings of the SIGEF Congress. Vol I. Buenos Aires, Argentina. Communication paper 2.11.

2 The concept of a neural graph was coined in the work of professors Kaufmann A and Gil-Aluja J (1995) Grafos Neuonrales para la economía y gestión de empresas (in Spanish). (Ed) Pirámide. Madrid. We follow these authors in the basic aspects.

Anna Maria Gil-Lafuente: *Fuzzy Logic in Financial Analysis*, StudFuzz **175**, 341–357 (2005)
www.springerlink.com

It is a well known fact that the use in the different fields of the principles that report on the functioning of artificial neural networks are no less than an attempt to simulate the operation of the nervous system of living beings. The principal element of this complex mechanism is the neuron. Disregarding the diversity existing between the different species and specific function in each one, the neuron is composed, in a very schematic way, of cellular body in which a nucleus exists. The cellular body is denominated "soma", which is prolonged by what is known as the "axon" and this is ramified giving rise to the "dendrites". By means of the dendrites information is transmitted from one neuron to another or others. This transmission, in the form of electro-chemical impulses takes place in the "synapses" and is more or less intense based on the electrical potential conducted through the dendrites, on the one hand, and the resistance made by a type of membrane hidden in the cellular body, which is known by the name of threshold, on the other. A representation, evidently much simplified of a neuron belonging to the nervous system of a living being could be as follows:

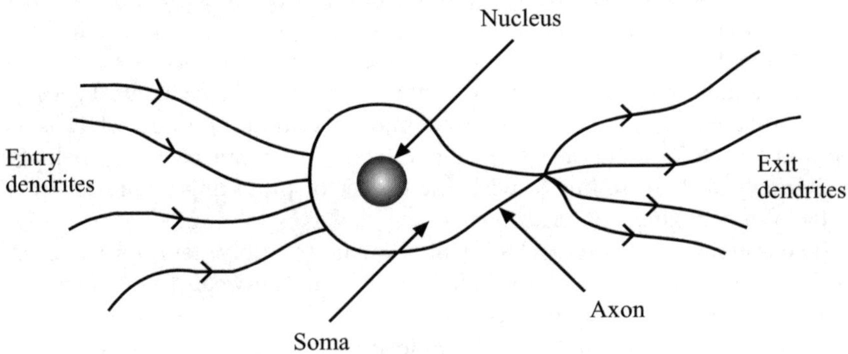

Fig. 19.1.

The advances that have taken place in the field of neurobiology have led to far greater knowledge of the brain of living beings and, having seen its perfection, there have been many scientists who have tackled the task of making a certain "reproduction" in an attempt to treat complex systems in different spheres of science. In this respect the task was embarked upon for drawing up certain schemes that, in one form or another, would allow for the reproduction of the operation of a natural neuron, initiating what would later become known as neuro-mimesis. The first of these schemes was drawn up by McCulloch and Pitts[3]. Its presentation is very simple and we have reproduced it below:

[3] McCulloch W S and Pitts N (1943) A logical calculus of the ideas immanent in the nervous system. Bulletin of Mathematical Biophysics, No. 5, pp. 115–133.

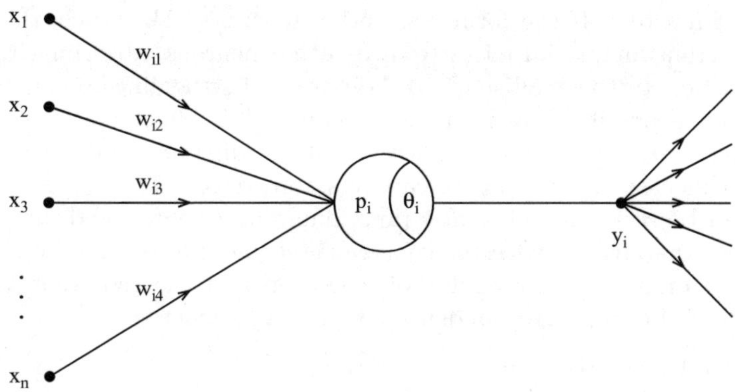

Fig. 19.2.

where:

x_i = entry stimulus.
w_{ij} = synaptic weighting of signal x_i for neuron i.
θ_i = threshold of neuron i.
p_i = potential struggling to exit neuron i.
y = actual potential exit.

If we analyse this scheme we can sense a process from which certain signals are transmitted proceeding from other neurons along the entry dendrites. These signals are subject to different processes (aggregation, compensation, stimulation, inhibition, etc.) depending on the function they must carry out. In this case the potential signals have an exit value of the previous neurons equal to x_i, $i = 1, 2, \ldots, n$. In their passage along the dendrites these values undergo modifications in the sense of reductions or amplifications. This function is exercised by certain "weights" which are designated by w_{ij}, $j = 1, 2, \ldots, n$. The product of $x_i . w_{ij}$ is the potential that reaches neuron i along dendrite j, $j = 1, 2, \ldots, n$. In this way the impulses, signals or potentials have been affected by the different weights assigned to each dendrite. When the different signals reach the central body of the cell, a wide range of possible operations takes place, from the most elementary to the most complex. In the simple case we have describe this is just a very easy sum. In this case an aggregation by addition takes place equal to:

$$p_i = \sum_{j=1}^{n} x_i . w_{ij}$$

and it is precisely potential p_i that is struggling to leave. Provided this potential p_i is capable of overcoming the threshold of neuron θ_i an exit potential will take place, which will go to other neurons by means of the exit dendrites (axonal dendrites). If the former situation does not take place, the neuron will stop transmitting impulses towards other neurons, absorbing the stimuli it receives. In the "collision" of the potential struggling to exit from the threshold it is possible that it might be damaged, in which case the potential expanding along the axonal dendrites will be different (and nearly always less). In this case an exit $y_i \neq p_i$ can be assumed.

As can be seen, the schematic functioning of a neuron and its dendrites maintains a parallelism with the concepts associated to the theory of graphs. Therefore comparing the scheme of a neuron and the representation of a graph, the following correspondence could be constructed:

Entry dendrites: arcs arriving at a vertex.
Cellular body and axon: vertex of a graph.
Exit dendrites: arcs leaving a vertex.

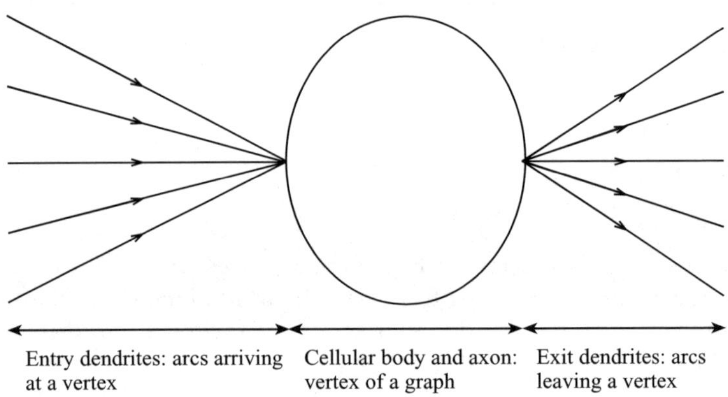

Fig. 19.3.

We now move on to describe the neural function assigned to the different potential values in [0; 1] based on the intensity of the impulses. In this way we could construct a semantic scale, which would show the level at which the electro-chemical impulses are transmitted:

0	no intensity
0,1	practically no intensity
0,2	nearly no intensity
0,3	little intensity
0,4	moderate intensity
0,5	average intensity
0,6	quite a lot of intensity
0,7	a lot of intensity
0,8	nearly total intensity
0,9	practically total intensity
1	total intensity

In this way, if a neuron emits an impulse of 0,7 it would mean that it has "a lot of intensity". But on the other hand if it emits an impulse of 0,1 then this would have "practically no intensity". As we pointed out when describing the McCulloch and Pitts model, the impulses or potentials leaving a neuron are affected in their passage along the dendrite by a weight which we are going to assume is included between [0; 1], in such a way that, the close to 0, the more will the intensity of the potential moving towards the cellular body of the arrival neuron be reduced. On the contrary, the closer to 1, the impulse moving towards the following neuron will be less reduced.

If what is wanted is that the sum of the product $x_i.w_{ij}$ is maintained in [0; 1] we must resort to the use of certain weights that give rise to a convex weighting (sum of the weights equal to the unit). For this it is sufficient to add the weights

$$\sum_{j=1}^{n} w_{ij} = W_j$$

and divide each product by this sum:

$$\frac{x_i.w_{ij}}{W_j}$$

which is equivalent to multiplying each potential x_i by

$$\frac{w_{ij}}{W_j}$$

The aggregation of these impulses or potentials by means of the sum gives rise to the potential struggling to leave the neuron, which we are sure, is to be found in [0; 1].

$$p'_i = \sum_{j=1}^{n} \frac{x_i.w_{ij}}{W_j}$$

When this potential equals or exceeds the value of the threshold of the neuron, this impulse is transmitted by the exit dendrites to other neurons. That is:

$$p'_i \succeq \theta_i, \ y_i > 0$$

$$p'_i \prec \theta_i, \ y_i = 0$$

The effect of the threshold on the potential can be expressed by a function of the same $y_i = f(p'_i)$ where if it is also desired that y_i be a value in [0; 1] it will be necessary for the function $f(p'_i)$ to represent a τ-conorm or an inference. Likewise, in a more simple case we could do $p'_i = y_i$ which indicates the non-existence of modifications in the potential.

19.2 Approach to the Problem

The objective of the majority of businesses that survive in the environment marked by the continuous and at times accelerated evolution of events, is to have available instruments to permit them to take decisions in order to arrive at the permanent adaptation of their structures.

In the financial context, one of the commonly sought after objectives is to obtain from among the different financial resources available, the one that is most suited to the specific needs of the business at a given moment.

We are conscious of the fact that the treatment of this problem by means of determinist or probability schemes has given rise to different models that on occasions have provided acceptable results. Nevertheless, we feel that the treatment of this problem by means of sequential instruments[4], or as in the case we are concerned with, of neural graphs, will allow us to arrive at suitable results, above all in surroundings undergoing permanent change.

In order to make it easier to follow the model we are going to develop, we shall start out with the existence of a business seeking a specific objective, which, in our case, consist of an increase of its presence in the market.

In order to carry out this objective, the persons responsible for finances in this business have considered four possible channels of financing:

get new contributions from the shareholders by means of a capital increase for the business.
request a long-term credit from some financial entity.
increase payment terms to suppliers.
request a short-term credit from some financial entity.

The materialisation of this project, based on the increase of market share relative to competition, requires obtaining, over n periods (in our example we will assume 10), M monetary units (in our case 5.000 m.u..) additional to existing financial means. As we have pointed out, this financing may come from either internal or external sources. Among all the possible options, we have chosen the four mentioned above which appear to be the most suitable after a fist prior study for carrying out the proposals of the business. The model we are presenting intends to bring to light a differentiating aspect of the works based on the Bellman principle of optimisation, informer of dynamic programming. This is perhaps the all too frequent fact that in a process such as we have suggested the total M (5.000 m.u.) is not required at one and the same time but rhythmically over the economic horizon under consideration. This circumstance, which we consider as fundamental as a start out point for the study, conditions the possibilities of combining the different sources selected, because once a source or several sources have been chosen

[4] Gil Lafuente A M (November 1996) Estrategias secuenciales para la captación de medios financieros (in Spanish), proceedings of the III SIGEF Congress, vol. I Buenos Aires, Argentina, Communication paper 2.11.

for a determined amount of money (and we will assume that this selection was optimum at the time it was considered) the combination selected now forms a part of future combinations, because it has already been applied. In this way to optimums to be attained over coming periods have to be found by combining the previous optimum or optimums with the possibilities at the analysed time. All this conditions the architecture of the network, in the sense that in the neurons of a certain level will always exist, at least one entry dendrite proceeding from one or more neurons of the previous level (the dendrite representing the optimum). This outline, in a certain way, does not fail to be based on dynamic programming, although from a different perspective to that of the better-known models. For the effects of simplification in the calculations (and for no other reason) we are going to assume that both the rhythm of the financial needs and the necessary amounts are constant and for the same amount. Therefore, if requirements are M and periods n, what is required for each period is M/n (in our case 500 m.u.).

It is quite obvious that each of these sources of financing will mean a different cost according to the nature of each one, also depending on the amount requested by the business.

On the other hand, in order to generalise the scheme and bring into play situations normally taking place in reality, we are going to introduce a series of limitations.

In this respect we would like to point out that the least expensive sources of financing are also the most difficult to obtain and it is not always possible to obtain the desired amount. Therefore let us assume that there are a series of conditioners at the time of making our requests to the different sources of financing:

Contributions form shareholders via an increase in capital can only be made by means of bands of 2.000 monetary units. Therefore, if it were decided to acquire financing from this source, the minimum would be 2.000 monetary units and the maximum 4.000 monetary units. The cost for the use of this source of financing would be located somewhere in the interval [0,08; 0,11]. The first band has been affected with a coefficient $w1 = 1$; but to the second band, taking into consideration that all the studies necessary for getting to know the solvency and liquidity of the business were already done at the time of requesting the first band of financing, and that investors are now prepared to place their financial resources at more reduced prices, a coefficient of $w_2 = 0{,}8$ is affected.

The sense of the coefficients, w_{ij}, $j = 1,\ 2$ is the percentage reduction of the cost of availability of the financial means.

With respect to the request for a long-term credit, this can be done in bands of 500 monetary units. In this way, in the event of using said source of financing in its minimum expression, only 500 monetary units would be taken; in the event of basing the total necessary financing on this long-term credit then the whole ten bands would be taken, that is 5.000 monetary units.

The cost arising from the use of this source of financing would be located in the interval [0,07; 0,12] based on the situation of the financial markets.

Relative to this then certain coefficients would be affected to the different possible financing bands. We then would arrive at:

Bands	1º	2º	3º	4º	5º	6º	7º	8º	9º	10º
Coefficients	0,90	0,90	0,90	0,90	0,90	0,90	0,85	0,85	0,80	0,80

With regard to the financing by deferring payment to suppliers, this amount would be limited only to 1.000 monetary units, because it is not easy to arrange for suppliers to defer payments for an ever-longer period. Obviously this type of financing has zero costs because while the payment deferment lasts the business has available resources with which to finance activities over a short period.

In the case we are occupied with, if the business manages to delay payments for its supplies for 15 days it can obtain, as we have said, up to 1.000 monetary units of resources for the short-term.

Finally there exists the possibility of requesting a short-term credit from a banking entity. This type of financing can be done in bands of 500 monetary units. In the case of using this source of funds exclusively up to ten bands would be taken which would mean 5.000 monetary units. The resulting cost is estimated as contained within the interval [0,08; 0,10]. Likewise, relative to the coefficients that affect the different bands, these have been considered as follows:

Bands	1º	2º	3º	4º	5º	6º	7º	8º	9º	10º
Coefficients	1	1	1	0,90	0,85	0,85	0,85	0,80	0,75	0,7

Having reached this point we can now consider the parallelism existing between the operation of a neural network and the financial process that is being studied in our work.

If we look at Fig. 19.3 the elements described with a series of concepts relative to the field of neuro-mimesis can be easily appreciated. Therefore, each one of the possible sources of financing affected by their respective restrictions is represented by neurons that are interlaced in the form of a network thanks to the impulses they generate. These impulses are conditioned by the thresholds that act as a form of barrier so that the business can avoid those monetary bands that are too expensive in their financial process. Only those impulses are transmitted, that is to say, they will only move on to form part of the financial structure of the business, that form the monetary bands that are equal to or less than the tolerated cost.

Let us assume that the following denominations are assigned to the concepts listed below:

A = Source of financing from an increase in payment terms to suppliers.
B = Source of financing via an increase in capital from contributions from shareholders.
C = Source of financing by requesting a long-term loan
D = Source of financing by requesting a short-term loan.

$x_A = 0$ (cost of financing source A).
$x_B = [0{,}08;\ 0{,}11]$ (cost of financing source B).
$x_C = [0{,}07;\ 0{,}12]$ (cost of financing source C.
$x_D = [0{,}08;\ 0{,}10]$ (cost of financing source D).

On the other hand the coefficients that affect each band of each of the sources of financing are summarised below:

Bands	1º	2º	3º	4º	5º	6º	7º	8º	9º	10º
w_A	1	1	—	—	—	—	—	—	—	—
w_B	1	1	1	1	0,80	0,80	0,80	0,80	—	—
w_C	0,90	0,90	0,90	0,90	0,90	0,90	0,85	0,85	0,80	0,80
w_D	1	1	1	0,90	0,85	0,85	0,85	0,80	0,75	0,70

Following on with this scheme, it is now necessary to propose a maximum acceptable threshold for each monetary band according to which the business is prepared to assume the corresponding financial cost.

In this sense, and merely as a didactic example, the following may be established:

Bands (m.u.)	Threshold (θ) (Maximum cost)
500	[0,030; 0,050]
1.000	[0,040; 0,060]
1.500	[0,070; 0,110]
2.000	[0,145; 0,220]
2.500	[0,190; 0,330]
3.000	[0,270; 0,440]
3.500	[0,320; 0,540]
4.000	[0,380; 0,650]
4.500	[0,440; 0,750]
5.000	[0,580; 0,860]

These are accumulated financial costs that the business is not prepared, or cannot exceed. With this information we are now in a position to commence the sequential process by incorporating it to a neural structure.

19.3 Introduction of Neural Structures to the Financial Problem

In the first place we will move on to calculate the different costs, affected by their corresponding coefficients that are arrived at by the use of the different bands form each of the sources of financing. Of all the combinations possible in the first place we will take the one or ones the cost of which is equal to or less than the established threshold, and later to take the optimum (in our case the lowest).

If the process is initiated with the first level of neurons, that is the first monetary band corresponding to each source of financing, it will be necessary to find the one (or ones) the cost of which is equal to or lower than the threshold existing for the first band, this is [0,03; 0,05]. In this way if what is intended is to arrive at:

$$\underset{\sim}{\mathbf{p}}_{ij} = (w_{ij} \cdot \underset{\sim}{\mathbf{x}}_j) \leq \underset{\sim}{\boldsymbol{\theta}}_i \quad \text{for } i = \{1, 2, 3, \ldots, 10\}, j = \{A, B, C, D\}$$

We will arrive at:

$$\begin{aligned}
\underset{\sim}{\mathbf{p}}_{1j} &= \mathrm{Min}\{(w_{1A}(\cdot)\underset{\sim}{\mathbf{x}}_A),\ [w_{1B}(+)w_{2B}(+)w_{3B}(+)w_{4B}](\cdot)\underset{\sim}{\mathbf{x}}_B, (w_{1C}(\cdot)\underset{\sim}{\mathbf{x}}_C),\\
&\quad (w_{1D}(\cdot)\underset{\sim}{\mathbf{x}}_D)\} \leq \underset{\sim}{\boldsymbol{\theta}}_1\\
&= \mathrm{Min}\{(1 \cdot 0),\ (4 \cdot 1(\cdot)[0{,}08; 0{,}11]),\ (0{,}90(\cdot)[0{,}07; 0{,}12]),\ (1(\cdot)\\
&\quad [0{,}08; 0{,}10])\} \leq [0{,}03; 0{,}05]\\
&= \mathrm{Min}\{(0),\ [0{,}32; 0{,}44],\ [0{,}063; 0{,}108],\ [0{,}08; 0{,}10]\} \leq [0{,}03; 0{,}05]\\
&= [0; 0] = 0
\end{aligned}$$

As can be seen, what has been included is the first group of 4 monetary bands from source B, since its possible use is limited to considering blocks of 4 bands, that is, 2.000 m.u.

In the event of opting to use this financial source, the business would be obliged to assume the cost of 4 bands (2.000 m.u.) in order to proceed to use only 500 m.u.

Based on the calculations made it can be seen that the only monetary band that remains below the established threshold is the one relative to source A. Therefore, the first band will be the only one subject to consideration at the time of constructing the necessary financial structure for the business to be able to attain its objective of increasing its market share.

This will be as follows:

$$\underset{\sim}{\mathbf{p}}_{1A} = [0; 0] = 0$$

and in the structure, the only active neuron is the first of segment A.

Therefore in the first period financing source A is used in the necessary band $N/n = 500$. On considering the second period, this band, already "applied" will form a part of the corresponding combination.

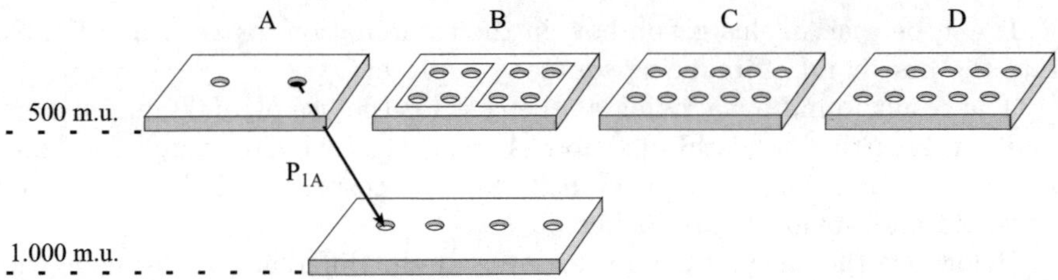

Fig. 19.4.

Once the first band at the lowest cost is arrived at, we move on to obtain financing of 1.000 m.u. (two bands) on the condition that their cost will be below the established threshold. In this way we arrive at:

$$\begin{aligned}
\underset{\sim}{\mathbf{p}}_{2j} &= \text{Min}\{\underset{\sim}{\mathbf{p}}_{1A}(+)w_{2A}(\cdot)\underset{\sim}{\mathbf{x}}_A),\ (\underset{\sim}{\mathbf{p}}_{1A}(+)[w_{1B}(+)w_{2B}(+)w_{3B}(+)w_{4B}](\cdot)\underset{\sim}{\mathbf{x}}_B,\\
&\quad (\underset{\sim}{\mathbf{p}}_{1A}(+)w_{1C}(\cdot)\underset{\sim}{\mathbf{x}}_C),\ \underset{\sim}{\mathbf{p}}_{1A}(+)w_{1D}(\cdot)\underset{\sim}{\mathbf{x}}_D)\} \leq \underset{\sim}{\boldsymbol{\theta}}_2\\
&= \text{Min}\{(1\cdot 0),\ (4\cdot 1(\cdot)[0{,}08;0{,}11]),\ (0{,}90(\cdot)[0{,}07;0{,}12])\ ,\\
&\quad (1(\cdot)[0{,}08;0{,}10])\} \leq [0{,}04;0{,}06]\\
&= \text{Min}\{(0), [0{,}32;0{,}44], [0{,}063;0{,}108], [0{,}08;0{,}10]\} \leq [0{,}04;0{,}06]\\
&= [0;0] = 0
\end{aligned}$$

Just as can be seen, the same considerations can be made for financing source B and as has been expressed above.

We arrive at:

$$\underset{\sim}{\mathbf{p}}_{2A} = [0;0] = 0$$

In the structure, even though potentials arrive from several neurons in the end only one of these will be active. Let us take a look at this in the following graph:

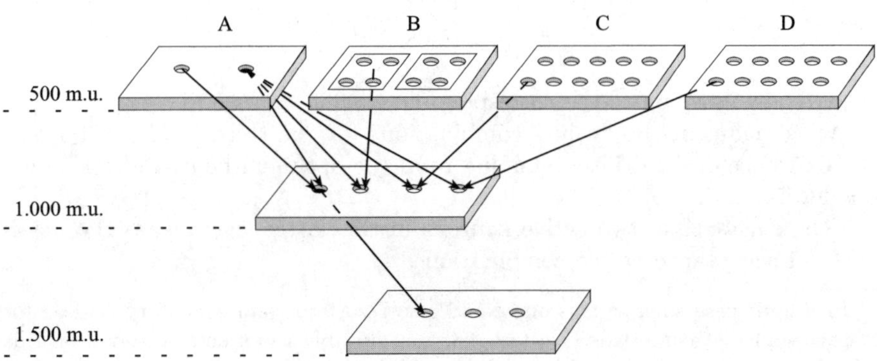

Fig. 19.5.

It can be seen in this graph how in the combination for arriving at 1.000 m.u. the first band of financing source A is present.

The result found for arriving at the cost for the use of 1.000 m.u. means, finally, taking two bands from source A the cost of which is zero. The other possible combinations lead to the fact that the costs exceed the established threshold and obviously are higher.

Below are the calculations for arriving at the different cost for using the remaining monetary bands. These are:

$$\begin{aligned}
\underset{\sim}{\mathbf{p}}_{3j} &= \mathrm{Min}\{(\underset{\sim}{\mathbf{p}}_{2A}(+)w_{1C}(\cdot)\underset{\sim}{\mathbf{x}}_C), (\underset{\sim}{\mathbf{p}}_{2A}(+)[w_{1B}(+)w_{2B}(+)w_{3B}(+)w_{4B}](\cdot)\underset{\sim}{\mathbf{x}}_B, \\
&\quad (\underset{\sim}{\mathbf{p}}_{2A}(+)w_{1D}(\cdot)\underset{\sim}{\mathbf{x}}_D)\} \le \underset{\sim}{\boldsymbol{\theta}}_3 \\
&= \mathrm{Min}\{(1\cdot 0), (4\cdot 1(\cdot)[0{,}08; 0{,}11]), (0, 90(\cdot)[0{,}07; 0{,}12]), (1(\cdot) \\
&\quad [0{,}08; 0{,}10])\} \le [0{,}04; 0{,}06] \\
&= \mathrm{Min}\{[0{,}32; 0{,}44], [0{,}063; 0{,}108], [0{,}08; 0{,}10]\} \le [0{,}07; 0{,}11] \\
&= [0{,}063; 0{,}108]
\end{aligned}$$

This result, allows us to highlight the fact that one neuron provokes a cost $\underset{\sim}{\mathbf{p}}_{3AC}$ that is clearly lower [0,07; 0,11] and therefore is active, while there exists another with a cost of [0,08; 0,10] the comparison of which with the threshold warrants consideration.

In fact, although the lower extreme $\underset{\sim}{\boldsymbol{\theta}}_3$ is less than that of $\underset{\sim}{\mathbf{p}}_{3AD}$ the opposite occurs with the upper extreme. From here stems the need to resort to a criterion that can be the middle points of the intervals. Thus:

$$\overline{\underset{\sim}{\boldsymbol{\theta}}_3} = \frac{0{,}07 + 0{,}11}{2} = 0{,}09 \qquad \overline{\underset{\sim}{\mathbf{p}}_{3AD}} = \frac{0{,}08 + 0{,}10}{2} = 0{,}09$$

This equality would appear to authorise the activity of the neuron, as if we accepted it[5]. In this way the following can be accepted.

$$\begin{aligned}
\underset{\sim}{\mathbf{p}}_{3AC} &= [0{,}063; 0, 108] \\
\underset{\sim}{\mathbf{p}}_{3AD} &= [0{,}08; 0{,}10]
\end{aligned}$$

We now move on to the corresponding graph:

Once again in the "valid" combinations for the third level can be seen the existence of dendrites stemming from the optimum neuron of the previous level.

The existence of two active neurons in this latter level means the possibility of acceptance of two combinations:

[5] In a limit case such as this one could also resort to complementary criteria for reviewing both intervals. We feel it is not suitable to extend ourselves on this subject at this juncture as it has been widely treated on other occasions.

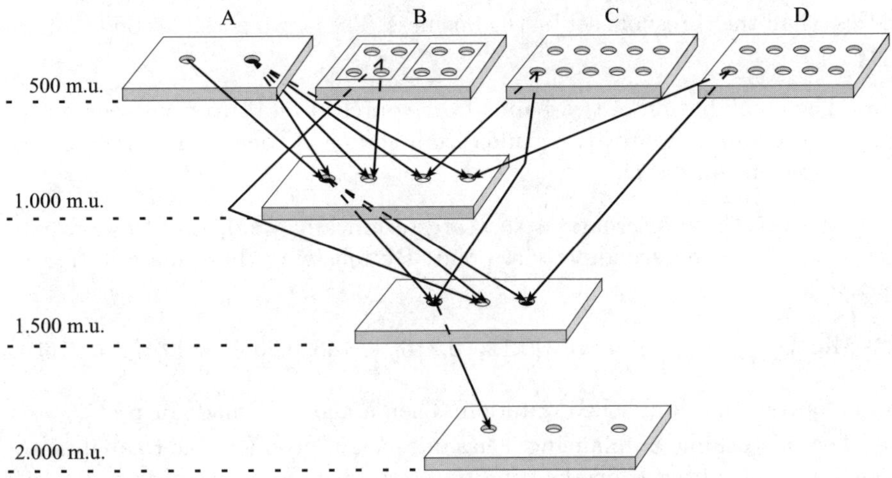

Fig. 19.6.

(a) Use of 1.000 m.u. from source A and 500 m.u. from source C.
(b) Use of 1.000 m.u. from source A and 500 m.u. from source D.

If the criterion of minimisation is used obviously at this level of resources combination (a) would be chosen. Even though technically the activity of the neuron representing combination (b) is accepted, this will have no interest from a financial point of view, since any combination with other sources will always provide amounts that are higher than those proceeding from (a), except where the use of combination (b) were to give some compensation that would advise this alternative, even at a higher cost.

Once again resort has been made to the use of source B due to the fact that it is necessary to take at least 2.000 m.u. of resources in one block.

Just as was done previously, we are going to calculate the cost arising from the use of 2.000 m.u. (four bands) in order to get to know which combination remains less than the established threshold. In this way:

$$\begin{aligned}
\underset{\sim}{\mathbf{p}}_{4j} &= \mathrm{Min}\{(\underset{\sim}{\mathbf{p}}_{3AC}(+)[w_{1B}(+)w_{2B}(+)w_{3B}(+)w_{4B}(\cdot)\underset{\sim}{\mathbf{x}}_B, (\underset{\sim}{\mathbf{p}}_{3AC}(+)w_{2C}(\cdot)\underset{\sim}{\mathbf{x}}_C),\\
&\quad (\underset{\sim}{\mathbf{p}}_{3AC}(+)w_{1D}(\cdot)\underset{\sim}{\mathbf{x}}_D)\} \le \underset{\sim}{\boldsymbol{\theta}}_4\\
&= \mathrm{Min}[0{,}063; 0{,}108](+)(4 \cdot 1(\cdot)[0{,}08; 0{,}11]), [0{,}063; 0{,}108]\\
&\quad (+)(0{,}90(\cdot)[0{,}07; 0{,}12]), [0{,}063; 0{,}108](+)(1(\cdot)\\
&\quad [0{,}08; 0{,}10]) \le [0{,}145; , 0{,}22]\\
&= \mathrm{Min}\{[0{,}383; 0{,}548], [0{,}126; 0{,}216], [0{,}143; 0{,}208]\} \le [0{,}145; 0{,}22]\\
&= [0{,}126; 0{,}216]
\end{aligned}$$

From the result arrived at the fact can be seen that for obtaining 2000 m.u. there are two different combinations of sources of financing the cost of which is less than the threshold set by the business. These two possible combinations are:

(a) The combination of two bands from source A and two from source C.
(b) The combination of two bands from source A, one from source C and one from source D.

Between these alternatives that are technically valid, the decision must be taken from the economic viewpoint. By following the same criterion we arrive at:

$$\text{Min}\{\underset{\sim}{\mathbf{p}}_{4\text{AC}}, \underset{\sim}{\mathbf{p}}_{4\text{ACD}}\} = \text{Min}\{[0{,}126; 0{,}216], [0{,}143; 0{,}208]\} = [0{,}126; 0{,}216]$$

on adopting the established criterion. Then a choice is made of $\underset{\sim}{\mathbf{p}}_{4AD}$.

For this volume of financing it has once again been feasible to analyse the possibility of source B on the condition of taking the four bands at one and the same time. Nevertheless its high cost in relation to the threshold makes it necessary to reject this financing alternative as happened with the previous situations.

In the following graph we now incorporate the corresponding level and its respective connections.

We continue to calculate the combinations that give rise to the different costs for the use of 2.500 m.u. For this we will continue to take into account the block formed by the four bands of source B in combination with the optimum of the previous level, even though it exceeds the figure required for 2.500 m.u. Now, as can be seen in the calculations, only the cost of this band already exceeds the threshold. For this the use of this source of financing will once more be avoided, even in combination with any of the others. In this way we arrive at:

$$\begin{aligned}
\underset{\sim}{\mathbf{p}}_{5j} &= \text{Min}\{(\underset{\sim}{\mathbf{p}}_{4AC}(+)w_{3C}(\cdot)\underset{\sim}{\mathbf{x}}_C), (\underset{\sim}{\mathbf{p}}_{4AC}(+)w_{1D}(\cdot)\underset{\sim}{\mathbf{x}}_D), \underset{\sim}{\mathbf{p}}_{4AC} \\
&\quad (+)[w_{1B}(+)w_{2B}(+)w_{3B}(+)w_{4B}](\cdot)\underset{\sim}{\mathbf{x}}_B\} \leq \underset{\sim}{\boldsymbol{\theta}}_5 \\
&= \text{Min}\{[0{,}126; 0{,}216](+)(0,90(\cdot)[0{,}07; 0{,}12]), [0{,}126; 0{,}216] \\
&\quad (+)(1(\cdot)[0{,}08; 0{,}10]), [0{,}126; 0{,}216](+)[0{,}32; 0{,}44])\} \leq [0{,}19; 0{,}33] \\
&= \text{Min}\{[0{,}189; 0{,}324], [0{,}206; 0{,}316], [0{,}446; 0{,}656]\} \leq [0{,}19; 0{,}33] \\
&= [0{,}189; 0{,}324]
\end{aligned}$$

It will be seen that the interval $[0{,}206; 0{,}316]$ could be considered as within the threshold $[0{,}19; 0{,}33]$ for the same reasoning stated above. Nevertheless, for strictly financing effects we are opting for the minimum possible cost. In this way, the result arrived at shows us that the only combination of monetary bands, which includes all the characteristics, previously established is the one that considers 1.000 m.u. from source A and 1.500 m.u. from source C. Any other combination means a higher cost than that established for this level.

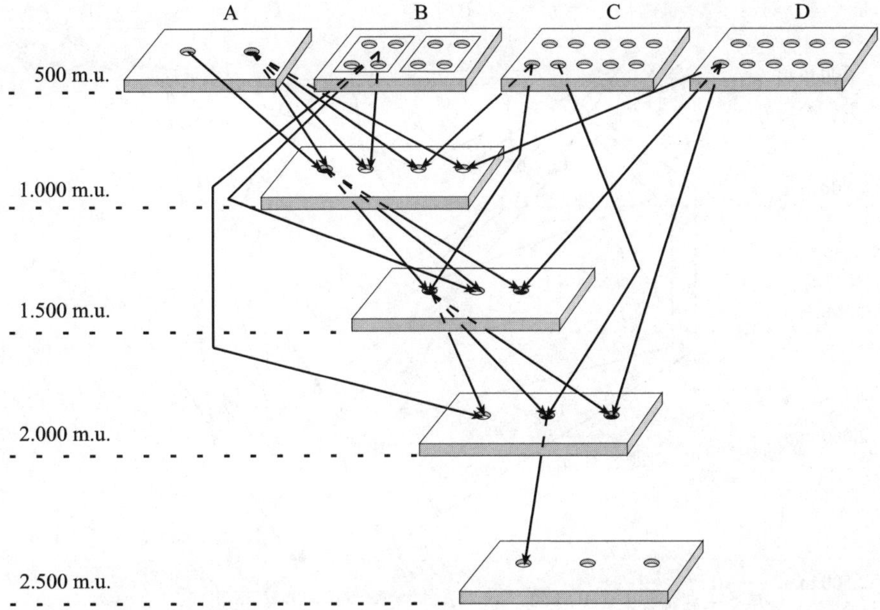

Fig. 19.7.

Therefore we have as the chosen combination the one with the following cost:

$$\underset{\sim}{\mathbf{p}}_{5j} = [0{,}189; 0{,}324]$$

We can see the neural architecture up to the fifth level (see Fig. 19.8).

In order to be able to see this in a better way the process has been represented from the previous figures that show the different levels of financing in "layers" of neurons, which become interconnected based on the impulses that leave each one, on the weighting affecting them and on the thresholds that act as barriers.

We could go on developing the scheme by passing on to the next levels and depleting the process. However convinced that by reiterating the calculations we would only fill more pages without contributing anything new, we feel it is suitable to stop at this point and make certain interesting considerations.

19.4 Final Considerations

The first of these refers to the hypothesis underlying in the reasoning that has been followed. It will not have escaped the notice of the reader that when one has arrived at a single possible combination or the optimum combination for each level, this combination becomes obligatory in the following levels,

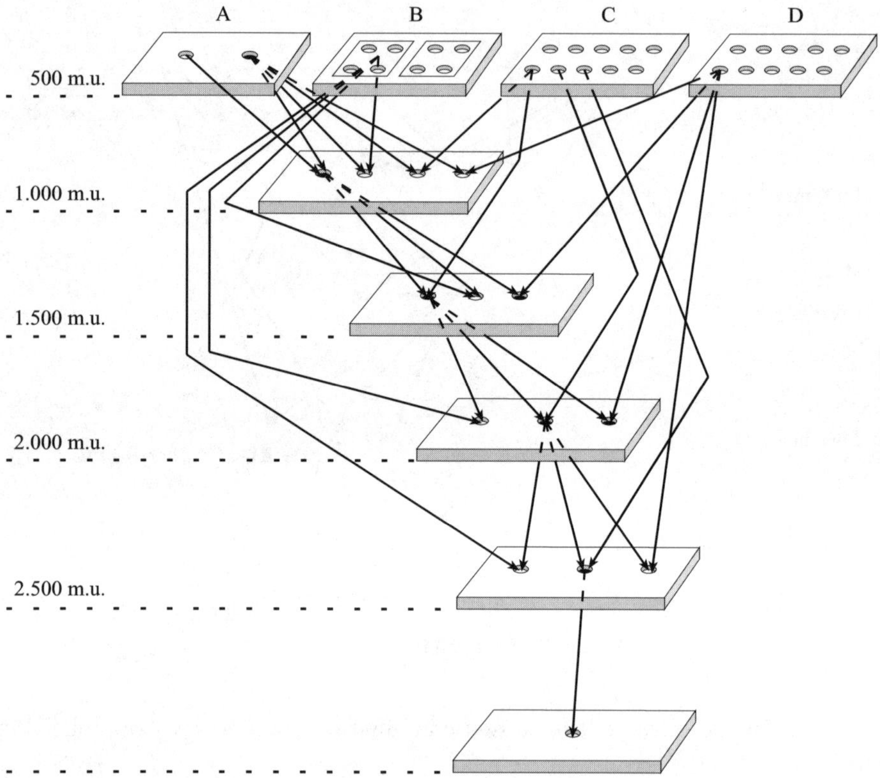

Fig. 19.8.

without any possibility of abandoning the same in order to seek possible combinations from the previous levels. Thus, for example, on considering the fifth level, and independently of the results that could be arrived at, certain combinations have not been taken into consideration such as the following: 2 bands from A and 1 from D, with a cost of $\underset{\sim}{\mathbf{p}}_{3AD}$ (technically valid) and 2 bands from C. All of which is in accordance with the essence of the problem under consideration and included in the initial hypothesis.

The second, affects the difficulties of establishing automatisms in the process a part from those arising from the intervention of the thresholds and optimisations for each level. We are referring principally to facts such as the limitation to $2N/n$ bands in B. We feel that these difficulties are reasonably overcome in the model we have formulated.

In the third, we should like to state that, these and other aspects have not stopped us drawing up this process, by using, in a certain manner, Bellman's principle, informer of dynamic programming. Now then, if the problem that is subject to study were to allow for starting out from a different basis, the use of this important tool for taking decisions would acquire certain nuances that are substantially different and a path would be used that is parallel to the one used in the solution to other problems[6]. In fact, if the clause according to which the financial needs are not totally N in an obligatory manner but that N is the number necessary at any given time and the objective sought is to determine for each band N/n, $2N/n$, $3N/n, \ldots, (n-1)N/n$, N, which is the optimum combination of bands and sources, then we would require to elaborate or use another model[7]. The problem in this case would be essentially different.

Finally we have attempted to present a scheme, with the aid of basic concepts of neural graphs, that is suitable for the treatment of the all too frequent problem that arises in business and institutional financial activities in general emanating from acquiring from different and compatible sources, the financial means for covering the needs arising from the start up of a project that requires the availability of monetary means at certain predetermined moments in time. In order to explain our model in more colloquial terms we have resorted to a numerical example convinced that its generalisation represents no problem whatsoever.

---

[6] In this respect see: Kaufmann A and Gil Aluja J (1987), Técnicas operativass de gestión para el tratamiento de la incertidumbre (in Spanish). (Ed) Hispano-Europea, Barcelona, Spain, pp. 393–405.

[7] Gil Aluja J (November 1997), La colocación óptima de los recursos financieros. El modelo ISFUNE (Investment Selection for Fuzzy Neural Networks), Proceedings of the 1st SIGEF Congress, Reus, Spain, Vol 1 pp. 9–44.

20 Assignment of Financial Means in Business

20.1 Approach to the Problem

Let us assume that in a determined moment a business is faced with certain financial requirements that are represented by the following N set:

$$N = \{N_1, N_2, N_3, N_4, N_5\}$$

where:

N_1 = finance for the annual purchase of raw materials
N_2 = finance for the acquisition of a plot of land next to current facilities
N_3 = finance for start up of an in-house transport system and collection of materials and products
N_4 = special finance for exporting products to a non UE country
N_5 = short-term finance for amounts fluctuating between 50 and 200 million for 1 and 3 months for sales made and not collected

These financial requirements have the following implicit characteristics:

$$C = \{C_1, C_2, C_3, C_4\}$$

where:

C_1 = price
C_2 = possibilities for renewal
C_3 = payback period
C_4 = commissions for opening and no draw down

On the other hand the market offers 6 financial products that could be suitable for covering the requirements of the business:

$$P = \{P_1, P_2, P_3, P_4, P_5, P_6\}$$

where:

P_1 = commercial discount line at "Fort Fait" rate
P_2 = credit for financing working capital

Anna Maria Gil-Lafuente: *Fuzzy Logic in Financial Analysis*, StudFuzz **175**, 359–381 (2005)
www.springerlink.com

P_3 = medium term mortgage loan
P_4 = short-term loan
P_5 = commercial discount line at scaled rates
P_6 = documentary credit with financing at 1 year non renewable

Therefore the need arises for assigning the products offered by the market to the financing requirements of the business in the most suitable way.

With regard to each one of the characteristics and for the effects of being able to establish a relation between the financing requirements of the business and the products offered by the market, we should now define, the hendecagonal scale which we will be using from here on, semantically. Thus for characteristic C_1, defined as the price for each financial element, we will have the following semantic meanings:

0	inexpensive
0,1	practically inexpensive
0,2	nearly inexpensive
0,3	close to inexpensive
0,4	more inexpensive than expensive
0,5	as inexpensive as expensive
0,6	more expensive than inexpensive
0,7	close to expensive
0,8	nearly expensive
0,9	practically expensive
1	expensive

Likewise, for characteristic C_2, that represents the possibility for renewal, we have:

0	totally renewable
0,1	practically renewable
0,2	easily renewable
0,3	quite easy to renew
0,4	more renewable than not renewable
0,5	as renewable as not renewable
0,6	more non renewable than renewable
0,7	quite difficult to renew
0,8	difficult to renew
0,9	practically non renewable
1	impossible to renew

Likewise we proceed to establish the semantic scale for characteristic $C3$, which defines the possibilities relative to obtaining a payback period as long drawn out as possible in time. This then will be:

0	short-term payback
0,1	payback at practically short-term
0,2	payback nearly at the short-term
0,3	payback close to the short-term
0,4	payback more at medium than short-term
0,5	payback at the medium term
0,6	payback more at the medium than long-term
0,7	payback close to the long-term
0,8	payback nearly at long-term
0,9	payback practically at the long-term
1	payback at long-term

	C_1	C_2	C_3	C_4
$\underset{\sim}{N}_1 =$	0,1	0,6	0,1	0,8
$\underset{\sim}{N}_2 =$	0	0,2	0,2	0,1
$\underset{\sim}{N}_3 =$	0,2	0,4	0,1	0,2
$\underset{\sim}{N}_4 =$	0	0,1	0,4	0,4
$\underset{\sim}{N}_5 =$	0,1	0,3	0,1	0,1

This means that, for example, that for N_1, defined as finance for the annual purchase of raw materials what is required is:

1. Nearly inexpensive money. The internal analysis of the business states that, relative to this characteristic, the desired price is a basic element at the time of assigning a financial product from among those offered by the market. In order for it to be accepted by the business, the price of the financial product must be very low.
2. Relative to the second characteristic, that is, the possibilities for renewal, it should be pointed out that this is an element of little importance at the time of analysing the financial products that are most suitable to the requirements of the business. This is why a representative level of a product with more possibilities of not being renewed than of being renewed would be acceptable to the business.
3. Relative to the payback period, the third characteristic with which the products and requirements have been defined, this is also an important item. We find ourselves faced with the need of resources at the short-term and therefore it is necessary that the payback of the principle is also done at the short-term in order to avoid high interest rate payments.

4. Finally, with regard to the fourth characteristic, that is, the existence of commissions, it can be seen that we are here faced with an unimportant item. In this context, the business would be ready to accept a financial product from the market that would include nearly the maximum commissions.

These valuations assigned to N_1 represent the maximum levels acceptable to the business. Therefore, for example those financial products the valuation of which was to exceed 0,1 would not be accepted for N_1. Likewise those financial products with a level for renewal greater than 0,6, nor a payback level greater than 0,1 or, finally, a valuation for commissions greater than 0,8 will not be accepted.

These criteria are applied in like fashion for the other financial requirements of the business (N_2, N_3, N_4 and N_5) relative to the previously defined characteristics.

Following on a profile is determined for each one of the financial products offered by the market.

	C_1	C_2	C_3	C_4
$\underset{\sim}{P}_1 =$	0,1	0,6	0,1	0,8
$\underset{\sim}{P}_2 =$	0	0,2	0,2	0,1
$\underset{\sim}{P}_3 =$	0,2	0,4	0,1	0,2
$\underset{\sim}{P}_4 =$	0	0,1	0,4	0,4
$\underset{\sim}{P}_5 =$	0,1	0,3	0,1	0,1

If we analyse, for example, the characteristics relative to P_1, that is, the financial product that consists of a commercial discount of the Fort Fait type, we will observe the following:

1. relative to the price of the money we find a product that is "nearly inexpensive".
2. relative to the possibilities of renewal P_1 is more renewable than not renewable.
3. relative to the payback period, this is as near to the short-term as to the long-term.
4. finally, relative to the commissions applied to P_1, we can see that practically the maximum commissions are applied.

This analysis can be done following the same criteria for all the financial products P_2, P_3, P_4, P_5 and P_6.

In short, we have broken down both the financial needs and the financial products into the four selected characteristics (price, possibilities of renewal, payback period and application of commissions) in order to assign valuations for later getting to know which of the financial products is better suited to each one of the financial requirements of the business.

20.2 A First Approximation: The Coefficient of Qualification

Having outlined the problem it is now necessary to find a series of instruments that allow us to relate the products offered by the market with the financial requirements of the business, based on the degree of compliance of their characteristics.

Therefore, from among all the possibilities that are offered by multivalent analysis, we have opted in a first stage for the use of a coefficient that we have designated as the "coefficient of qualification". This is based on the progressive acceptance of those characteristics corresponding to the financial products the valuations of which remain within the threshold set by the financial needs of the business. Whenever any characteristic exceeds tolerable levels, we then proceed to gradually penalise the corresponding characteristic. Let us develop this idea. We will call:

$\mu_{C_b}(P_i) \in [0; 1]$ the value of the membership function relative to financial product P_i and characteristic C_b.

$\mu_{C_b}(N_j) \in [0; 1]$ the value of the membership function relative to financial requirement N_j and characteristic C_b.

In this way what we arrive at is:

1. When the characteristic of the financial product is located within the limits established for the financial requirement of the business, that is:

$$\mu_{C_b}(P_i) \preceq \mu_{C_b}(N_j)$$

Then the characteristic is accepted with a value determined by $(1 - \mu_{C_b}(N_j))$. We can see that, when the level of requirement is very high as is the case with N_1 with characteristic C_1, when a financial product satisfies the demand with a specific characteristic, we give this financial product greater importance than in other cases in which the level of requirement is lower.
2. If the characteristic of the financial product remains outside the threshold established for the financial requirement, that is:

$$\mu_{C_b}(P_i) \succ \mu_{C_b}(N_j)$$

We proceed to progressively penalise the corresponding financial product by assigning a valuation for this characteristic of:

$$\overline{1 \wedge (\mu_{c_b}(N_j) + \mu_{c_b}(P_i))}$$

This means penalising to a greater extent those financial products that do not reach the levels of the financial requirements for which the requirement is small and, on the other hand those financial products that do not exceed the levels established by certain financial requirements with a greater need are penalised less.

Let us now compare $\underset{\sim}{\mathbf{N}}_1$ and $\underset{\sim}{\mathbf{P}}_1$ relative to each one of the aforementioned characteristics:

For C_1:

$$\mu_{C_1}(P_1) \succ \mu_{C_1}(N_1)$$

because:

$$0{,}8 \succ 0{,}1$$

This is a case for penalisation therefore the following formula is applied:

$$\overline{1 \wedge (\mu_{C_1}(N_1) + \mu_{C_1}(P_1))}$$

which will give rise to:

$$\overline{1 \wedge (0{,}1 + 0{,}8)} = \overline{0{,}9} = 0{,}1$$

For C2:

$$\mu_{C_2}(P_1) \preceq \mu_{C_2}(N_1)$$

given that:

$$0{,}4 \preceq 0{,}6$$

therefore the following is taken:

$$(1 - \mu_{C_2}(N_1))$$

which will give rise to

$$(1 - 0{,}6) = 0{,}4$$

For C3:

$$\mu_{C_3}(P_1) \succ \mu_{C_3}(N_1)$$

given that:

$$0{,}5 \succ 0{,}1$$

here we proceed to penalise and therefore the following is applied:

$$\overline{1 \wedge (\mu_{C_3}(N_1) + \mu_{C_3}(P_1))}$$

which will give rise to:

$$\overline{1 \wedge (0{,}1 + 0{,}5)} = \overline{0{,}6} = 0{,}4$$

For C4:

$$\mu_{C_4}(P_1) \succ \mu_{C_4}(N_1)$$

given that:

$$0{,}9 \succ 0{,}8$$

the corresponding valuation is assigned, which is as a result of the following calculation:

$$\overline{1 \wedge (\mu_{C_4}(N_1) + \mu_{C_4}(P_1))}$$

giving rise to:

$$\overline{1 \wedge (0{,}8 + 0{,}9)} = \overline{1} = 0$$

Finally a calculation is made of the level of acceptance of financial product P_1 relative to financial requirement N_1, by carrying out a weighted addition of the values obtained for each characteristic. The result then is:

$$Q(\underset{\sim}{\mathbf{N}}_1, \underset{\sim}{\mathbf{P}}_1) = \frac{0{,}1(+)0{,}4(+)0{,}4(+)0}{4} = \frac{0{,}9}{4} = 0{,}225$$

$$\begin{array}{lll}
Q(\underset{\sim}{\mathbf{P}}_1, \underset{\sim}{\mathbf{N}}_1) = 0{,}225 & Q(\underset{\sim}{\mathbf{P}}_3, \underset{\sim}{\mathbf{N}}_1) = 0{,}200 & Q(\underset{\sim}{\mathbf{P}}_5, \underset{\sim}{\mathbf{N}}_1) = 0{,}300 \\
Q(\underset{\sim}{\mathbf{P}}_1, \underset{\sim}{\mathbf{N}}_2) = 0{,}225 & Q(\underset{\sim}{\mathbf{P}}_3, \underset{\sim}{\mathbf{N}}_2) = 0{,}300 & Q(\underset{\sim}{\mathbf{P}}_5, \underset{\sim}{\mathbf{N}}_2) = 0{,}300 \\
Q(\underset{\sim}{\mathbf{P}}_1, \underset{\sim}{\mathbf{N}}_3) = 0{,}250 & Q(\underset{\sim}{\mathbf{P}}_3, \underset{\sim}{\mathbf{N}}_3) = 0{,}225 & Q(\underset{\sim}{\mathbf{P}}_5, \underset{\sim}{\mathbf{N}}_3) = 0{,}325 \\
Q(\underset{\sim}{\mathbf{P}}_1, \underset{\sim}{\mathbf{N}}_4) = 0{,}200 & Q(\underset{\sim}{\mathbf{P}}_3, \underset{\sim}{\mathbf{N}}_4) = 0{,}250 & Q(\underset{\sim}{\mathbf{P}}_5, \underset{\sim}{\mathbf{N}}_4) = 0{,}375 \\
Q(\underset{\sim}{\mathbf{P}}_1, \underset{\sim}{\mathbf{N}}_5) = 0{,}200 & Q(\underset{\sim}{\mathbf{P}}_3, \underset{\sim}{\mathbf{N}}_5) = 0{,}275 & Q(\underset{\sim}{\mathbf{P}}_5, \underset{\sim}{\mathbf{N}}_5) = 0{,}275 \\
Q(\underset{\sim}{\mathbf{P}}_2, \underset{\sim}{\mathbf{N}}_1) = 0{,}200 & Q(\underset{\sim}{\mathbf{P}}_4, \underset{\sim}{\mathbf{N}}_1) = 0{,}325 & Q(\underset{\sim}{\mathbf{P}}_6, \underset{\sim}{\mathbf{N}}_1) = 0{,}300 \\
Q(\underset{\sim}{\mathbf{P}}_2, \underset{\sim}{\mathbf{N}}_2) = 0{,}250 & Q(\underset{\sim}{\mathbf{P}}_4, \underset{\sim}{\mathbf{N}}_2) = 0{,}425 & Q(\underset{\sim}{\mathbf{P}}_6, \underset{\sim}{\mathbf{N}}_2) = 0{,}450 \\
Q(\underset{\sim}{\mathbf{P}}_2, \underset{\sim}{\mathbf{N}}_3) = 0{,}200 & Q(\underset{\sim}{\mathbf{P}}_4, \underset{\sim}{\mathbf{N}}_3) = 0{,}325 & Q(\underset{\sim}{\mathbf{P}}_6, \underset{\sim}{\mathbf{N}}_3) = 0{,}350 \\
Q(\underset{\sim}{\mathbf{P}}_2, \underset{\sim}{\mathbf{N}}_4) = 0{,}225 & Q(\underset{\sim}{\mathbf{P}}_4, \underset{\sim}{\mathbf{N}}_4) = 0{,}375 & Q(\underset{\sim}{\mathbf{P}}_6, \underset{\sim}{\mathbf{N}}_4) = 0{,}450 \\
Q(\underset{\sim}{\mathbf{P}}_2, \underset{\sim}{\mathbf{N}}_5) = 0{,}225 & Q(\underset{\sim}{\mathbf{P}}_4, \underset{\sim}{\mathbf{N}}_5) = 0{,}375 & Q(\underset{\sim}{\mathbf{P}}_6, \underset{\sim}{\mathbf{N}}_5) = 0{,}425
\end{array}$$

Once the coefficient of qualification is found for each one of the financial requirements we proceed to the assignment of each product to each requirement in such a way that the higher the coefficient arrived at the greater their acceptance.

The main characteristic of the coefficient of qualification is that, on the contrary to the other methods normally based on the proportional valuation of the differences or distances, whichever the case, existing between financial products and requirements, it includes the double property expressed below:

1. When the levels required are reached or exceeded the characteristic that is the object of analysis is awarded with the particularity that a greater specific weight is assigned when the requirements are large and in spite of the fact that these are exceeded. On the other hand, when we are dealing with levels that are easy to exceed and in spite of this they are not reached, then a small specific weight is assigned.

2. When the required level is not reached, a system of penalisation is established that is not proportional but progressive, based on whether the objective is easy or difficult to attain. In the first case (easily attainable objective), the fact of not exceeding it would mean heavy penalising. On the other hand, for a difficult to attain objective the fact of not exceeding it would mean a very small penalty.

On the other hand and without trying to draw this work out too long it should be pointed out that other methods exist that are very useful for treating determined problems and among these we could mention the Hamming distance, the adequacy coefficient and the Hungarian algorithm which we will develop further on.

Thus, whilst the Hamming distance implies proportional penalisation both in the event the characteristic were not to reach the desired levels and if they were exceeded, preventing the analysis of certain elements within the problem of assignment, the use of the coefficient of adequacy allows for total admittance of the characteristics that reach the required levels without awarding in a greater or lesser degree the effort carried out in order to exceed the objectives.

20.3 Assignment of Valuations to the Characteristics of the Financial Elements

In order to continue with the process of assignment it is necessary to consider each characteristic in relation to each one of the financial requirements. For this it will be necessary to normalise, in a statistical sense[1], the established values in order later to carry out a convex weighting[2]. In this way we will have to divide each level by the sum of the established levels and this must be done for each of the financial requirements.

As a prior step to the calculation, and for effects of greater visualisation of the results, we show the initial data with its corresponding operations in the form of matrices using the coefficient of qualification.

For products P_1, P_2, P_3, P_4, P_5 and P_6 we will have:

$[\underset{\sim}{P}_1] =$

	C_1	C_2	C_3	C_4	
N_1	0,1	0,4	0,4	0	0,225
N_2	0,2	0,4	0,3	0	0,225
N_3	0	0,6	0,4	0	0,250
N_4	0,2	0,5	0,1	0	0,200
N_5	0,1	0,3	0,4	0	0,200

$[\underset{\sim}{P}_2] =$

	C_1	C_2	C_3	C_4	
N_1	0	0	0,6	0,2	0,200
N_2	0,1	0,1	0,5	0,3	0,250
N_3	0	0	0,6	0,2	0,200
N_4	0,1	0,2	0,6	0	0,225
N_5	0	0	0,6	0,3	0,225

[1] We should remember that in statistical normalisation the sum of all the elements of each row must be equal to the unit.

[2] In convex weighting a determined weight is assigned to each characteristic in order to give it a different relative importance in such a way that the sum of the weights is equal to the unit.

$[\underset{\sim}{P}_3] =$

	C_1	C_2	C_3	C_4	
N_1	0,2	0,4	0,2	0	0,200
N_2	0,3	0,8	0,1	0	0,300
N_3	0,1	0,6	0,2	0	0,225
N_4	0,3	0,7	0	0	0,250
N_5	0,2	0,7	0,2	0	0,275

$[\underset{\sim}{P}_4] =$

	C_1	C_2	C_3	C_4	
N_1	0	0,4	0,7	0,2	0,325
N_2	0	0,8	0,8	0,1	0,425
N_3	0	0,6	0,7	0	0,325
N_4	0	0,9	0,6	0	0,375
N_5	0	0,7	0,7	0,1	0,375

$[\underset{\sim}{P}_5] =$

	C_1	C_2	C_3	C_4	
N_1	0,3	0,4	0,5	0	0,300
N_2	0,4	0,4	0,4	0	0,300
N_3	0,2	0,6	0,5	0	0,325
N_4	0,4	0,5	0,6	0	0,375
N_5	0,3	0,3	0,5	0	0,275

$[\underset{\sim}{P}_6] =$

	C_1	C_2	C_3	C_4	
N_1	0,2	0,4	0,6	0	0,300
N_2	0,3	0,8	0,5	0,2	0,450
N_3	0,1	0,6	0,6	0,1	0,350
N_4	0,3	0,9	0,6	0	0,450
N_5	0,2	0,7	0,6	0,2	0,425

We can see that each matrix $[\underset{\sim}{\mathbf{P}}_i]$ shows the degree of closeness existing between the characteristic of the requirements and those required by product P_i itself.

In the simplest case, if it were to be considered that all the characteristics were equally important, the above matrices could be summarised into a single matrix in which the elements represented the coefficient of qualification[3]. For this it would be sufficient to add each row of matrices $[\underset{\sim}{\mathbf{P}}_j]$ and divide them by 4. The result has been placed in the columns to the left of each matrix. In this way each matrix is summarised in a vector, the joining of which will give rise to the following matrix $[\underset{\sim}{\mathbf{P}}]$.

Therefore we would arrive at:

$[\underset{\sim}{P}] =$

	P_1	P_2	P_3	P_4	P_5	P_6
N_1	0,225	0,200	0,200	0,325	0,300	0,300
N_2	0,225	0,250	0,300	0,425	0,300	0,450
N_3	0,250	0,200	0,225	0,325	0,325	0,350
N_4	0,200	0,225	0,250	0,375	0,375	0,450
N_5	0,200	0,225	0,275	0,375	0,275	0,425

This matrix[4] brings to light the level of adequacy of the financial products to the requirements of the business. It should be taken into account that the nearer to 1 that the valuations arrived at are, the greater the level of similarity

[3] What we have called the coefficient of qualification up to this point could have a denomination, which would not exactly adapt itself to its arithmetical significance, from now on we will to call it coefficient GIL.

[4] If matrix $[P]$ were to have been arrived at by the use of any method based on distances in stead of using the coefficient GIL we would have to find $\overline{[P]}$ in order to operate with levels of resemblance.

and, therefore, on the other hand, the nearer to 0 they are, the lower will be the level of resemblance.

In this case we can see that in the matrix that was arrived at before there is no element the valuation of which exceeds level 0,5. This brings to light the fact that the degree of compliance by the financial products of the required characteristics by the financial requirements (price, renewal, pay back period and commissions) is relatively low.

In order that in matrix $[\underset{\sim}{\mathbf{P}}]$ there were to appear elements that exceed level 0,5 it would be necessary for the characteristics of the financial products to reach or exceed the levels required by the financial requirements. In this respect we can see in the outline of this work that this requirement is not complied with on many occasions.

As an example we can analyse, starting out from the matrix of adequacy or qualifications $[\underset{\sim}{\mathbf{P}}]$, which is the financial product with a higher level relative to N_1. By taking a look at the results arrived at we will see, without a doubt that this is P_4, because the valuation representing its level of resemblance is 0,325. In this way we could successively assign the different financial products to the requirements by taking as a reference the highest level of qualification.

Just as this case has been outlined, it can be assumed that all the characteristics have the same importance (same weight) at the time when analysing the financial products and requirements. But it is all too often frequent that for a determined financial product certain characteristics are more relevant than others.

With the object of getting closer to reality, we are going to consider that the characteristics have a different importance at the time of evaluating the products for covering financial requirements. With this we will be giving a greater generalisation to the scheme we are presenting.

Merely as an indication we now establish certain weights within $[0; 1]$ for each characteristic, which will show the relative importance that they have for each one of the financial requirements. We therefore could establish, for example:

For $\underset{\sim}{\mathbf{N}}_1$:

$$C_1 = 0{,}5 \quad C_2 = 0{,}4 \quad C_3 = 0{,}8 \quad C_4 = 0{,}9$$

For $\underset{\sim}{\mathbf{N}}_2$:

$$C_1 = 0{,}9 \quad C_2 = 0{,}4 \quad C_3 = 0{,}3 \quad C_4 = 0{,}9$$

For $\underset{\sim}{\mathbf{N}}_3$:

$$C_1 = 0{,}8 \quad C_2 = 0{,}7 \quad C_3 = 0{,}2 \quad C_4 = 1$$

For $\underset{\sim}{\mathbf{N}}_4$:

$$C_1 = 0{,}9 \quad C_2 = 0{,}6 \quad C_3 = 0{,}4 \quad C_4 = 0{,}7$$

For $\underset{\sim}{\mathbf{N}}_5$:

$$C_1 = 1 \quad C_2 = 0{,}8 \quad C_3 = 0{,}4 \quad C_4 = 0{,}8$$

With this data we now proceed to normalise in the statistical sense, that is, we divide each estimate of the levels by the total sum relative to each financial requirement. Therefore for N1 we have $C_1 + C_2 + C_3 + C_4 = 2{,}6$ and therefore in box (N_1, C_1) we place $0{,}5/2{,}6 = 0{,}19$. We can now successively arrive at the values for all the elements and show the following matrix of weights, which we shall call $[\underset{\sim}{\mathbf{E}}]$.

$[\underset{\sim}{E}] =$

	C_1	C_2	C_3	C_4
N_1	0.19	0.15	0.31	0.35
N_2	0.36	0.16	0.12	0.36
N_3	0.30	0.26	0.07	0.37
N_4	0.35	0.23	0.15	0.27
N_5	0.33	0.27	0.13	0.27

This matrix of weights, will be used for converting each matrix $[\underset{\sim}{\mathbf{P}}_i]$ in a vector. For this it will be sufficient to calculate the product $[\underset{\sim}{\mathbf{E}}] \times [\underset{\sim}{\mathbf{P}}_i]$, for $i = 1,2,3,4,5,6$, by taking an element of $[\underset{\sim}{\mathbf{E}}]$ and multiplying it successively by an element of $[\underset{\sim}{\mathbf{P}}_i]$. Once all the products have been calculated, element by element, the ones corresponding to each row are added. In this way we will arrive at the objective of possessing a vector for each matrix $[\underset{\sim}{\mathbf{P}}_i]$.

The detail of all these operations is as follows:

$[\underset{\sim}{E}] \times [\underset{\sim}{P}_1] =$

	C_1	C_2	C_3	C_4
N_1	0,19	0,15	0,31	0,35
N_2	0,36	0,16	0,12	0,36
N_3	0,30	0,26	0,07	0,37
N_4	0,35	0,23	0,15	0,27
N_5	0,33	0,27	0,13	0,27

$\times$

	C_1	C_2	C_3	C_4
N_1	0,1	0,4	0,4	0
N_2	0,2	0,4	0,3	0
N_3	0	0,6	0,4	0
N_4	0,2	0,5	0,1	0
N_5	0,1	0,3	0,4	0

$=$

	C_1	C_2	C_3	C_4	
N_1	0,019	0,060	0,124	0	0,203
N_2	0,072	0,064	0,036	0	0,172
N_3	0	0,156	0,028	0	0,184
N_4	0,070	0,115	0,015	0	0,200
N_5	0,033	0,081	0,052	0	0,166

If we do the same operations for $[\underset{\sim}{\mathbf{P}}_1]$, $[\underset{\sim}{\mathbf{P}}_2]$,$[\underset{\sim}{\mathbf{P}}_3]$, $[\underset{\sim}{\mathbf{P}}_4]$,$[\underset{\sim}{\mathbf{P}}_5]$ and $[\underset{\sim}{\mathbf{P}}_6]$ we will arrive at the following results:

$$[\underset{\sim}{E}]\times[\underset{\sim}{P}_1] = \begin{array}{c|c|} N_1 & 0{,}203 \\ N_2 & 0{,}172 \\ N_3 & 0{,}184 \\ N_4 & 0{,}200 \\ N_5 & 0{,}166 \end{array} \quad [\underset{\sim}{E}]\times[\underset{\sim}{P}_3] = \begin{array}{c|c|} N_1 & 0{,}160 \\ N_2 & 0{,}248 \\ N_3 & 0{,}200 \\ N_4 & 0{,}266 \\ N_5 & 0{,}281 \end{array} \quad [\underset{\sim}{E}]\times[\underset{\sim}{P}_5] = \begin{array}{c|c|} N_1 & 0{,}272 \\ N_2 & 0{,}256 \\ N_3 & 0{,}251 \\ N_4 & 0{,}345 \\ N_5 & 0{,}245 \end{array}$$

$$[\underset{\sim}{E}]\times[\underset{\sim}{P}_2] = \begin{array}{c|c|} N_1 & 0{,}256 \\ N_2 & 0{,}220 \\ N_3 & 0{,}116 \\ N_4 & 0{,}171 \\ N_5 & 0{,}159 \end{array} \quad [\underset{\sim}{E}]\times[\underset{\sim}{P}_4] = \begin{array}{c|c|} N_1 & 0{,}347 \\ N_2 & 0{,}260 \\ N_3 & 0{,}205 \\ N_4 & 0{,}297 \\ N_5 & 0{,}307 \end{array} \quad [\underset{\sim}{E}]\times[\underset{\sim}{P}_6] = \begin{array}{c|c|} N_1 & 0{,}284 \\ N_2 & 0{,}368 \\ N_3 & 0{,}265 \\ N_4 & 0{,}402 \\ N_5 & 0{,}387 \end{array}$$

Once all the products have been arrived at, these results will be shown in a matrix, which we will call $[P']$, the results found previously.

In this way we will have:

$[\underset{\sim}{P}'] =$

	P_1	P_2	P_3	P_4	P_5	P_6
N_1	0,203	0,256	0,160	0,347	0,272	0,284
N_2	0,172	0,220	0,248	0,260	0,256	0,368
N_3	0,184	0,116	0,200	0,205	0,251	0,265
N_4	0,200	0,171	0,266	0,297	0,345	0,402
N_5	0,166	0,159	0,281	0,307	0,245	0,387

This matrix $[\underset{\sim}{\mathbf{P}}']$ is the result of a convex weighting of the initial matrix $[\underset{\sim}{\mathbf{P}}]$. From this point on and for effects of initiating the process of assignment, it is possible to take as the starting out point either matrix $[\underset{\sim}{\mathbf{P}}]$ or matrix. $[\underset{\sim}{\mathbf{P}}']$. The difference between these is in the fact that matrix $[\underset{\sim}{\mathbf{P}}']$ is affected by certain weights that show the relative importance that has been given to each one of the characteristics. In this respect, matrix $[\underset{\sim}{\mathbf{P}}]$ lacks this weighting and, therefore, its use means granting the same relative importance to all the characteristics. From now on and with the object of initiating the process of assignment, we will take matrix $[\underset{\sim}{\mathbf{P}}']$ as a basis for the development of the model we propose below.

20.4 Assignment Algorithms

A first approximation for resolving the problem of assignment can take place by means of a series of algorithms, which, although they do not always provide the optimum solution, they do help to find a good solution. From among these algorithms there is one[5] that is very useful for its simplicity in the application we are going to discover very succinctly below. We start out from matrix $[\underset{\sim}{\mathbf{P}}']$ and continue with the steps we have listed below:

1. Choose the element of matrix $[\underset{\sim}{\mathbf{P}}']$ possessing the highest value. This element represents, by means of its row and column respectively, the financial requirement of the business that is going to be covered by one of the products offered in the market.
2. Eliminate from the matrix the row and column that correspond to the financial requirement and the product arrived at in the previous point so that a matrix of a lower order is arrived at.
3. Successively repeat this process until all the financial requirements of the business are covered (until vacating the matrix).
4. In order to get to know the total degree of suitability of the assignment made it is sufficient to find the sum of the values of each element of the matrix corresponding to the assignments made.

In order to find the degree of suitability it is sufficient to divide the degree of absolute suitability by the number of assignments made.

This extraordinarily simple algorithm provides a good solution, although it is not always the optimum solution.

For this, on many occasions, it is recommended that another calculation procedure with the name of Hungarian[6] algorithm be used, the name for this algorithm deriving from the mathematician of than nationality, König[7], who was the ideologist behind it. In short, what it is a way of finding the affectations in such a manner that what is arrived at is a process of optimisation starting out from two sets of related elements by means of a supposedly known matrix.

---

[5] For greater detail see Kaufmann A and Gil Aluja J (1986) Introducción a la teoría de los subconjuntos borrosos a la gestión de las empresas (in Spanish). (Ed) Milladoiro, Santiago de Compostela, Spain, 1st ed. Chap. VII, pp. 151–152.

[6] This algorithm has been applied in an interesting work by Gil Aluja J (November 1995) Modelos no numéricos de asignación en la gestión de personal (in Spanish), paper presented at the II SIGEF Congress. Santiago de Compostela, Spain, vol. 1 Congress Proceedings pp. 458–466.

[7] König D (1916) Théorie der endlichen und unendlichen graphen, later reprinted by (Ed) Chelsea Publ. Co. New York, 1950. This work was made known by Kuhn H.W. in an article The Hungarian method for the assignment problem. Naval Research Quarterly. Vol. 2 No. 1–2 March-June, 1955, pp. 83–98.

For developing this process it is necessary to start out from a matrix in which, as happens the greater part of the time, the number of rows (financial requirements) does not coincide with the number of columns (financial products), so that it is necessary to work with rectangular matrices.

However, with the object of making the algorithm more operative it is customary to transform the rectangular matrix into a square matrix in which, therefore, the number of rows will be equal to the number of columns. For this it is necessary to add, in this case, the additional fictitious row or rows that show the valuations for which no possibility exists of any financial product adapting to this financial requirement or vice-versa. If we consider the matrix of the previous heading we will have a matrix $[\underset{\sim}{\mathbf{P}}'']$ adopting the following form:

$[\underset{\sim}{P}''] =$

	P_1	P_2	P_3	P_4	P_5	P_6
N_1	0,797	0,744	0,840	0,653	0,728	0,716
N_2	0,828	0,780	0,752	0,740	0,744	0,632
N_3	0,816	0,884	0,800	0,795	0,749	0,735
N_4	0,800	0,829	0,734	0,703	0,655	0,598
N_5	0,834	0,841	0,719	0,693	0,755	0,613
F	1	1	1	1	1	1

There are other algorithms that can be used for the solution of this problem (the Branch and Bound algorithms being one of them) For this case we are going to use the Hungarian algorithm.

20.5 The Process of Assignment

Starting out from matrix $[\underset{\sim}{\mathbf{P}}'']$ it is possible to initiate the process detailed by the following steps:

1. We take the first column ($P1$) and subtract the smallest element from each element of the same.
   In each box we will have:

$$P_{ij}^{(1)} = P_{ij} - \min_i P_{ij}$$

   This process is repeated for each of the remaining columns so that one zero at least appears in each one of them.
   We then do the same operations for each one of the rows so that once again at least one zero appears in each one. We then arrive at:

$$P_{ij}^{(1)} = P_{ij} - \min_j P_{ij}$$

which gives rise to the matrix represented below:

$[\underset{\sim}{A}] =$

	P_1	P_2	P_3	P_4	P_5	P_6
N_1	0	0	0,121	0	0,073	0,118
N_2	0,031	0,036	0,033	0,087	0,089	0,034
N_3	0,019	0,140	0,081	0,142	0,094	0,137
N_4	0,003	0,085	0,015	0,050	0	0
N_5	0,037	0,097	0	0,040	0,100	0,015
F	0,203	0,256	0,281	0,347	0,345	0,402
	0,797	0,744	0,719	0,653	0,655	0,598

$[\underset{\sim}{B}] =$

	P_1	P_2	P_3	P_4	P_5	P_6	
N_1	0	0	0,121	0	0,073	0,118	0
N_2	0	0,005	0,002	0,056	0,058	0,003	0,031
N_3	0	0,121	0,062	0,123	0,075	0,118	0,019
N_4	0,003	0,085	0,015	0,050	0	0	0
N_5	0,037	0,097	0	0,040	0,100	0,015	0
F	0	0,053	0,078	0,144	0,142	0,199	0,203

2. Once we have arrived at matrix $[\underset{\sim}{B}]$ we check to see if it is possible to make an assignment in which the values $P_{ij}^{(1)}$ of the solution are all zeros. If this is so then we have arrived at an optimum. On the contrary the process is continued by means of the following steps:
   (a) The rows containing the least zeros are taken one by one.
   (b) One of the zeros of the considered row is framed and the remaining zeros that are members of the same are crossed out, as well as the zeros in the column of which the framed zero is a member.
   (c) This process is repeated for each row until there are no zeros to frame.
   If we consider matrix $[\underset{\sim}{B}]$ that we previously arrived at the result would be the following:

$[\underset{\sim}{B}] =$

	P_1	P_2	P_3	P_4	P_5	P_6
N_1	~~0~~	~~0~~	0,121	[0]	0,073	0,118
N_2	~~0~~	0,005	0,002	0,056	0,058	0,003
N_3	~~0~~	0,121	0,062	0,123	0,075	0,118
N_4	0,003	0,085	0,015	0,05	[0]	~~0~~
N_5	0,037	0,097	[0]	0,04	0,1	0,015
F	[0]	0,053	0,078	0,144	0,142	0,199

From this matrix we represent the graph associated to the same in which the areas corresponding to the framed zeros appear with a thick line and the zeros that have been crossed out with a discontinuous line:

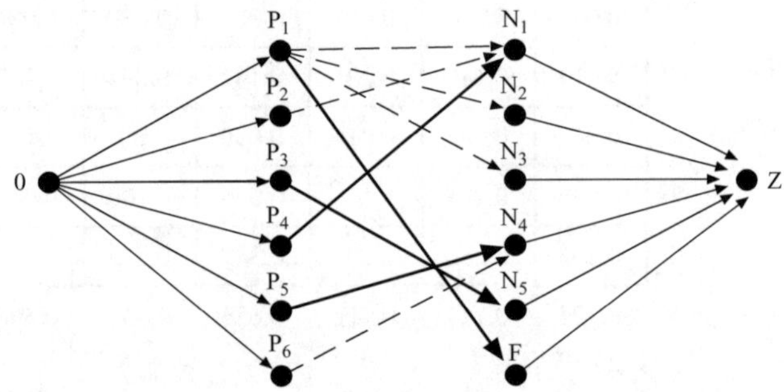

Fig. 20.1.

On establishing the thick and discontinuous lines, that is having framed and crossed out zeros in the previous matrix, what we have done is establish the flow of value 1, which leaves from 0 and arrives at Z, through P_i and N_j. In the graph shown it can be seen that there are two arcs N_2Z and N_3Z that do not end up at Z along which there is no flow.
On taking the decisions with reference to which zeros are to be framed and which crossed out there has been an incidence on the following aspects:

Send flow from P_1 to F to saturate arc FZ with value 1.
Send flow from P_3 to N_5 to saturate arc N_5Z with value 1.
Send flow from P_4 to N_1 to saturate arc N_1Z with value 1.
Send flow from P_5 to N_4 to saturate arc N_4Z with value 1.

We can see, however, that there are two financial products P_2 and P_6 that are not assigned, and in the same way, two financial requirements have not been covered.

3. Obtain the least number of rows and columns that contain all the zeros; in order to do this we must proceed with the following steps:
   (a) All the rows in which there are not framed zeros are signalled with an arrow.
   (b) The columns in which a crossed out zero does exist in a row with an arrow are signalled.
   (c) The rows in which a squared zero does exist in a column signalled with an arrow are likewise signalled. This process is repeated until no further rows or columns can be signalled.
   (d) Finally a line is drawn through the rows that are not signalled by arrows and through the columns that are signalled with arrows. These rows and columns constitute the least number that posses framed or crossed out zeros.

By following this process we arrive at:

	P_1	P_2	P_3	P_4	P_5	P_6	
N_1	0	0		[0]			
N_2	0	0,005	0,002	0,056	0,058	0,003	←
N_3	0	0,121	0,062	0,123	0,075	0,118	←
N_4					[0]	0	
N_5			[0]				
F	[0]	0,053	0,078	0,144	0,142	0,199	←
	↑						

The rows and columns through which a line has been drawn constitute the so-called "minimum support" of matrix $[\underset{\sim}{\mathbf{B}}]$. In the above matrix we have placed the numbers in those boxes that are not crossed with a line. In this case there are 3 rows and 1 column the sum of which gives us a result of 4. We can see that in the initial matrix $[\underset{\sim}{\mathbf{B}}]$ there were already 4 framed zeros which as these are the maximum possible number they are called the "framed index". This framed index is designated, in this case by $Q[\underset{\sim}{\mathbf{B}}]$. The representation by means of a graph does not permit the analysis of certain questions. If we consider that $N+$ is the set of rows that are members of the "minimum support" and $P+$ the number of columns, we arrive at:

$$N+ = \{N_1, N_4, N_5\}$$
$$P+ = \{P_1\}$$

the arcs of which are represented by thick lines

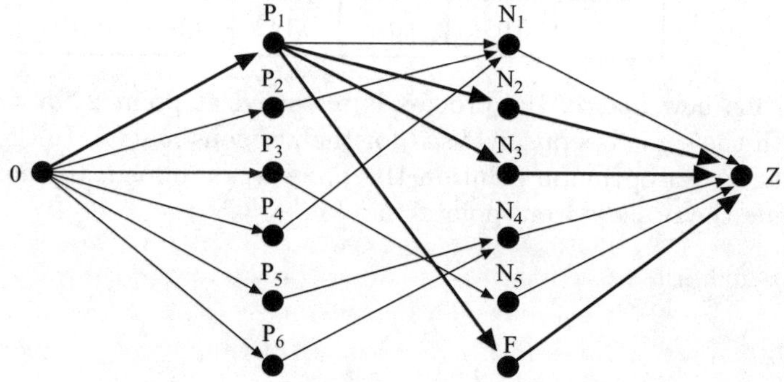

Fig. 20.2.

It will be seen in this graph that no arc exists that joins vertices of which one is a member of the minimum support and the other is not a member of the same.

4. The lowest value is taken from among the elements of the matrix that have not been lined. The figure that is taken is subtracted from the elements of the columns that are not lined and this is added to the rows that are lined. In this way we arrive at a matrix with elements $P_{ij}^{(2)}$ such as we have shown below:

	P_1	P_2	P_3	P_4	P_5	P_6	
N_1	0	0		0			
N_2	0	0,005	0,002	0,056	0,058	0,003	←
N_3	0	0,121	0,062	0,123	0,075	0,118	←
N_4					0	0	
N_5			0				
F	0	0,053	0,078	0,144	0,142	0,199	←
	↑						

The smallest element of the matrix in this case is 0,002.

	P_1	P_2	P_3	P_4	P_5	P_6
N_1	0,002	0	0,021	0	0,073	0,118
N_2	0	0,003	0	0,054	0,056	0,001
N_3	0	0,119	0,060	0,121	0,073	0,116
N_4	0,005	0,085	0,015	0,050	0	0
N_5	0,039	0,097	0	0,040	0,100	0,015
F	0	0,051	0,076	0,142	0,140	0,197

5. With this new matrix the process is restarted at point 2, by continuing on with the same operations used for the previous matrix. In the event of arriving at an optimum solution the process is stopped; to the contrary continue the process from points 3 and 4.

	P_1	P_2	P_3	P_4	P_5	P_6
N_1	0,002	0	0,021	0	0,073	0,118
N_2	0	0,003	0	0,054	0,056	0,001
N_3	0	0,119	0,060	0,121	0,073	0,116
N_4	0,005	0,085	0,015	0,050	0	0
N_5	0,039	0,097	0	0,040	0,100	0,015
F	0	0,051	0,076	0,142	0,140	0,197

From this matrix we arrive at the graph associated to the same.

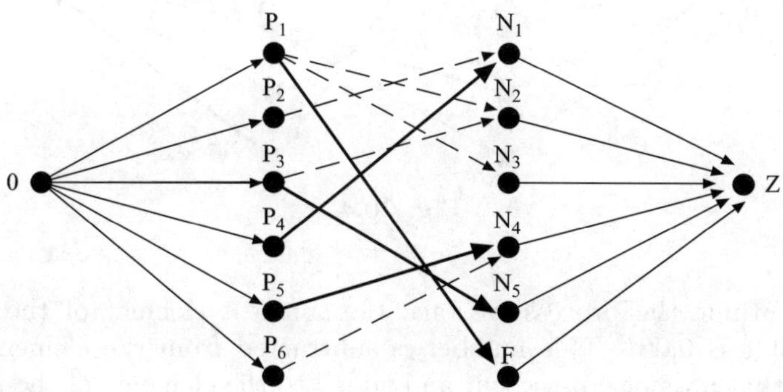

Fig. 20.3.

In this way we arrive at the matrix possessing the minimum number of rows and columns containing zeros. We will have:

	P_1	P_2	P_3	P_4	P_5	P_6	
N_1		0		0			
N_2	0	0,003	0	0,054	0,056	0,001	←
N_3	0	0,119		0,121	0,073	0,116	←
N_4					0	0	
N_5		0,097	0	0,040	0,100	0,015	←
F	0	0,051		0,142	0,140	0,197	←
	↑		↑				

The minimum support of this matrix is formed by 2 rows and 2 columns (the rows and columns crossed with a line).

The framing index will be $Q(\underset{\sim}{\mathbf{B}}) = 4$ in such a way that:

$$N+ = \{N_1, N_4\}$$
$$P+ = \{P_1, P_3\}$$

The corresponding arcs are represented in the following graph by thick lines:

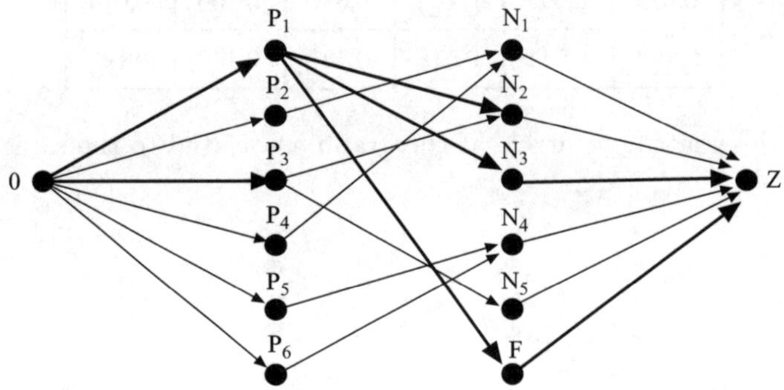

Fig. 20.4.

To continue the process we take the smallest element of the previous matrix that is 0,001. This number is subtracted from the elements of the columns that are not crossed out and added to the elements of the rows that are crossed out. We now arrive at:

	P_1	P_2	P_3	P_4	P_5	P_6
N_1	0,003	0	0,022	0	0,073	0,118
N_2	0	0,002	0	0,053	0,055	0
N_3	0	0,118	0,060	0,12	0,072	0,115
N_4	0,006	0,085	0,016	0,050	0	0
N_5	0,039	0,096	0	0,039	0,099	0,014
F	0	0,05	0,076	0,141	0,139	0,196

Since we have not arrived at an optimum solution we continue with the process of framing and crossing out zeros:

	P_1	P_2	P_3	P_4	P_5	P_6
N_1	0,003	0	0,022	0	0,073	0,118
N_2	0	0,002	0	0,053	0,055	0
N_3	0	0,118	0,060	0,12	0,072	0,115
N_4	0,006	0,085	0,016	0,050	0	0
N_5	0,039	0,096	0	0,039	0,099	0,014
F	0	0,05	0,076	0,141	0,139	0,196

Once again we will see that there is a financial product, P_2, that has no assignment as well as a financial requirement, N_3, that has not been covered. This is why we renew the process once again arriving at the least number of rows and columns that contain all the zeros:

	P_1	P_2	P_3	P_4	P_5	P_6	
N_1							
N_2							
N_3		0,118	0,060	0,120	0,072	0,115	←
N_4							
N_5							
F		0,050	0,076	0,141	0,139	0,196	←
	↑						

The lowest element from among those that are not lined is 0,050. This number is subtracted from the elements of the not lined columns and added to the elements of the rows that are crossed out. The following matrix is the result:

	P_1	P_2	P_3	P_4	P_5	P_6
N_1	0,053	0	0,022	0	0,073	0,118
N_2	0,050	0,002	0	0,053	0,055	0
N_3	0	0,068	0,010	0,070	0,022	0,065
N_4	0,056	0,085	0,016	0,050	0	0
N_5	0,089	0,096	0	0,039	0,099	0,014
F	0	0	0,026	0,091	0,089	0,146

Below we show the graph associated to this matrix:

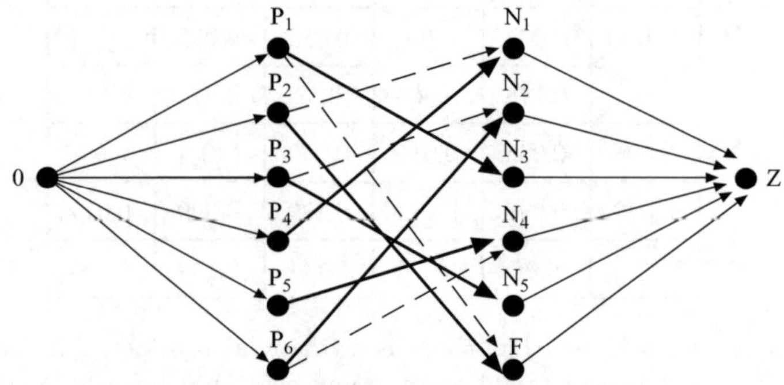

Fig. 20.5.

Finally we can see that all the financial products have been assigned to all the financial requirements, therefore, arriving at an optimum situation. The process we have followed has concluded as follows:

1. Product P_1 is assigned to N_3.
2. Product P_2 is assigned to F.
3. Product P_3 is assigned to N_5.
4. Product P_4 is assigned to N_1.
5. Product P_5 is assigned to N_4.
6. Product P_6 is assigned to N_2.

Considering that F represents a fictitious financial requirement which was introduced into the process in order to simplify the algorithm which at the same time has allowed us to make an optimum assignment, it can be seen that financial product P_2 is rejected for effects of its use in the business.

In this way, and taking the initial matrix again, we arrive at the minimum possible distance, provided by the assignment we arrived at:

$$d(P_j, N_i) = d(P_1, N_3) + d(P_3, N_5) + d(P_4, N_1) + d(P_5, N_4) + d(P_6, N_2)$$

$$\delta(P_j, N_i) = \frac{d(P_j, N_i)}{5} = \frac{1{,}675}{5} = 0{,}355$$

20.6 Final Considerations

In the work we have presented the existence of certain algorithms has been brought to light that allow for resolving, among others, the important problem of the assignment of financial resources to the demands made by businesses. The models we have used are characterised, apart from their operating simplicity, for being based on multivalent logic, key aspect for the development of techniques that permit the analysis of future situations in an eminently changing economic-financial environment.

We have developed the Hungarian algorithm in a certain amount of detail due to its undoubted interest, which is both technical as well as practical. Obviously many other algorithms and models exist in financial literature the object of which is arriving at solutions as suitable as possible for the treatment of assignment in the widest possible sense, but these differ from the ones used in this work as they do not consider the effort necessary for adapting themselves to the requirements sought by businesses. Therefore, the greater or lesser success of the results, or the better or worse decision in the assignments obtained depends both on the methodology used, as on the origin and process of information and from obtaining the initial data. It is in this context when studies based on operating techniques for treating uncertainty acquire a certain significance.

21 The Economic-Financial Value of a Business

21.1 Classical Criteria of Valuation

Particular unanimity exists in the statement that economic interchanges take place in areas that are all the time more extended and that financial relations between businesses in different countries are increasingly getting closer and closer. We are immersed in a process of interconnection and integration of economic systems into large blocks that, at the same time maintain interests that at times are common and at other opposite, but always interrelated.

In this complex mosaic of interests, ever more frequently, processes of acquisition, purchasing, absorption, concentration or merger of businesses appear that require the establishment of a valuation that allows for setting a purchase and sale price that is acceptable to the eventual purchasers and sellers of the same.

One of the more outstanding characteristics of the problem of valuation arises from the fact that it is difficult to arrive at a standard for fixing the price, due to the fact that each purchase and sale operation for a business is different from all the rest. And this for reasons so diverse as the particularity of the business, its economic-financial structures, its organisation, the markets in which it is located, its political surroundings and a long etcetera.

On the other hand, the fixing of a price that allows for the transfer of ownership of the business (or part of the business) from one financial group to another, is not, generally speaking, the specific result of a technical study, but the consequence of laborious negotiations in which the parties commence discussions starting out from positions that are more or less opposite. The mission of financial analysis should consist in making life easier for the negotiators, by supplying data that is sufficiently trustworthy and, by means of available techniques, propitiate convergence up to the point that an agreement is possible.

In previous chapters a certain meaning of the "value of the business" has been given linked to the consideration given by shareholders and creditors to the securities they posses (shares and liabilities). But financial analysis assumes another notion of value when it refers to the position adopted by the eventual sellers and purchasers of a business or a more or less important part of the same.

Anna Maria Gil-Lafuente: *Fuzzy Logic in Financial Analysis*, StudFuzz **175**, 383–409 (2005)
www.springerlink.com

In the first of these cases attention is fixed on the "value" of the shares (capital) and debentures (liabilities), elements, therefore, that belong to the liabilities side of the balance sheet or financial structure, while the other places special emphasis on the "value" of the assets that are necessary for generating profit and eventually on the value these may reach in the more or less immediate future.

We are now going to pay attention to the latter concept of the value of the business, bringing to light certain techniques that are useful in a context of uncertainty. Right from the outset, we should point out that the valuation of a business includes elements with a high content of subjectivity inherent in every process in which the notion of value is present. It should not come as a surprise, then, that certain economic aspects are incorporated, which will be considered in monetary terms, in which both subjective and objective elements appear. Subjectivity arises not only due to the difficulty of estimating values relative to the future, but as a consequence of the need to value the "importance" that the exploitation of a business acquires for an external entity.

As is well known, a business, in order to reach its objectives, requires certain production factors and an organisation. The first are materialised in the patrimonial volume of assets, but the organisation, without which no activity whatsoever can take place and therefore no profit obtained, is not specifically included in the economic structure of the balance sheet.

This is why, only in a first approximation, it can be accepted that the value of a business is given by the set of assets and rights that it possesses, encompassed under the term of "patrimony". In a certain manner, and only in a certain manner, it can be said that the value of a business is expressed by the set of elements of the functional assets it possesses. The value of assets only brings to light the "possibility" of generating profit, that constitutes one of the aspects to be considered in the determination of the value of a business.

When we delve into the specialised literature, it will be seen that there are many and varied proposals that are carried out for the process of valuation. Without the pretence of the following being exhaustive we could point out the following criteria, as an example:

1. Criterion of historical or retrospective price.

   For each one of the elements of assets consideration is given to the price that was paid at the time of their incorporation into the business. Independently of these aspects, the inefficiency of this criterion is brought to light merely by pointing out the difference existing between the purchase value and economic value, that is to say the suitability for generating profit.

   In this sense, Pourbaix points out "that total acquisition costs (adjusted if necessary to take into account wear and tear) correspond to the real value of an economic entity. Nevertheless, the coherence of this reasoning is destroyed by continuous inflation, and above all by the extraordinary dynamism of scientific and technical progress"[1]. In spite of this fact, this valuation criterion is frequently taken into account in order to value determined patrimonial elements. Indeed, in relation to stocks, it may be interesting to consider, by taking one of the following procedures as a basis:

- FIFO (first-in-first-out), according to which the latest prices for raw materials existing in the warehouse are assigned.
- LIFO (last-in-first-out) assigning the prices of the oldest raw materials to total stocks.
- "Average price", that consists in weighting the prices by the amounts entering the warehouse over a determined period.

Now the problems caused by current economic systems, characterised by very rapid changes that give rise to important inflationary processes and accelerated technical progress, are the reason why there is a proposal for valuation in which certain correction indices are taken into account that modify the dimension of the patrimonial masses. This practice gives rise to different appreciations by the parties taking part in the valuation and, therefore, a lack of consensus. This problem springs to light so frequently that, from several decades back, specific procedures have been proposed that attempt to alleviate the inconveniences caused by this.

Merely as an indication we mention the procedure called "useful stock" which is established, on the one hand, "based on the quantities that are necessary to feed the machinery of the transformation cycle and, on the other hand, in order to maintain stop-gap stock the importance of which is linked to the period required for its replacement, taking into account the random elements that can have an influence on this period"[2]. The difficulties that this type of procedure entails highlights the advantages of using those models that take as their basis the replacement value.

2. Criterion of replacement value.

The basis of this criterion is to be found in the estimate of the prices of each of the patrimonial masses of assets according to the market situation at the time of the hypothetical sale of the business and their economic effectiveness. This is also known as the "fundamental value".

In this sense, the concept of value arises from the existence of a patrimony (fundamental value) and in a certain way from its capacity for generating profit. In the "fundamental value" an important role is played by total assets, which is brought into play in order to arrive at the set objectives. Now the, valuation of the elements of the assets is not always easy,

[1] Pourbaix Claude (1969) Valeur de l'entreprise. (Ed) Dunod, Paris, France, p. 37.

[2] Hanon de Louvet Ch (1955) Nouveau traité d'analyse et discussion de bilans. (Ed) Editions Comptables, Commerciales et Financières, Brussels, Belgium, p. 149.

for this reason specialists on the subject establish certain considerations relative to the patrimonial volume that makes up the economic structure. "Determination of the "fundamental value" is done in two stages that consist, in the first place, in taking a quantitative and qualitative inventory of the volume of assets; then, in second place, in valuing the same".[3]

It should come as no surprise then, that in the process of valuation of business the fundamental value has occupied a position of privilege. Pourbaix states that "the fundamental value makes reference to the set of assets and rights possessed by a business that form a homogeneous economic structure. That is to say that external elements are excluded strictly speaking from exploitation. Let us specify that we are only dealing with asset values. In this phase of the calculations there is no deduction for liability funds"[4].

The European Union of Accounting Experts considers that the elements that intervene in assets of a business can be classified into the following groups:

- Values that are susceptible to being included in the balance sheet as individualised elements:
  (1) Fixed assets and stocks
  (2) Participation and credit rights
  (3) Liquid assets and bill portfolio
  (4) Patents, licenses and "acquired" goodwill.
- Values susceptible to be included in the balance sheet during a limited period (organisation, research and project expenses)
- Values not susceptible to be included in the balance sheet and difficult to value (goodwill created by the business, client value, market position, etc.)

The elements that make up the economic structure of the balance sheet pass through different situations relative to the elaboration process. In this sense, accounting literature establishes a distinction between functional, extra-functional and anti-functional elements, the consideration of which is important when attempting to determine the fundamental value of the business. Indeed, given that the anti-functional and extra-functional elements do not take part in obtaining profit, their amount acquires a very limited sense and can only be incorporated into the process as a complementary element to the value.

Merely as an indication, we will illustrate the treatment of some of the patrimonial masses, such as buildings and industrial equipment.

With regard to the buildings, it would seem adequate to start out from the prices current at the moment of valuation of the constructed square

[3] European Union of Accounting Experts, (1962) Evaluación de empresas y partes de Empresa. (Ed) Deusto, Bilbao, Spain.

[4] Pourbaix Claude, op. cit., p. 233.

meter, deducting depreciation. Obviously in this process a fact that constitutes the counterpart of the valuation done at current prices should be taken into account: depreciation, which can be considered as an increasing function of time.

In this sense, Koller[5] proposes the following expression:

$$D = \frac{(n+20)^2}{140} = 2{,}86$$

where

$D =$ percentage depreciation
$n =$ time transpired since building construction

Which can be transformed, approximately, into the following:

$$D = \frac{n^2}{140} + \frac{4}{14^n}$$

In relation to the valuation of industrial equipment at replacement value, the European Union of Accounting Experts proposes the following formula[6]:

$$\text{Replacement value} = A \cdot \left(1 + \frac{100}{M}\right) \cdot \frac{D^r}{D}$$

where:

$A =$ initial value;
$M =$ percentage increase in acquisition cost;
$D^r =$ residual use duration;
$D =$ total use duration.

In short, the criterion for the replacement value or fundamental value is obtained through the estimate of the different patrimonial masses calculated at the price of the increase in the valuation, just as if one were setting up a business that was a twin of the one in existence.

3. Criterion of capitalised profit.

   This is based on the fact, as has already been pointed out, that the business as an organic unit strives to obtain profit, for which objective it is not just sufficient to have a patrimony (fundamental value), but also an "organisation". It is precisely these profits, which are the axis on which the calculation of the value of the business is based.

In fact, according to this criterion, valuation is done by means of the estimate of profits for certain future periods, discounted by means of the corresponding interest rates, at the time when the valuation takes place. From

[5] Koller (1959) Die Ermittlung von Gesanmtwert der Unternimens, (2nd Ed.), Dusseldorf, Germany, pp. 48–49.

[6] European Union of Accounting Experts, op. cit., pp. 30–31.

this point of view, the value of the business depends on the following variables: profit for each period, rates of interest and the time span covering the profits under consideration.

The difficulties existing for estimating the future profits of a business will not escape the reader. In this respect, two positions can be adopted: extrapolation of the profits obtained in previous accounting periods and estimating by means of the opinion of experts, without any direct relation with historical data.

It is obvious that "profit" normally undergoes fluctuations over time, giving rise to serious considerations as a consequence of the uncertainty in which the outcome of the activity of the business is submerged. There are many elements that condition the profit of a business: its production organisation, supply channels, qualifications of its personnel, distribution network, financial and commercial image, etc.

In classical studies on valuation it has been established that in order to determine this profit, in such a way that it generates a certain amount of acceptance and that its quantity can be considered valid in the short and medium term, the balance sheets of the business relative to a period of time that fluctuates between three and six accounting periods are taken as the basis, as well as the balance sheets from other businesses with similar characteristics, that can be suitable for making comparisons.

With all this, an attempt is made to obtain a "price" for the business independently of the position of the current owners of the same, possible purchasers, as well as the elements that contribute to the current political-economic situation. In this way what it means is to find a value that could be attributed by any person to any particular object.

It is obvious that this "value for the business" will be subject to fluctuations if complementary subjective factors exercise a specific influence on the hypothetical purchase and sale decision.

Some authors consider that, taking into account the elements that influence in the accounting determination of the profit, it is more convenient to estimate in its place the cash-flow generated, that is, the profit on the exploitation without deducting depreciation, that expresses, therefore, the capacity that the business has for generating its own financing. The advantage of incorporating the cash-flow in stead of profit is due, principally to the elimination of the influence that depreciation has on the results of each accounting period.

Neither is it easy to estimate the interest rates to be considered for each period. If in a stable economic system and even in a stationary situation the rates considered are subject to serious argument, in a situation as changing as the current one the problem is aggravated. Only the possibility for using uncertain values can lead these estimates to a position that allows for acceptance with a certain degree of consensus.

Finally, the third element to be considered in the valuation by means of the criterion of up-dated profit is the estimate of the number of periods in which future profits should be included. We hasten to say that unanimity does not exist in this respect on the part of the specialists on the subject and proposals fluctuate between very different limits according to the type of business, dimensions of the same, etc. We are not going to linger now on this problem, but this will be tackled immediately under the section on the determination of goodwill.

There are many other criteria for the determination of the value of a business. Some have a particular interest for a specific case, as occurs with the "liquidation value", valid for when it is foreseen that a business will be liquidated, the basis of which can be found in the addition of the values obtained from the sale of the patrimony deducting payments made for debts and liquidation expenses.

To these criteria, which could be called "pure" we have to add the "mixed" criteria in the calculation of which several of the previously described criteria take part in some way.

Therefore, the complex task of arriving at a value for the business is increased by the existence of several criteria that are susceptible to be taken into consideration. On the other hand, different criteria are frequently used in practice in order to quantify the different patrimonial masses that comprise the economic structure of the balance sheet. In this way the liquidation value for the anti and extra-functional elements is used, although it can be said that the replacement value constitutes the most frequently used valuation criterion for patrimonial elements.

21.2 Determination of the Value of the Business

In the previous section certain criteria, called pure criteria, among which one stands out that takes into consideration the result as a fundamental part of the calculation. The basis for this can be found in the determination of profit over a certain number of periods, to be determined, following on the time of the valuation, and their justification in the acceptance that the value of a business does not depend solely on its patrimony, but on the degree of suitability it posses for generating profit in the future.

In general lines it would appear that this criterion is the one most generally accepted although, as we will see below, it is subjected to certain modifications.

In the first place we would like to point out that a business, the activity of which can be qualified as normal, should have a value calculated on capitalised profit higher than the value obtained by means of the criterion of replacement value or fundamental value. This difference is known by the name of goodwill.

The concept of goodwill arises in this way because profits exceed a figure considered as the norm, and are the result of determined intangible values that the business posses independently of the tangible value of the assets of the same.

The calculation of the goodwill by means of the profit hoped for brings up important technical difficulties. Specialised works on the subject generally present two paths as desirable: the "direct method", that consists in estimating certain possible profit and certain normal profits and updating the difference, and in the "indirect method", that consists in updating the estimate of profit and deducting the fundamental value. Notwithstanding the problems arising from its use in reality, these methods have a certain acceptance by professionals who have to apply them.

Now, even in the case that it were to be possible to estimate certain future profits that can be accepted both by the selling party and the purchasing party, the number of periods that should be taken into consideration in the calculation process would still have to be resolved. The positions in this respect do not coincide. Thus, as is included in the work by Pourbaix, "in practice resort is made, it would appear by tradition, to the product of multiplying the average profit by three[7]. This would correspond to the start-up period. The lack of profits in general would be prolonged over a period of three years"[8]. Obviously this practical rule can only be taken into consideration as valid for determined types of activities and even among them merely as an indication.

In a more general perspective, the analysis of a large number of actual cases appears to bring to light the fact that this multiplier of the annual profit fluctuates in the majority of cases between 1 and 5. In this respect it is said that within the commercial sphere, for the food industry this factor is placed between 3 and 4, and for pharmacies 4 is admitted.

It should come as no surprise, that in the light of these difficulties, the notion of uncertainty arises relative to the data it is necessary to estimate. For this reason, in the more advanced works, techniques are incorporated that are based on the theory of fuzzy sub-sets and their multiple variations.

But before delving into the use of valid techniques for the treatment of uncertainty let us see what forms are taken by the criteria of capitalised profit when this is qualified with that of the fundamental value, giving rise to what are called mixed methods. Traditionally what is known as the "German method", "indirect method" or "Schmalenbach method" and the "Saxon method" or "direct method" have been considered as the most important.

The German method starts out from the idea that the capitalised value of profits, conceptually the best criteria, must be submitted to a qualification as a consequence of the uncertainty of future data. In the absence of a valid technique, classical writers have placed themselves in a position of caution consisting in adding half of the goodwill to the fundamental value or, what amounts to the same, deduct half of the goodwill from the capitalised value.

---

[7] Gauchey report to the 6th Annual Congress of the Chambre Des Experts Judiciares Evaluateurs, (November 5, 1960), Aix-an-Provence, France.

[8] Pourbaix C, op. cit., p. 60.

Analytically we would arrive at:

$$\begin{aligned} V_e &= V_s + \frac{1}{2} \cdot GW \\ &= V_c - \frac{1}{2} \cdot GW \\ &= V_c - \frac{1}{2} \cdot (V_c - V_s) \\ &= \frac{V_c - V_s}{2} \end{aligned}$$

as

$$\begin{aligned} V_e &= \text{value of the business;} \\ V_s &= \text{fundamental value;} \\ V_c &= \text{capitalised value;} \\ GW &= \text{goodwill} = V_c - V_s \end{aligned}$$

with which the criterion is limited to assigning the median between capitalised value and fundamental value as the value of the business.

We feel there is no need to insist on the difficulties of accepting this calculation method, the simplicity of which is accompanied initially by adopting an arbitrary position, which assigns the same weight to the fundamental value as it does to the capitalised value.

We now move on then to the second of these two methods, known under the name of "Anglo Saxon method". Calculation for this method is done by adding the total goodwill to the fundamental value. We can see that, independently of the part of the goodwill that is considered, both methods coincide in essence, the only variation being the calculation method. Indeed, while in the German method the goodwill is arrived at as the difference between the capitalised value and the fundamental value (indirectly), in the Anglo Saxon method goodwill is calculated by up dating the difference between expected profit and "normal" yields of the fundamental value (directly).

If I is the interest rate considered as "normal" and B future profit "stable" expected in the future, the value of the business will be:[9]

$$V_e = V_s + (B - i \cdot V_e) \cdot \frac{(1+i)^n - 1}{(1+i)^n \cdot i}$$

From this we start out from the principle according to which normal yields must be calculated on the "value of the business", which is what any eventual purchasers will pay, instead of doing it on the fundamental value.

[9] It can be seen that in this case it has been considered that the interest rate "normal" i coincides with the rate taken for the up dating basis. In certain cases it may be convenient to use different rates, for which it would be sufficient in the formula to substitute $i \cdot V_s$ by $r \cdot V_s$. When $i > r$ future profit is penalised in this way reducing goodwill.

In the event that the estimate of profits and interest rates for future periods were to give rise to different values in each of the years, we would have:

$$V_e = V_s + (B_1 - i_1 \cdot V_e) \cdot (1+i_1)^{-1} + (B_2 - i_2 \cdot V_e) \cdot (1+i_1)^{-1} \cdot (1+i_2)^{-1} + \cdots + (B_n - i_n \cdot V_e) \cdot (1+i_n)^{-1} \cdot (1+i_2)^{-1} \cdot \cdots \cdot (1+i_n)^{-1}$$

Based on this expression, we are now going to the introduce uncertainty that is inherent in all valuation process, when it is assumed that both future profits and interest rates cannot be estimated in terms of certainty.

21.3 Estimating Interest Rates in Uncertainty

In order to develop the scheme we are proposing, we are going to use a numerical example. In this case we are going to consider that the most important problem arises as a consequence of the estimate of future profit and interest rates. On the other hand, both sellers and purchasers have agreed on the number of periods over which the profit should be considered, specifically three years. Therefore, $n = 3$.

Relative to the estimate for the interest rates, initially official estimates have been taken and these establish for the future periods rates of $[i_1^{(1)}, i_1^{(2)}] = [0{,}10; 0{,}12]$, $[i_2^{(1)}, i_2^{(2)}] = [0{,}12; 0{,}14]$ $[i_3^{(1)}, i_3^{(2)}] = [0{,}08; 0{,}10]$, but purchasers and sellers are not totally in agreement with the figures and they have consulted a group of m experts, in this case $m = 7$, to express their opinions within a hendecadaire scale[10]. The answers are included in the Table 21.1.

Table 21.1.

Expert	[0,10; 0,12]	Expert	[0,12; 0,14]	Expert	[0,08; 0,10]
1	[0,8; 0,9]	1	[0,6; 0,8]	1	0,4
2	0,7	2	[0,9; 1]	2	[0,3; 0,5]
3	[0,6; 0,8]	3	[0,4; 0,6]	3	[0,7; 0,8]
4	[0,9; 1]	4	[0,5; 0,9]	4	0,4
5	[0,5; 0,8]	5	[0,7; 0,8]	5	[0,6; 0,9]
6	[0,7; 1]	6	0,7	6	[0,4; 0,7]
7	[0,6; 0,7]	7	[0,6; 0,9]	7	[0,8; 0,9]

[10] In relation to this subject, see Kaufmann A and Gil Aluja J (1992) Técnicas de gestión de empresa. Previsiones, decisiones y estrategias (in spanish). (Ed) Pirámide, Madrid, Spain, p. 17, where different ways of expressing subjectivity are shown.

From this data the following statistics are arrived at:

Table 21.2.

	[0,10; 0,12]		[0,12; 0,14]		[0,08; 0,10]	
0						
0,1						
0,2						
0,3					1	
0,4			1		3	2
0,5	1		1			1
0,6	2		2	1	1	
0,7	2	2	2	1	1	1
0,8	1	2		2	1	1
0,9	1	1	1	2		2
1		2		1		

and the following expertons:

Table 21.3.

<table>
<tr><th></th><th colspan="2">[0,10; 0,12]</th><th colspan="2">[0,12; 0,14]</th><th colspan="2">[0,08; 0,10]</th></tr>
<tr><td>0</td><td colspan="2">1</td><td colspan="2">1</td><td colspan="2">1</td></tr>
<tr><td>0,1</td><td colspan="2">1</td><td colspan="2">1</td><td colspan="2">1</td></tr>
<tr><td>0,2</td><td colspan="2">1</td><td colspan="2">1</td><td colspan="2">1</td></tr>
<tr><td>0,3</td><td colspan="2">1</td><td colspan="2">1</td><td colspan="2">1</td></tr>
<tr><td>0,4</td><td colspan="2">1</td><td colspan="2">1</td><td>0,857</td><td>1</td></tr>
<tr><td>0,5</td><td colspan="2">1</td><td>0,857</td><td>1</td><td>0,428</td><td>0,714</td></tr>
<tr><td>0,6</td><td>0,857</td><td>1</td><td>0,714</td><td>1</td><td>0,428</td><td>0,571</td></tr>
<tr><td>0,7</td><td>0,571</td><td>1</td><td>0,428</td><td>0,857</td><td>0,285</td><td>0,571</td></tr>
<tr><td>0,8</td><td>0,285</td><td>0,714</td><td>0,142</td><td>0,714</td><td>0,142</td><td>0,428</td></tr>
<tr><td>0,9</td><td>0,142</td><td>0,428</td><td>0,142</td><td>0,428</td><td>0</td><td>0,285</td></tr>
<tr><td>1</td><td>0</td><td>0,285</td><td>0</td><td>0,142</td><td>0</td><td>0</td></tr>
</table>

We then arrive at the R^+-expertons. For the interest rate of the first year we have:

Table 21.4.

	α	Experton			α	R^+ – Experton	
	0	1			0	0,1200	
	0,1	1			0,1	0,1200	
	0,2	1			0,2	0,1200	
	0,3	1			0,3	0,1200	
	0,4	1			0,4	0,1200	
0,10+(0,12-0,10) (·)	0,5	1		=	0,5	0,1200	
	0,6	0,857	1		0,6	0,1171	0,1200
	0,7	0,571	1		0,7	0,1114	0,1200
	0,8	0,285	0,714		0,8	0,1057	0,1142
	0,9	0,142	0,428		0,9	0,1028	0,1085
	1	0	0,285		1	0,1000	0,1057

The mathematically expected value will be:

$$(\underset{\sim}{\mathbf{i}}_1) = [0{,}1137; 0{,}1168]$$

Thus we arrive at the fact that the aggregate opinion of the experts leads us to an estimate for the interest rate for the first year of between 11,37% and 11,68%.

For the interest rate for the second year:

Table 21.5.

	α	Experton			α	R^+ – Experton	
	0	1			0	0,1400	
	0,1	1			0,1	0,1400	
	0,2	1			0,2	0,1400	
	0,3	1			0,3	0,1400	
	0,4	1			0,4	0,1400	
0,12+(0,14-0,12) (·)	0,5	0,857	1	=	0,5	0,1371	0,1400
	0,6	0,714	1		0,6	0,1342	0,1400
	0,7	0,428	0,857		0,7	1,1285	0,1371
	0,8	0,142	0,714		0,8	0,1228	0,1342
	0,9	0,142	0,428		0,9	0,1228	0,1285
	1	0	0,142		1	0,1200	0,1228

The mathematically expected value will be:

$$(\underset{\sim}{\mathbf{i}}_2) = [0{,}1325; 0{,}1362]$$

The aggregate opinion of the experts leads to an estimate for the interest rate during the second year of between 13,25% and 13,62%.

For the interest rate for the third year we have (see Table 21.6):

Table 21.6.

$$(\underset{\sim}{\mathbf{i}}_3) = [0{,}0902;\ 0{,}0931]$$

		Experton				R⁺ – Experton	
	0	1			0	0,1000	
	0,1	1			0,1	0,1000	
	0,2	1			0,2	0,1000	
	0,3	1			0,3	0,1000	
	0,4	0,857	1		0,4	0,0971	0,1000
0,08+(0,10-0,08) (·)	0,5	0,428	0,714	=	0,5	0,0885	0,0942
	0,6	0,428	0,571		0,6	0,0885	0,0914
	0,7	0,285	0,571		0,7	0,0857	0,0914
	0,8	0,142	0,428		0,8	0,0828	0,0885
	0,9	0	0,285		0,9	0,0800	0,0857
	1	0	0		1	0,0800	0,0800

The mathematically expected value will be

$$(\underset{\sim}{\mathbf{i}}_3) = [0{,}0902; 0{,}0931]$$

Finally the aggregate opinion of the experts is an estimate of the interest rate for the third year of between 9,02% and 9,31%.

To summarise, the rates of interest that the experts propose, having seen and studied the official figures, are the following:

$$(\underset{\sim}{\mathbf{i}}_1) = [0{,}1137; 0{,}1168]$$
$$(\underset{\sim}{\mathbf{i}}_2) = [0{,}1325; 0{,}1362]$$
$$(\underset{\sim}{\mathbf{i}}_3) = [0{,}0902; 0{,}0931]$$

With this process one of the objectives that were set has been arrived at, and this consists in "thinning down" the extension of the interval in order to allow a more adequate margin for negotiation, which is equivalent to reducing the uncertainty.

Now then, what has also cropped up is an eventual problem that appears quite clearly in this example, and this consists of the excessive translation of the interval towards determined positions in the case we are occupied with, getting much close to the upper extremes. If this agrees with the opinion of the experts and above all, is accepted by the eventual sellers and buyers, the problem no longer exists. However, it is possible, and enters into what quite frequently occurs, that having obtained the results, these do not authentically represent the personal opinions of the imagined future reality. In these circumstances we feel that it would be suitable if we proceeded to do a parameterisation of the data provided by the experts.

For this, we take the parameterised identification of Sugeno[11] as the basis, adapting the same to the requirements of the specific problem:

$$i^{(\lambda)} = \frac{i}{1+(1+i)\cdot\lambda}, \quad \lambda \in]-1;\infty]$$

This is a "perfect" parameterised identification since it complies with the three conditions:

1. $(i=0) \Rightarrow (i^{(\lambda)}=0), \quad (i=1) \Rightarrow (i^{(\lambda)}=1)$
2. $\left(i^{(\lambda)}\right)^{(\lambda)} > i, \quad \lambda < 0$
   $\left(i^{(\lambda)}\right)^{(\lambda)} = i, \quad \lambda = 0$
   $\left(i^{(\lambda)}\right)^{(\lambda)} < i, \quad \lambda < 0$
3. $(i > j) \Rightarrow \left(i^{(\lambda)} > j^{(\lambda)}\right)$

It is obvious that one of the important elements in the solution of this problem is the selection of the value of parameter λ. For our effects, we feel it is sufficient to consider a $\lambda \in [0;1]$.[12] Let us assume the experts all agree in making $\lambda = 0{,}5$.

The parameterisation for obtaining $\varepsilon(\underset{\sim}{\mathbf{i}}^{(\lambda=0,5)})$ commences at level 1, that is at $[0;0{,}285]$.

$\alpha = 1:$

$$[0;0{,}285]^{(\lambda=0,5)} = \left[\frac{0}{1+1\times 0{,}5}; \frac{0{,}285}{1+0{,}715\times 0{,}5}\right] = [0;0{,}209]$$

$\alpha = 0{,}9:$

$$[0{,}142;0{,}428]^{(\lambda=0,5)} = \left[\frac{0{,}142}{1+0{,}858\times 0{,}5}; \frac{0{,}428}{1+0{,}572\times 0{,}5}\right] = [0{,}099;0{,}332]$$

---

11 Sugeno N (1974) Theory of fuzzy integrals and its applications, Doctoral thesis in Engineering, Institute of Technology, Tokyo, Japan.

12 To consider $\lambda = 0$ would mean relinquishing the parameterisation.

$\alpha = 0{,}8$:

$$[0{,}285; 0{,}714]^{(\lambda=0{,}5)} = \left[\frac{0{,}285}{1 + 0{,}715 \times 0{,}5}; \frac{0{,}714}{1 + 0{,}286 \times 0{,}5}\right] = [0{,}209; 0{,}555]$$

$\alpha = 0{,}7$:

$$[0{,}571; 1]^{(\lambda=0{,}5)} = \left[\frac{0{,}571}{1 + 0{,}429 \times 0{,}5}; \frac{1}{1 + 0 \times 0{,}5}\right] = [0{,}470; 1]$$

$\alpha = 0{,}6$:

$$[0{,}857; 1]^{(\lambda=0{,}5)} = \left[\frac{0{,}857}{1 + 0{,}143 \times 0{,}5}; \frac{1}{1 + 0 \times 0{,}5}\right] = [0{,}779; 1]$$

For the values of $\alpha = 0{,}5$ and lower, we will obviously arrive at a result that is always equal to the unit.

And finally we can write (see Table 21.7).

The mathematically expected value of the parameterised experton will be equal to $[0{,}657; 0{,}809]$. In this way we have an estimate of the interest to be considered in year 1:

$$\underset{\sim}{\mathbf{i}}^{(\lambda=0{,}5)} = 0{,}10 + (0{,}12 - 0{,}10)(\cdot)\,[0{,}657; 0{,}809] = [0{,}1131; 0{,}1161]$$

It will be seen that the results mean a very reduced displacement to the left. This can be considered as normal, if we take into account the fact that

Table 21.7.

α				α		
0	1			0	1	
0,1	1			0,1	1	
0,2	1			0,2	1	
0,3	1			0,3	1	
0,4	1			0,4	1	
0,5	1		=	0,5	1	
0,6	0,857	1		0,6	0,7990	1
0,7	0,571	1		0,7	0,4700	1
0,8	0,285	0,714		0,8	0,2090	0,5520
0,9	0,142	0,428		0,9	0,0990	0,3320
1	0	0,285		1	0,0000	0,2090

The left experton is raised to the power $(\lambda=0{,}5)$.

[0,657; 0,809]

the experts, with their respective opinions, have given rise to the appearance of the great quantity of 1s in the experton, which indicates a notable tendency to placing themselves in the upper extreme of the interval. Parameterisation, as is known, does not exercise (nor must it exercise) any influence whatsoever on this position of convergence of opinions. However, a value of $\lambda > 0{,}5$ would give rise to a increased displacement to the left.

In the parameterisation done with the object of arriving at $\varepsilon(\underset{\sim}{\mathbf{i}}_2^{(\lambda=0,5)})$ we are going to omit the detail of the calculations. We will have in this case (see Table 21.8).

Nevertheless it could happen that in spite of having resorted to all the stages we have pointed out, those responsible for the purchase and sale operation of the business were not to have been in agreement on any of the three proposals for the interest rates. In this case the process should be repeated, starting out with other initial provisions, different from those established officially by the monetary authorities, in which the opinions, suggestions and criticisms that have arisen in the process followed up to this point would be included. And this as many times as necessary. The computer, fortunately resolves the problem of speed in calculations.

For operative effects, we are going to assume that estimates $\underset{\sim}{\mathbf{i}}_1^{(\lambda=0,5)}$, $\underset{\sim}{\mathbf{i}}_2^{(\lambda=0,5)}$ and $\underset{\sim}{\mathbf{i}}_3^{(\lambda=0,5)}$ have been accepted. We move on then to estimate the profits for the three years.

Table 21.8.

$[0{,}12;\ 0{,}14]^{(\lambda=0,5)}$

α				α		
0	1			0	1	
0,1	1			0,1	1	
0,2	1			0,2	1	
0,3	1			0,3	1	
0,4	1			0,4	1	
0,5	0,857	1	=	0,5	0,7990	1
0,6	0,714	1		0,6	0,5550	1
0,7	0,428	0,857		0,7	0,3320	0,7990
0,8	0,142	0,714		0,8	0,0990	0,5550
0,9	0,142	0,428		0,9	0,0990	0,3320
1	0	0,142		1	0	0,0990

$[0{,}588;\ 0{,}778]$

Table 21.9.

Purchaser's experts	[4.000; 6.000]	Seller's experts	[4.000; 6.000]
1	[0,1; 0,4]	1	[0,7; 0,9]
2	[0,2; 0,3]	2	1
3	0,3	3	[0,6; 0,8]
4	[0,3; 0,6]	4	[0,5; 0,7]
5	0,5	5	[0,4; 0,8]

21.4 Estimating Profits in Uncertainty

We will start out from certain initial estimates, expressed by means of a confidence interval as wide as is necessary, but that is accepted by both parties, purchasers and sellers. Therefore a joint meeting will establish a level below which the profit for one year will not descend and another above which there will be no profit in the corresponding year.

We continue with a numerical example and will assume that an initial agreement has been reached with the following estimates:

$$\underset{\sim}{\mathbf{B}}_1 = [4.000; 6.000]\,, \quad \underset{\sim}{\mathbf{B}}_2 = [3.000; 6.000] \quad \text{and} \quad \underset{\sim}{\mathbf{B}}_3 = [2.000; 5.000]$$

With the object of placing the profit for each one of the three years to be considered within each of the intervals $\underset{\sim}{\mathbf{B}}_1$, $\underset{\sim}{\mathbf{B}}_2$, and $\underset{\sim}{\mathbf{B}}_3$, each one of the parties resorts to a group of n experts (in this case we will assume five) who will give their opinion, also by means of intervals in $[0; 1]$.

Therefore for the profit estimated for the first year, the experts appointed by the purchasers and those appointed by the sellers give the following opinions:

The corresponding statistics are arrived at within the hendecadaire system (see Table 21.10).

And from there we arrive at the expertons and R^+-expertons (see Table 21.11).

Given that the mathematically expected values of the respective expertons are [0,28; 0, 42] and [0,64; 0, 82] we arrive at:

$$\varepsilon\left(\underset{\approx}{\mathbf{B}}{}_1^{(C)}\right) = 4.000 + (6.000 - 4.000)(\cdot)[0{,}28; 0{,}42] = [4.560; 4.840]$$

$$\varepsilon\left(\underset{\approx}{\mathbf{B}}{}_1^{(V)}\right) = 4.000 + (6.000 - 4.000)(\cdot)[0{,}64; 0{,}82] = [5.280; 5.640]$$

In this first phase we can already see an approximation between the positions of the purchasing experts and the selling experts, since the distance

Table 21.10.

Purchasers

	[4.000; 6.000]	
0		
0,1	1	
0,2	1	
0,3	2	
0,4		1
0,5		2
0,6	1	1
0,7		1
0,8		
0,9		
1		

Sellers

	[4000,00; 6000,00	
0		
0,1		
0,2		
0,3		
0,4	1	
0,5	1	
0,6	1	
0,7	1	1
0,8		2
0,9		1
1	1	1

Table 21.11.

	[4.000; 6.000]		
0		1	
0,1		1	
0,2	0,8		1
0,3	0,6		1
0,4	0		0,6
0,5	0,2		0,4
0,6	0		0,2
0,7	0		0
0,8	0		0
0,9	0		0
1	0		0

[0,28; 0,42]

Purchaser's experton

	[4.000; 6.000]		
0		1	
0,1		1	
0,2		1	
0,3		1	
0,4		1	
0,5	0,2		1
0,6	0,6		1
0,7	0,4		1
0,8	0,2		0,8
0,9	0,2		0,2
1	0,2		0,2

[0,64; 0,82]

Seller's experton

that separated both extremes of the interval 4.000 and 6.000 has been reduced, by something more than half. In fact the purchasing position has established a minimum of 4.560 while the seller has established a maximum of 5.640.

When both position are considered sufficiently close, this result can serve as the basis for the corresponding negotiation. When these circumstances are arrived at the technical process gives way to the political decision. In the event that the interval were to be considered as too wide, and, therefore, excessive uncertainty has been incurred, a revision should be made of the expert's opinions on the basis of the new interval $[4.560; 5.640]$. This can either be done separately, as was done before, or as is our criterion, grouping the new opinions of the purchasing and selling experts in order to arrive at a single R^+-expertons for each period. On this occasion we will use the second procedure.

After first consulting the ten experts on the positioning of the interval $[4.560; 5.640]$ we arrive at (see Table 21.12).

Table 21.12.

Expert	[4.560; 6.000]
1	[0,3; 0,4]
2	[0,2; 0,4]
3	0,1
4	[0,1; 0,3]
5	0,2
6	[0,7; 0,9]
7	[0,6; 0,8]
8	[0,9; 1]
9	0,8
10	[0,5; 0,8]

Data

[4.560; 5.640]		
0		
0,1	2	1
0,2	2	1
0,3	1	1
0,4		2
0,5	1	
0,6	1	
0,7	1	
0,8	1	3
0,9	1	1
1		1

Statistics

[4.560; 5.640]		
0	1	
0,1	1	
0,2	0,8	0,9
0,3	0,6	0,8
0,4	0,5	0,7
0,5	0,5	0,5
0,6	0,4	0,5
0,7	0,3	0,5
0,8	0,2	0,5
0,9	0,1	0,2
1	0	0,1

Experton

From the experton that has been arrived at we find the R^+-experton:

The mathematically expected value of the R^+-experton will be:

$$\varepsilon(\underset{\sim}{\mathbf{B}}_1) = [5.052; 5.198]$$

In this way we arrive at the corresponding profit for the first year that can be incorporated into later calculations on the value of the business.

Table 21.13.

	0	1			0	5.640	
	0,1	1			0,1	5.640	
	0,2	0,8	0,9		0,2	5.424	5.532
	0,3	0,6	0,8		0,3	5.208	5.424
	0,4	0,5	0,7		0,4	5.100	5.316
4.560+(5.640-4.560) (·)	0,5	0,5	0,5	=	0,5	5.100	5.100
	0,6	0,4	0,5		0,6	4.992	5.100
	0,7	0,3	0,5		0,7	4.884	5.100
	0,8	0,2	0,5		0,8	4.776	5.100
	0,9	0,1	0,2		0,9	4.668	4.776
	1	0	0,1		1	4.560	4.668
		Experton				R^+ – Experton	

Table 21.14.

Purchaser's experts	[3.000; 6.000]	Seller's experts	[3.000; 6.000]
1	[0,1; 0,4]	1	[0,7; 0,9]
2	0,4	2	0,7
3	[0,3; 0,6]	3	[0,7; 0,9]
4	[0,5; 0,7]	4	[0,5; 0,8]
5	0,3	5	[0,8; 0,1]

It is evident that, in the event that it were to be considered that the result arrived at should be the object of a transfer, it is possible to use, just as was done in the case of interest rates, a process of parameterisation. We will leave this process here as it would only give rise to unnecessary repetition, without providing anything further.

We will now move on to the estimate of profit for the second year. The experts establish their positions:

Table 21.15.

Expert	[4.140; 5.520]	
1	0,5	0,6
2	0,4	
3	0,3	0,5
4	0,2	0,6
5	0,3	
6	0,6	0,8
7	0,7	
8	0,8	0,9
9	0,8	
10	0,9	1

Data

Who provide the following expertons (see Table 21.15).
With the following mathematically expected values for the R^+-expertons:

$$\varepsilon\left(\underset{\sim}{\mathbf{B}}_2^{(C)}\right) = [4.140; 4.560]$$
$$\varepsilon\left(\underset{\sim}{\mathbf{B}}_2^{(V)}\right) = [4.980; 5.520]$$

The experts are asked to give their opinion relative to the interval [4.140; 5.520]. These opinions are shown in the Table 21.16.

Once the R^+-experton is arrived at, the following mathematically expected value is found:

$$\varepsilon(\underset{\sim}{\mathbf{B}}_2) = [4.899; 5.050]$$

that constitutes an estimate of profit for the second year.

Finally we move on to the estimate for profit for the third year.

The experts give the following positions in the Table 21.17.
which give rise to the following expertons in the Table 21.18.

We then immediately find the mathematically expected value of the R^+-expertons:

$$\varepsilon\left(\underset{\sim}{\mathbf{B}}_3^{(C)}\right) = [2.960; 3.560]$$
$$\varepsilon\left(\underset{\sim}{\mathbf{B}}_3^{(V)}\right) = [4.220; 5.400]$$

The position of the ten experts relative to interval [2.960; 4.400] is expressed in the Table 21.19.

Table 21.16.

	[3.000; 6.000]			[3.000; 6.000]	
0	1		0	1	
0,1	1		0,1	1	
0,2	0,8	1	0,2	1	
0,3	0,8	1	0,3	1	
0,4	0,8		0,4	1	
0,5	0,4	0,6	0,5	1	
0,6	0	0,6	0,6	0,8	1
0,7	0	0,2	0,7	0,6	1
0,8	0	0	0,8	0,2	0,8
0,9	0	0	0,9	0	0,4
1	0	0	1	0	0,2
	[0,38; 0,52]			[0,66; 0,84]	
	Purchaser's experton			Seller's experton	

Table 21.17.

Purchaser's experts	[2.000; 5.000]	Seller's experts	[2.000; 5.000]
1	[0,3; 0,4]	1	[0,6; 0,7]
2	[0,3; 0,6]	2	0,7
3	[0,2; 0,6]	3	[0,7; 0,9]
4	0,4	4	0,8
5	[0,4; 0,5]	5	0,9

Having found the corresponding experton and R^+-experton we arrive at the following mathematically expected value of the latter:

$$\varepsilon(\underset{\sim}{\mathbf{B}}_3) = [3.608; 3.723]$$

which can be taken as the estimate of profit for the third year.

In this way we now have available the necessary uncertain data for moving on to calculate the value of the business in the sphere of uncertainty.

Table 21.18.

α	[2.000; 5.000]	
0	1	
0,1	1	
0,2	1	
0,3	0,8	1
0,4	0,4	1
0,5	0	0,8
0,6	0	0,4
0,7	0	0
0,8	0	0
0,9	0	0
1	0	0
	[0,32; 0,52]	

Purchaser's experton

α	[2.000; 5.000]	
0	1	
0,1	1	
0,2	1	
0,3	1	
0,4	1	
0,5	1	
0,6	1	
0,7	0,8	1
0,8	0,4	0,6
0,9	0,2	0,4
1	0	0
	[0,74; 0,80]	

Seller's experton

Table 21.19.

Expert	[2.960; 4.400]	
1	0,3	0,6
2	0,2	0,4
3	0,4	
4	0,1	0,2
5	0,3	
6	0,6	0,8
7	0,7	0,9
8	0,5	
9	0,8	
10	0,8	0,9

Data

21.5 Arriving at the Value of the Business

Once the estimates arrived at by the process we have described are accepted as valid, we now move on to select the criterion for valuation that will be taken as the basis for calculations. If the following formula is accepted:

$$V_e = V_s + (B_1 - i_1 \cdot V_e) \cdot (1+i_1)^{-1}$$
$$+ \cdots + (B_n - i_n \cdot V_e) \cdot (1+i_n)^{-1} \cdot (1+i_2)^{-1} \cdot \cdots \cdot (1+i_n)^{-1}$$

the following nomenclature can be adopted for reasons of simplicity:

$$I_1^{-1} = (1+i_1)^{-1}$$
$$I_2^{-1} = (1+i_1)^{-1} \cdot (1+i_2)^{-1}$$
$$I_3^{-1} = (1+i_1)^{-1} \cdot (1+i_2)^{-1} \cdot (1+i_3)^{-1}$$

which gives rise to an expression that is very simple and suitable for immediate application:

$$V_e = \frac{V_s + B_1 \cdot I_1^{-1} + B_2 \cdot I_2^{-1} + B_3 \cdot I_3^{-1}}{1 + i_1 \cdot I_1^{-1} + i_2 \cdot I_2^{-1} + i_3 \cdot I_3^{-1}}$$

In order to arrive at V_e, it is necessary to have available, apart from the estimates for interest rates and profits for each period that have already been arrived at (uncertain variables expressed by confidence intervals), the fundamental value V_s that must be calculated according to commonly accepted criteria of which mention has been made in the first section of this chapter. Normally there should be no insoluble problems that prevent the consideration of this calculation element in certain terms[13]. For this we will start out from the case that a fundamental value has been arrived at that is accepted by both parties. In the event there were not to be acceptance a revision should be made of the mass or masses or patrimonial element that is the object of discussion.

In our case, the experts have established and purchasers and sellers accepted, a fundamental value $V_s = 28.000$.

Below is a summary of the estimates:

$$V_e = 28.000$$
$$\underset{\sim}{\mathbf{i}}_1 = [11{,}31; 11{,}61], \quad \underset{\sim}{\mathbf{i}}_2 = [13{,}17; 13{,}55] \quad \text{and} \quad \underset{\sim}{\mathbf{i}}_3 = [8{,}95; 9{,}20] \quad \text{in } \%.$$
$$\underset{\sim}{\mathbf{B}}_1 = [5.035; 5.173], \quad \underset{\sim}{\mathbf{B}}_2 = [4.899; 5.050] \quad \text{and} \quad \underset{\sim}{\mathbf{B}}_3 = [3.068; 3.723]$$

[13] The incorporation of the fundamental value in the uncertain form in no way alters or complicates the process that is followed from here on.

From the interest rates we arrive at the corresponding updating rates:

$$(1(+)\underset{\sim}{\mathbf{i}}_1)^{-1} = (1(+)\,[0{,}1131; 0{,}1161])^{-1} = \left[\frac{1}{1{,}1161}, \frac{1}{1{,}1131}\right] = [0{,}8959; 0{,}8983]$$

$$(1(+)\underset{\sim}{\mathbf{i}}_2)^{-1} = (1(+)\,[0{,}1317; 0{,}1355])^{-1} = \left[\frac{1}{1{,}1355}, \frac{1}{1{,}1317}\right] = [0{,}8806; 0{,}8836]$$

$$(1(+)\underset{\sim}{\mathbf{i}}_3)^{-1} = (1(+)\,[0{,}0895; 0{,}0920])^{-1} = \left[\frac{1}{1{,}0920}, \frac{1}{1{,}0895}\right] = [0{,}9157; 0{,}9178]$$

where:

$$\underset{\sim}{\mathbf{I}}_1^{-1} = (1 + \underset{\sim}{\mathbf{i}}_1)^{-1} = [0{,}8959; 0{,}8983]$$
$$\underset{\sim}{\mathbf{I}}_2^{-1} = (1 + \underset{\sim}{\mathbf{i}}_1)^{-1} \cdot (1 + \underset{\sim}{\mathbf{i}}_2)^{-1} = [0{,}7889; 0{,}7937]$$
$$\underset{\sim}{\mathbf{I}}_3^{-1} = (1 + \underset{\sim}{\mathbf{i}}_1)^{-1} \cdot (1 + \underset{\sim}{\mathbf{i}}_2)^{-1} \cdot (1 + \underset{\sim}{\mathbf{i}}_3)^{-1} = [0{,}7224; 0{,}7284]$$

From here the updated values for profits and interest rates will be:

$$\underset{\sim}{\mathbf{B}}_1(\cdot)\underset{\sim}{\mathbf{I}}_1^{-1} = [5.035; 5.173]\,(\cdot)\,[0{,}8959; 0{,}8983] = [4.510; 4.646]$$
$$\underset{\sim}{\mathbf{B}}_2(\cdot)\underset{\sim}{\mathbf{I}}_2^{-1} = [4.899; 5.050]\,(\cdot)\,[0{,}7889; 0{,}7937] = [3.864; 4.008]$$
$$\underset{\sim}{\mathbf{B}}_3(\cdot)\underset{\sim}{\mathbf{I}}_3^{-1} = [3.608; 3.723]\,(\cdot)\,[0, 7224; 0{,}7284] = [2.606; 2.711]$$

$$\underset{\sim}{\mathbf{i}}_1(\cdot)\underset{\sim}{\mathbf{I}}_1^{-1} = [0{,}11315; 0{,}1161]\,(\cdot)\,[0{,}8959; 0{,}8983] = [0{,}1013; 0{,}1042]$$
$$\underset{\sim}{\mathbf{i}}_2(\cdot)\underset{\sim}{\mathbf{I}}_2^{-1} = [0{,}1317; 0{,}1355]\,(\cdot)\,[0{,}7889; 0{,}7937] = [0{,}1038; 0{,}1075]$$
$$\underset{\sim}{\mathbf{i}}_3(\cdot)\underset{\sim}{\mathbf{I}}_3^{-1} = [0{,}0895; 0{,}0920]\,(\cdot)\,[0{,}7224; 0{,}7284] = [0{,}0646; 0{,}0670]$$

And finally we arrive at:

$$V_e = \frac{28.000(+)\,[10.980; 11.365]}{1(+)\,[0{,}2697; 0, 2787]} = \left[\frac{38.980}{1{,}2787}, \frac{39.365}{1{,}2697}\right] = [30.484; 31.003]$$

Therefore, the value of the business that has been estimated in an environment of uncertainty will be shown by the confidence interval [30.484; 31.003], which is equivalent to stating that it will be no less than 30.484 monetary units nor higher than 31.003 monetary units. There is therefore a margin for manoeuvre of 519 monetary units, which will be the object of final negotiation between purchasers and sellers.

At this point we could consider what could be done in the event that the confidence interval were too wide (the value too uncertain) for it to constitute a firm basis for agreement in the negotiation. We feel that the answer could be arrived at in the same sense of the use of the case for interest rates and

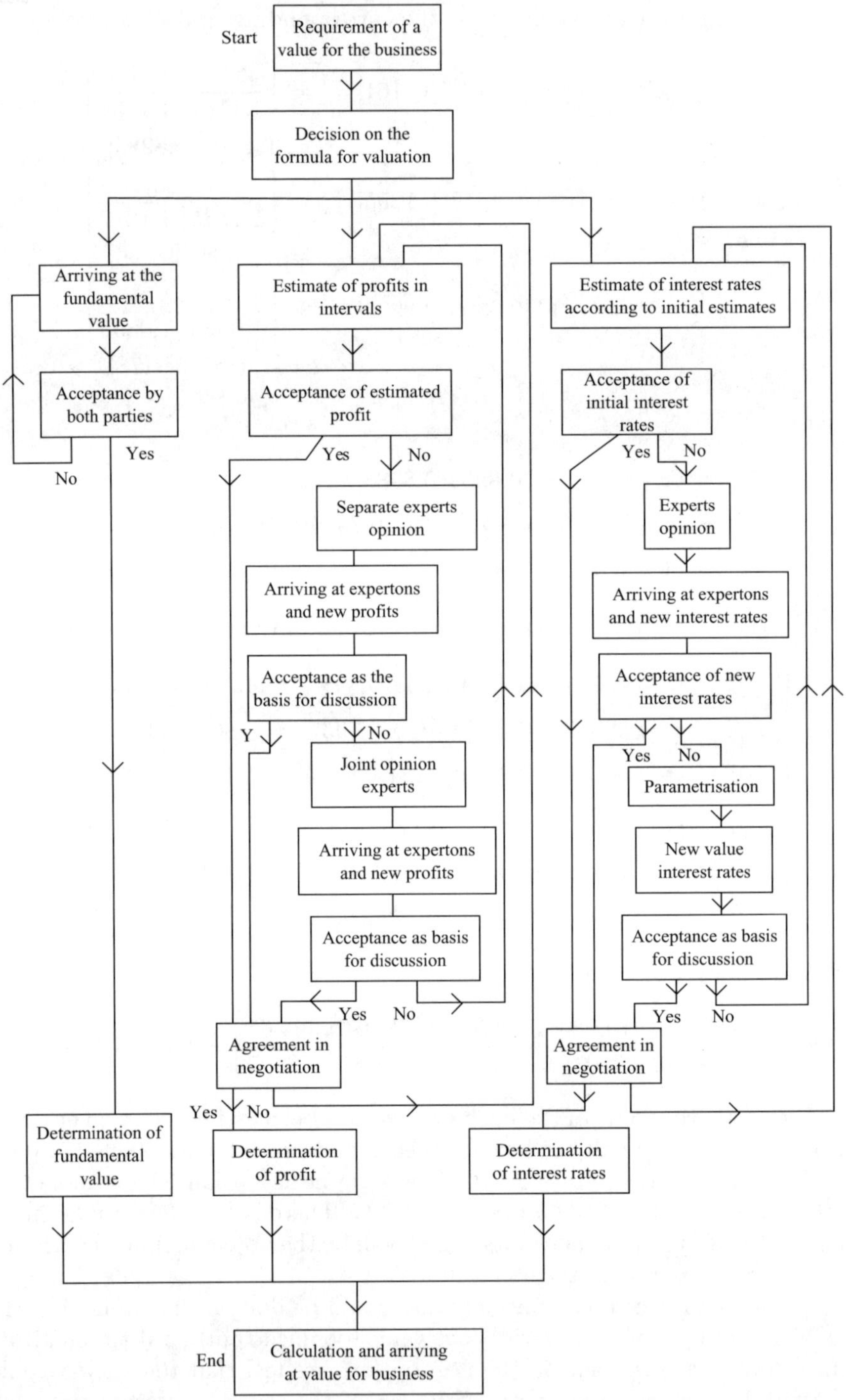
Start
Requirement of a value for the business
Decision on the formula for valuation
Arriving at the fundamental value
Acceptance by both parties
Yes
No
Estimate of profits in intervals
Acceptance of estimated profit
Yes
No
Separate experts opinion
Arriving at expertons and new profits
Acceptance as the basis for discussion
Y
No
Joint opinion experts
Arriving at expertons and new profits
Acceptance as basis for discussion
Yes
No
Agreement in negotiation
Yes
No
Estimate of interest rates according to initial estimates
Acceptance of initial interest rates
Yes
No
Experts opinion
Arriving at expertons and new interest rates
Acceptance of new interest rates
Yes
No
Parametrisation
New value interest rates
Acceptance as basis for discussion
Yes
No
Agreement in negotiation
Determination of fundamental value
Determination of profit
Determination of interest rates
End
Calculation and arriving at value for business

Fig. 21.1.

profits: by means of counter experts, the construction of expertons and R^+-expertons. The use of this technique constitutes an important aid, as we have seen, for a large number of problems arising in the financial reality of our times.

As a summary and with the object of systemising the model we have explained, we have drawn the following Fig. 21.1 in which all actions required for arriving at the value of the business are established.

It will be seen that in its construction the same steps have been followed as those existing for the numerical example. It is obvious that this organisation chart can be modified if the phases that comprise the proposed process are extended or reduced. In this respect the significant fact is the incorporation of parameterisation in the estimate for interest rates, while this has not been used for estimating profits. The inclusion of this phase would obviously modify the organisation chart. We have attempted to maintain the chart with the same basic assumptions as used in developing the model.

Finally we would like to point out that the model we have proposed is susceptible to modification if the hypotheses that have been used throughout this development are modified. Our only intention has been to open up the way for carrying out new works in which the use of expertons and parameterisation may give rise to fruitful results when a problem arises such as valuation, in which negotiation occupies a position of importance, where subjectivity plays a fundamental role.

22 The Setting up of a Financial Product

22.1 Introduction

The economic and financial context of our day and age is characterised by a succession of situations, in which facts and phenomena that affect business, at all levels, take place with a speed and intensity without precedent in history. It is for this reason that economic entities, the object of which is to render services or launch financial products, carry out their activities in an increasingly complex framework. But this complexity is also aggravated by the enormous concurrence existing in the sector, as a consequence of the continued and "apparent" varied offer of financial products and services that can be found in the market, the "real" characteristic of which is its high degree of homogeneity. These fact make it necessary, if not essential, for financial institutions to set up and strengthen a "reserve" or "bank" of ideas[1] destined to maintain a stock of products that are suitable for launching in the market when required.

In this sense, the efforts which business executive have to make for continued adaptation to new situations is important, but at the same time essential if what is sought is survival in the age of extreme, even fierce competition.

Therefore, in summary, what this is, is an attempt to have available new products and/or services for launching at the precise moment onto the market that requires the same, in this way anticipating concurrent institutions which must initiate a long path that stretches from the birth of the idea right up to placing the product or service on the market.

In order to make the birth of new ideas possible for launching new financial products or services on the market, or with the object of improving already existing products, we have at our disposal several different procedures.

[1] This subject has been included in a work: Kaufmann A, Gil Aluja J and Gil Lafuente AM (1994) La creatividad en la gestión de las empresas, (Ed) Pirámide, Madrid, Spain. the technical elements of which, basically pp. 59–61, 81–91 and 150–170 constitute the support for this work.

Anna Maria Gil-Lafuente: *Fuzzy Logic in Financial Analysis*, StudFuzz **175**, 411–426 (2005)
www.springerlink.com

22.2 Elements for Forming an Idea Bank

When a permanent mutability exists of economic and financial phenomena, specialists on the subject bring into operation a series of management instruments which will allow the executives of financial institutions to bring into practice the setting up and maintenance of an idea bank.

One of these instruments is the creativity circle with which a creative dynamism can be attained within financial institutions without which it would be difficult to maintain the necessary levels of competitiveness required by current day economic systems.

The organisation and development of these creativity circles is made easy, given the current status of data processing technology, with the use of computers. Although it is man who thinks, imagines, creates it will be the computer, which will make calculations more agile with the processing of information, in accordance with certain parameters that must be provided, by doing the calculations in fractions of seconds.

The efficiency of understanding between man and machine can be found in specialisation; the human being must carry out the activities in those tasks that the machine cannot do in the current state of technical knowledge: imagine, create, in short think for itself. But this is no obstacle for leaving to computers all those repetitive tasks and based on complying with orders, tasks that they can do far more efficiently since electronic equipment does not by itself commit errors nor disobey orders. It is in this sense that the computer is becoming a very close ally in the creative process.

The interest for creative activity has acquired certain nuances and given rise to different paths of research. We will now move on to give a brief description of those we consider the most interesting.

22.3 The Tree Method

We will commence this brief description by mentioning the so-called tree method. This is a set of rules that have their origin in what is known as decision trees, presented by Fustier[2] and applied on different occasions in the field of inventive research.

The process commences with the definition of the objective of the research. In a second stage the main guidelines are established, and finally for each of the latter the known theoretical means are brought to light in order to reach the desired objective.

When the tree method is used for creative activities it is also called the "tree of discovery". The use of discovery trees is susceptible to application

[2] Fustier M (1980) La logique de l'arbre (in French) Ecully-Rhône, ALGOE.

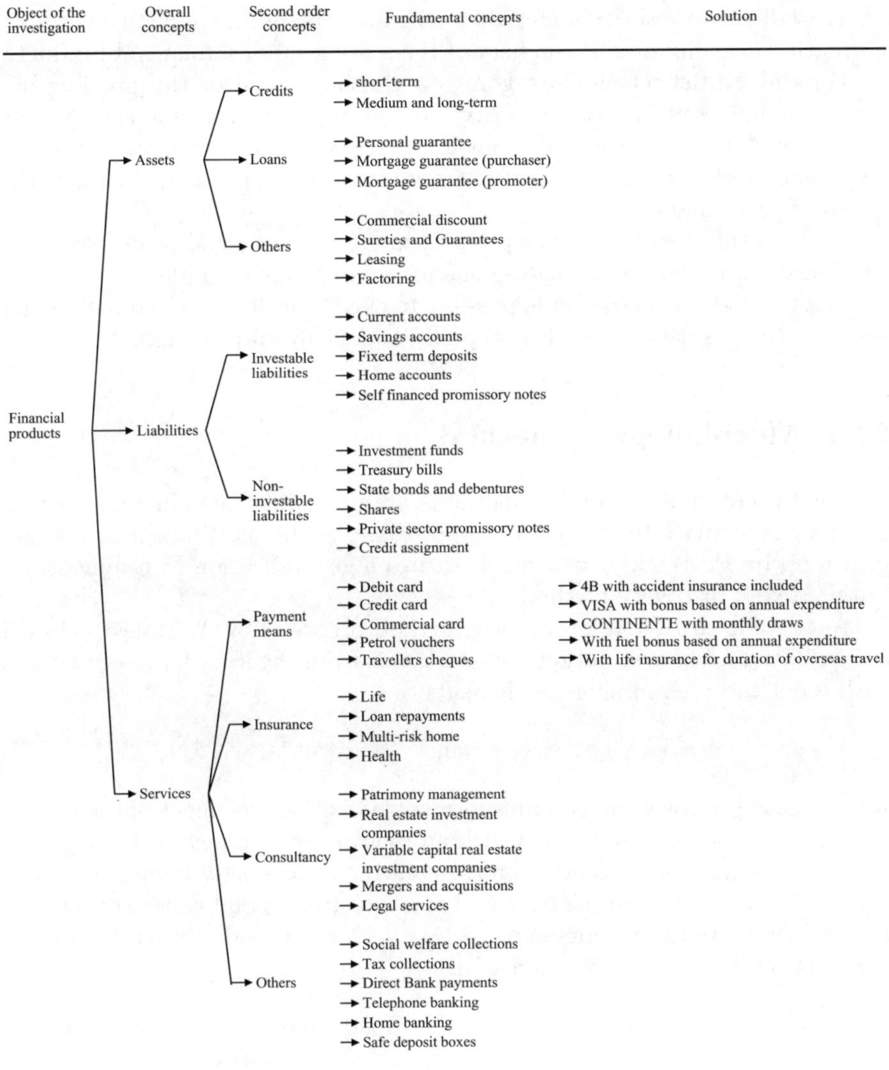

Fig. 22.1.

to any creative problem it is desired to resolve or investigate, both if it is normative research[3] or non normative.

[3] Research is called normative when the object of the investigation is established beforehand and the purpose is centered on the creation of models and instruments for attaining the set objective. On the other hand, we talk about non normative research when the object of the investigation is not specified.

We will now use the tree method as the instrument for carrying out a non-normative investigation in the field of the setting up of a financial product.

For didactic effects we are going to consider some of the possible approaches that could be done relative to the object of investigation, taking into account that the concepts that arise, as well as the different solutions, may vary based on the scope of the study of the financial institution, the group of participants, etc.

On the other hand, of all the possible solutions that could be derived from the concepts, we will only analyse one for them as an example.

With this three method it is possible to carry out different types of analyses from the casuistry offered by the different financial products.

22.4 Morphological Method

Among the creativity analytical methods, standing out due to its repercussion in research media into this area of knowledge, is the morphological method drawn up by F. Zwicky[4], which constitutes a generalisation to n-dimensions of the Moles[5] discovery method.

As is sufficiently well known, the method proposed by A. Moles[6] is based on the notion of bi-association and starts out from the knowledge of two sets with a not too high number of elements:

$$A = \{a_1, a_2, \ldots, a_n\} \quad \text{and} \quad B = \{b_1, b_2, \ldots, b_n\}$$

which, placed as rows and columns form a matrix. Each element of the matrix constitutes a bi-association for which an innovating significance is sought.

Concentrating our attention on the creativity process for a financial product and only as an example we could consider two second order concepts of the tree presented in the previous section such as A: "non investable liabilities" and B: insurance". We arrive at:

A = {Investment funds, Treasury bills, Bonds and Treasury debentures, Shares, Private Sector promissory notes, credit assignment}.

B = {life, loan pay-back, home multi-risk, health}

[4] Zwicky F Reference taken from Kaufmann A, Gil Aluja J and Gil Lafuente AM (1994) La creatividad en la gestión de las empresas (in Spanish). (Ed) Pirámide, Madrid, Spain.

[5] Moles A and Caude R (1972) Creativité et méthodes d'innovation (in French). Paris, France. (Ed) Fayard, Mane Collection Management.

[6] Moles A and Caude R, op. cit.

Non investable liabilities \ Insurance		Life b_1	Loan Pay-back b_2	Home Multi-risk b_3	Health b_4
Investment funds	a_1	[0,8; 0,9]	0	0	0
Treasury bills	a_2	[0,9; 1]	0	0	0
Treasury Bonds and Debentures	a_3	[0,9; 1]	0	0	0
Shares	a_4	[0,3; 0,5]	1	[0,7; 0,9]	0
Private sector promissory notes	a_5	[0,7; 0,9]	0	0	0
Credit assignment	a_6	[0,5; 0,6]	[0,4; 0,7]	[0,3; 0,6]	0

Each one of the pairs is valued in $[0; 1]$ according to their possibilities of being taken into account for the effects of constituting a new financial product. In this way we could consider:

$\{a_1, b_1\}$: life insurance associated with holding an investment fund.
$\{a_2, b_1\}$: life insurance associated with holding treasury bills.
$\{a_3, b_1\}$: life insurance associated with holding treasury bonds and debentures.
$\{a_4, b_1\}$: life insurance associated with holding shares.
$\{a_4, b_2\}$: insurance of loan repayment associated when financing is available for purchasing shares.
$\{a_4, b_3\}$: multi risk insurance associated when the shares belong to real estate company.
$\{a_5, b_1\}$: life insurance associated to holding company promissory notes.
$\{a_6, b_1\}$: life insurance associated with holding a credit assignment.
$\{a_6, b_2\}$: insurance for loan repayment associated to a credit assignment for covering possible contingencies.
$\{a_6, b_3\}$: multi risk home insurance associated to when the credit assignment proceeds from real estate companies.

The Zwicky method takes a further step forward, in order to decompose the object of thought, in our case a financial product, in all the parts in which it is formed, independently of its possibility of being carried out.

In this way, the different combinations of these primary and fundamental elements that constitute a financial product are successively grouped. In this way hypothetical products are arrived at that may be known or unknown, useful or useless, realisable or not realisable, etc., according to the combination considered.

With this preamble, we are now ready to use this method in an attempt to create a new financial product, in this case for example a credit. For greater simplicity we will only take 6 characteristics, each of which may adopt several different forms. In this way we have:

A. Price
- a_1 : EURIBOR rate
- a_2 : EURIBOR + 1%
- a_3 : EURIBOR + 2%

B. Amount
- b_1 : less than 10 million
- b_2 : between 10 and 50 million
- b_3 : between 50 and 200 million
- b_4 : more than 200 million

C. Payback period
- c_1 : less than 12 months
- c_2 : between 12 and 18 months
- c_3 : over 18 months

D. Speed in granting
- d_1 : before 24 hours
- d_2 : between 24 and 48 hours
- d_3 : between 48 hours and one week
- d_4 : more than a week

E. Possibilities for renewal
- e_1 : slender
- e_2 : average
- e_3 : high

F. application commissions
- f_1 : none
- f_2 : opening
- f_3 : opening and no draw down

We now move on to assign the generally accepted denominations to the case we are presenting. Therefore the elements, that are mutually substitutable, corresponding to a determined characteristic give rise to the concept of "formative set".

The gathering together of all the formative groups gives rise to a "general morphology". In our case, the general morphology is:

$$
\begin{aligned}
A &= \{a_1, a_2, a_3\} \\
B &= \{b_1, b_2, b_3, b_4\} \\
C &= \{c_1, c_2, c_3\} \\
D &= \{d_1, d_2, d_3, d_4\} \\
E &= \{e_1, e_2, e_3\} \\
F &= \{f_1, f_2, f_3\}
\end{aligned}
$$

In order to commence the creativity process for a financial product, a random choice is made of an element of each of the formative groups of the previously described morphology[7]. On this occasion the following sextuplet has appeared as the combination:

$$(a_3,\ b_3,\ c_3,\ d_4,\ e_1,\ f_3)$$

In which, by analysing the significance of each of the elements, we will attempt to find a new financial product that contains, in a determined degree, the characteristics that have been pointed out.

For this, the new product should contain:

a_3: price EURIBOR + 2%
b_3: between 50 and 200 million
c_3: payback more than 18 months
d_4: granted in a period of more than 1 week
e_1: slender possibilities for renewal
f_3: application of opening and non-draw down commissions

A quick glance will tell us that this possible mode of credit will have a high price and be granted in large amounts. Its study by the financial institution will be done in great and scrupulous detail, although from the outset it will be felt that from it a credit mode may arise that is designed for medium to large sized businesses in the incorporation phase, that require a large amount of resources in order to start up their activities. Its viability is justified since we are dealing with incipient businesses in the market and, therefore, their capacity for negotiating prices and commissions is very limited. Also for this reason it can be imagined that these resources will be required by the business over a period that is longer than an accounting period, but when the borrowing business has been capitalised it will no longer require this financial product. This would justify the characteristic why the possibilities for renewal are slender.

We are now in a position to commence the process of inventive creativity. But before this we are going to look at, although very briefly, how a creativity group is formed.

22.5 Setting up a Creativity Group

With the object of allowing ideas to emerge from the different combinations of elements of each one of the formative sets, what is required is the setting up of a creative group. In this case, we are going to assume that eight people take part, of whom only three are experts; the other five should not be specialists but have a similar social and cultural level in order that communications should not be interrupted.

[7] In this very elementary case there already appear 1,296 possible combinations.

As coordinator of this group the central element of the process will be selected: the moderator or leader, whose task consists in stimulating interchange, summarising ideas, provoking the imagination, discovering elements from the unconscious, etc. Among the tasks is that of proposing the acceptance of new ideas after discussion and the consequent valuation relative to 3 criteria: coherence, innovation and possibility of realisation.

Finally forming the group will be an "operator" whose mission consists in introducing the data that appear in a computer for treatment of the data and in this way be able to have the results available quickly.

Parallel to the forming of the creativity group, the organisation of the sessions that are to be held should be established in detail. In this sense, it is generally accepted as ideal that the creativity meetings should have a duration of two days split over 4 sessions.

Meetings should take place in a room set out in such a way that the operator and computer are located at one end; in the centre there should be a table for five of the participants; the moderator should occupy a place of privilege; and finally at the other end of the room another table that constitutes what we call the "purgatory" at which will be seated three of the participants in shifts, who have a vote but no voice and can only give their opinions by written messages. Relegated to this purgatory are those participants who monopolise the conversations and in this way, momentarily separated, other members are allowed to voice an opinion and participate more actively in the discussion.

22.6 Development of the Activity of the Group

Once the role to be played by each participant in the creativity group is known, the sessions commence by bringing to light the ideas and associations that arise from the same.

When the moderator accepts one of these, a valuation is given from 3 different points of view, which correspond to the already mentioned criteria:

- The criterion of coherence of all the formative elements of the new idea (c).
- The criterion of technical innovation and commercial innovation (i_t and i_c).
- The criterion of its possible realisation (r).

These criteria will later be aggregated. We propose "expertons" as an instrument that will permit us to do the aggregation of all the valuation provided by the eight participants in the creativity group. As is normal, each valuation will be expressed in a confidence interval in $[0; 1]$.

In the first place we introduce the previously established criterion of coherence.

Obviously the three specialists in the group play a more significant role when valuating the criteria of innovation and realisation, as a consequence of their profound knowledge on the subject, so that they can contribute, in this sense, comments and clarification of a high technical value.

We are now ready to commence the development of the process that should lead to the introduction into the bank of new financial products in the credit mode for immediate release in the market, using the STIM 6 method for inventive stimulation.

22.7 Coherence in the Association of Ideas

We start out from the previously described morphology, which brings to light the different characteristics of a credit, as well as the first random extraction arrived at from each one of the formative sets. Taking these elements, a square matrix is drawn up in which in each box appears, that is for each bi-association, the valuation given relative to the criterion of coherence. Below we have shown the corresponding matrix:

	a_3	b_3	c_3	d_4	e_1	f_3
a_3	1	[0,7; 0,9]	[0,6; 0,9]	[0,4; 0,5]	[0,8; 0,9]	[0,7; 0,8]
b_3		1	1	[0,9; 1]	[0,6; 0,8]	[0,6; 0,7]
c_3			1	0,8	[0,8; 0,9]	[0,7; 0,9]
d_4				1	[0,9; 1]	[0,6; 0,7]
e_1					1	[0,8; 0,9]
f_3						1

The valuation corresponding to each bi-association has been arrived at with the aid of the R^+-expertons. For this the opinion has been sought of a group of experts from the research department for the launching of new products for the financial entity, who have expressed the same by means of valuations in $[0; 1]$, using the hendecadaire system.

As an example and for the effects of not drawing this work out too much, we will analyse some of these bi-associations used in this case.

(a) Bi-association $\{a_3, b_3\}$: {EURIBOR + 2%, between 50 and 200 million}.

These two characteristics reveal a credit of a high value at a relatively expensive price. The ideas proposed led towards the possibility of awarding punctuality in payments relative to the payback of the principal and interest payments, as only those businesses with a low level of negotiation regarding the price would be in a position to take on a debt at such a high interest rate.

Therefore among the proposals the following appeared: credits negotiated at EURIBOR + 2%, between 50 and 200 million for which, if the businesses pay punctually at the time of renewal or in the event of renegotiating the interest rates, the same would be decreased by 0,25% successively until reaching a minimum of EURIBOR + 0,75%.

(b) Bi-association $\{b_3,\ c_3\}$: {between 50 and 200 million, payback longer than 18 months}

The proposals from the group relative to these two characteristics revolve around: increasing the pay-back period in 6 additional months if payment has been punctual at all times; in the event of renewal, decrease the interest rate by 0.25%, or eliminate commissions, provided there has been no delay in paying interest.

(c) Bi-association $\{c_3,\ d_4\}$: payback longer than 18 months, granting in over a week}

These two characteristics highlight the need for studying in certain detail those credits payback period, which exceeds 12 months. The idea presented by the group is centred around awarding early payback of the credit, as well as punctuality in interest payments. In this way pay-back of 25% of the credit every 6 months will give rise to a reduction in the interest rate of between 0,1% and 0,25% of the initially agreed upon interest rate and of the interest rates current in the market at the time of revision.

(d) Bi-association $\{d_4,\ e_1\}$: {granting in over 1 week, slender possibilities for renewal}

These characteristics bring to light the need for a credit by a business, either for starting up its production activity or for restructuring a certain aspect or section of the business.

Therefore, here we are dealing with the need to analyse granting financial resources for a specific situation in a business.

The proposals arising from the group of participants specifically took into account the fact of the continuity of the relationship commencing between the financial entity and the business. Here there were several possibilities: if the payments corresponding to interest and return of the principal take place in a strictly punctual manner and it is seen that the activity of the business, by means of its collections and payments is carried on quite normally in accordance with forecasts, the business could be rendered certain services free of charge (handling receipts, direct debit for salaries, a reduction in certain tariffs such as commissions applied to transfers, issuing cheques, etc.).

(e) Bi-association $\{e_1,\ f_3\}$: {possibilities for renewal slender, commissions for opening and no draw down}

With regard to these characteristics the participants are unanimous in the need to encourage, on the one hand, the undertaking to pay interest and, in second place, in gaining the loyalty of the client by offering a whole range of financial products under better conditions than those currently offered by the market. Therefore the ideas put forward were mainly the following: if there has been punctuality in payments, new alternative products could be offered at the same interest rate but charging 0,5% less for commissions. In this way, each time the business initiates any financial operation with its financial entity it would see commissions decreasing until they totally disappeared. The borrower would find advantages in this by the mere fact of carrying out its operations with one and the same lending institution, and the latter would have gained the clients loyalty.

Broadly speaking these could be, among many others, some of the ideas that could be extracted from an analysis of the different bi-associations.

22.8 From Tri-Associations to Coherent n-Associations

Having reached this point it is now necessary to establish a coherence threshold. In this sense we are only going to consider those bi-associations the mathematically expected value of which reaches a level of 0,8 relative to the degree of coherence, which would be translated in the hendecadaire scale into the fact that the ideas proposed by the group of participants should be "nearly coherent". The Φ-fuzzy relation, would then be converted into the following Boolean matrix.

	a_3	b_3	c_3	d_4	e_1	f_3
a_3	1	1	0	0	1	0
b_3		1	1	1	0	0
c_3			1	1	1	1
d_4				1	1	0
e_1					1	1
f_3						1

For obtaining each one of the elements of this Boolean matrix we have assigned a 1 to those boxes the valuation of which reaches the threshold of coherence and a 0 in the remainder.

When the valuation assigned to a determined bi-association is expressed by means of a confidence interval, in order to become aware as to whether it reaches the threshold of coherence we have opted for making the entropy fall by adding the two extremes and dividing the result by 2.

In this way, by means of a graph, we could represent the relations that arise with the application of the threshold of coherence:

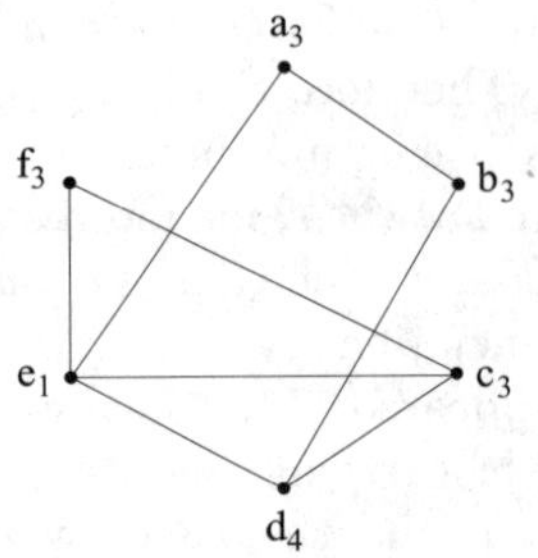

Fig. 22.2.

A simple glance shows that, for this level of coherence that there are two tri-associations:

(a) $\{b_3, c_3, d_4\}$: between 50 and 200 million. Payback longer than 18 months, granting over one week)

From this tri-association the following idea was put forward by the members of the group: taking as the basis the award for prompt payment of interest and early pay-back of the principal, when 25% of the credit is paid-back every 6 months there is an automatic reduction in the interest rate of between 0,1 and 0,25% based on the initial price and market conditions.

(b) $\{c_3, e_1, f_3\}$: {payback longer than 18 months, slender possibilities of renewal, opening and no draw down commissions}

Proposals made by the group are based on establishing an award for early payback and gaining the loyalty of the client. The ideas are to reduce interest rates from between 0,1 and 0,25% provided there is a 25% payback of the credit every 6 months, as well as a decrease in commissions of 0,5% for successive operations with the same financial entity, provided there is at least one operation a year.

Finally an analysis is made of the penta-association arising as a consequence of the established coherence threshold:

(c) $\{a_3, b_3, c_3, d_4, e_1\}$: {EURIBOR + 2%, between 50 and 200 million, payback longer than 18 months, granted in over 1 week, slender possibilities of renewal}

The opinion of the participants is directed towards awarding punctual payment, early payback and gaining the loyalty of the client.

Among the ideas that are proposed we can mention the following: if the business pays punctually, it will have the option of paying back 25% of the loan every 6 months in exchange for a reduction in the interest rate of 0,25% as well as a reduction of 0,50% in commissions on successive operations that it may do with the same financial entity, with the additional possibility that these loans may be granted in a minimum period.

22.9 Criterion of Innovation

Once the ideas have been analysed under the criterion of coherence it will be necessary to submit the proposals that have exceeded the established threshold to the criteria of technological innovation and commercial innovation. For analysing these criteria the opinion and experience of the experts that form the creativity group is essential.

The case of technological innovation in this case is not important, because the meeting of the group has as its objective the offer of services that are not present in the market. The technology for which they are designed is, therefore, irrelevant.

By following the guidelines set for the criterion of coherence, those ideas are analysed that attain a certain level of innovation, which it is necessary to predetermine. Therefore consideration could be given for example that the minimum level for commercial innovation is 0,9 (practically innovative).

On analysing the valuations corresponding to the criteria of commercial innovation from the previously obtained bi-associations, which exceeded the threshold of coherence, we arrive at a Φ-fuzzy relation, which could be as follows:

	a_3	b_3	c_3	d_4	e_1	f_3
a_3	1	[0,9; 0,1]	0	0	1	0
b_3		1	[0,4; 0,5]	0,9	0	0
c_3			1	[0,9; 1]	[0,7; 0,9]	0,8
d_4				1	1	0
e_1					1	0,9
f_3						1

This Φ-fuzzy relation is converted into the following Boolean matrix if an innovation threshold of $\alpha = 0{,}9$ is considered. For this we use the same procedure as for the previous criterion:

	a_3	b_3	c_3	d_4	e_1	f_3
a_3	1	1	0	0	1	0
b_3		1	0	1	0	0
c_3			1	1	0	0
d_4				1	1	0
e_1					1	1
f_3						1

From the associated graph we can see the n-associations that appear under the criterion of commercial innovation:

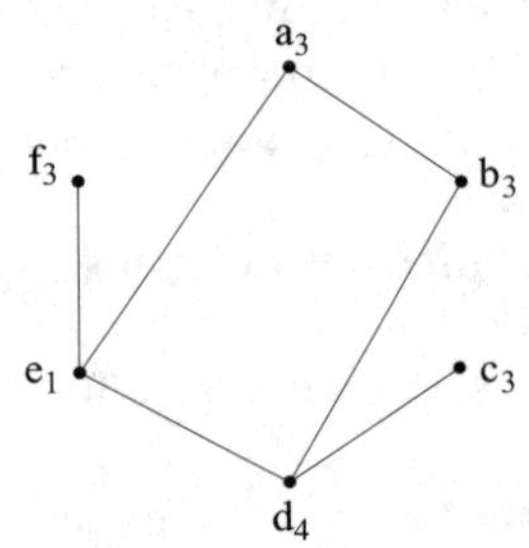

Fig. 22.3.

Immediately we can see the presence of a cuatri-association formed, in this case, by elements $\{a_3, b_3, d_4, e_1\}$ corresponding to characteristics {EURIBOR+ 2%, between 50 and 200 million, granted in over one week, slender possibilities for renewal}. It will be seen that this corresponds to the same previous combination of elements in which c_3 has been eliminated.

Relative to this association of characteristics, the group proposes the following idea: provided the business pays the corresponding interest expiry dates the borrower will be awarded with a half yearly reduction of the interest rates of 0,25%. Also, provided it carries out more than one operation per year with this financial entity. The business will be benefited by the immediate granting of the operations it may do in a standard manner and provided the volume is between 50 and 200 million}

22.10 Possibility of Realisation

Having analysed the previous proposal of the members of the group we must now take a look at the criterion of feasibility. What has to be done now is to analyse the possibilities that this financial product being put into practice. The threshold of possibilities of its realisation is established at 0,9 (practically feasible).

If we analyse each of the bi-associations that have exceeded the criterion of commercial innovation we can construct the following Φ-fuzzy relation:

	a_3	b_3	c_3	d_4	e_1	f_3
a_3	1	1	0	0	0,9	0
b_3		1	0	[0,9; 1]	0	0
c_3			1	1	0	0
d_4				1	[0,9; 1]	0
e_1					1	1
f_3						1

Transforming this Φ-fuzzy relation into a Boolean matrix, based on the feasibility criterion established we arrive at:

	a_3	b_3	c_3	d_4	e_1	f_3
a_3	1	1	0	0	1	0
b_3		1	0	1	0	0
c_3			1	1	0	0
d_4				1	1	0
e_1					1	1
f_3						1

The associated graph of the above Boolean relation is as follows:

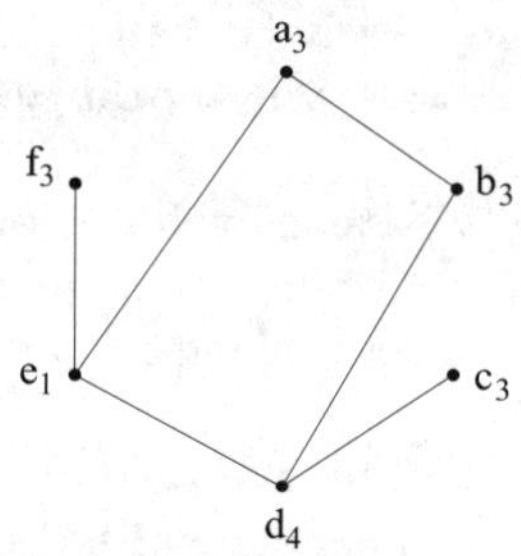

Fig. 22.4.

From the same it can be seen that the same cuatri-association exists as the previous one formed by elements $\{a_3, b_3, d_4, e_1\}$ corresponding to characteristics {EURIBOR + 2%, between 50 and 200 million, granted in over one week, slender possibilities for renewal}.

The proposal of the members of the group relative to the idea that is presented to the bank and which could give rise to a new financial product, is based on the following peculiarities:

(a) It is a credit the payments of which corresponding to interest must be made on the specifically established dates with no delay whatsoever.
(b) Punctuality of a credit (relatively expensive for the volume of the same) is awarded by reducing the interest rate by 0,25% every six months period.
(c) As the credit is conceived with slender possibilities for renewal, the second objective is to retain the loyalty of the borrower with the financial entity by carrying out, at least, one operation per annum of between 50 and 200 million.
(d) The previous characteristic is made possible by offering reductions of 0.50% in commissions for new operations that are formalised and by means of the immediate granting of the credit within the framework of standard operations.

This financial product will pass over therefore, to the idea "bank" of the financial entity for launching on the market at the time when a potential demand is detected.

22.11 Conclusions

The ideas that have arisen could be different if the members of the group were persons of a different profile to those that were chosen, as well as the time and place chosen for holding the sessions.

It should be considered that with the mere arrival at an "idea" this process does not stop, since it is necessary to consider it again in order to initiate the tasks that commence a different combination of elements chosen at random.

In this way, at each of the creativity sessions carried out by the financial entity few or many new proposals will appear all of which will fed into the idea "bank".

The frequency of creativity sessions will obviously depend on specific circumstances. In any event, two or three sessions should be held per annum with the object of arriving at a convenient up dating of the ideas forming the idea "bank".

Only in this way will the objective of stimulating and redirecting the human team that makes up the financial entity be achieved, and at the same time stimulating production.

And it is only in this context that it is possible to overcome the competition and break traditional moulds of products and services that have become outdated due to their excessive homogeneity.

23 Effects of Economic Surroundings on Financial Activity

23.1 Relations of Causality in Financial Activity

Businesses, in the exercise of their financial activity are affected by a high number of interactions that proceed from the economic sphere as well as from the social, cultural, political etc. spheres. As a result of this, modifications occur in the framework in which they carry on their activities.

As financial systems develop, and a process of grouping of the economies of different countries takes place, forming a large international financial market, the need arises for serious study of the intensity of the incidence of certain elements of the surroundings on the principal variables or indicators of the financial health of businesses. In short, it is becoming increasingly more important to be aware of the scope of the steps taken by attempting to show the consequences that these have on the internal financial environment.

Now, as is particularly well known, the relations of causality do not always end up in the direct binomial cause and effect, but in many cases an effect is at one and the same time the cause of another effect.

In this way the network of interconnections can be so thick that the human mind, on its own, is totally incapable of covering it all. Students of the subject have attempted to find certain procedures that allow for the evaluation of accumulated incidents on determined elements that are considered fundamental in financial analysis.

In this way certain instruments have been created in an attempt to totally or partially minimise any possible errors, forgetfulness or negligence that can quite often cause malfunctions in the sphere of financial management.

In traditional studies, information available is generally treated and elaborated only taking into account the direct relations of cause and effect. This form of action is, in the majority of cases, incomplete and can cause quite a few upsets, given the importance acquired by indirect or second generation effects that frequently act as intermediaries.

Recently, Professors Kaufmann and Gil Aluja have drawn up a whole range of models[1] that allow for getting closer to the objective of globalisation of the direct and indirect incidents existing between a group of causes

[1] Kaufmann A and Gil Aluja J (1989) Modelos para la investigación de efectos olvidados (in Spanish). Santiago de Compostela, Spain. (Ed) Milladoiro.

Anna Maria Gil-Lafuente: *Fuzzy Logic in Financial Analysis*, StudFuzz **175**, 427–444 (2005)
www.springerlink.com

and a group of effects. We are referring to the so called theory of forgotten effects, the initial support of which is constituted by "matrices of qualitative consequence" and the treatment of which brings to light a large number of mechanisms that would be very difficult if not impossible to discover by means of mere intuition or experience.

An analysis, however superficial it may be, of the life of businesses immersed in the economic-financial concert of this day and age, would bring to light the fact that whatever modification to the surroundings, however slight it may appear to be, causes internal effects which on may occasions condition actions in the financial sphere.

Increasingly, then, it is becoming more important for business executives not only to arrive at the network of interconnections that relates any change in the surroundings of the business, whatever its nature may be, with the financial phenomena produced in the same, but also to estimate the intensity of any repercussions caused by these external movements. Also, a valuable element for decision taking is constituted by becoming aware of those relations of causality that are normally forgotten by the experts and which, in short, often play an important role in those mechanisms that determine financial equilibrium. We are referring to the so called forgotten effects which are a consequence of those implicit but not evident relations between elements located outside the direct limits of influence of the business (social, political, macroeconomic, etc.) and its financial set up.

With the object of developing this original idea, we are going to start out from the case of a business the objective of which is to attain or maintain a good financial position by means of its profits, share price quotation, market share, etc.

Without trying to draw up an exhaustive list of the external causes that have an incidence on the financial phenomena of the business, which would make the description of the proposed technique difficult, we are going to take into consideration the following causes:

$a =$ inflationary process;
$b =$ entry of capital from abroad;
$c =$ external armed conflict of a certain seriousness;
$d =$ generalised economic recession;
$e =$ tax increases;
$f =$ strict labour laws;
$g =$ change in cash parity;
$h =$ modifications in inter-bank interest rates
$i =$ technical progress with economic repercussions.

When variations in one or several of these elements are detected by those responsible in the businesses, an attempt is made in some way to estimate their incidence on the main economic-financial variables, with the object of articulating measures that lead to minimising the negative effects or take advantage, wherever possible and in the best way, of the positive effects. These actions include many diverse aspects that affect both the human factor (continuous training, introduction of creative circles, increase in the degree of communication of personnel with the outside, etc.) and the material elements (improvement or setting up of laboratories for better quality control, acquisition of new production equipment, etc.).

Whatever the measures adopted, there will inevitably be repercussions, to a greater or lesser degree, in the financial variables of the business.

To list all the financial elements that may be affected by one or several of these external elements would be an arduous, if not impossible, task. For our study we have taken the following into consideration:

A = volume of external financing;
B = cost of outside capital;
C = market share;
D = sales prices;
E = financial solvency
F = financial risk;
G = production costs;
H = self financing
I = ratio of indebtedness;
J = profits;
K = share price quotation.

The first observation to be made on this list of concepts and instruments of financial analysis is that, also between them all, there exist relations of causality. Just as an example, it can be stated that profits have a consequence on the share quotation, or that sales prices and production costs affect profits. But this is quite normal and the same could also be said about those elements considered as causes.

23.2 A First Approximation to a Solution of Incidences

With the object of estimating the incidences of each of the causes proceeding from the macro-economic field on each one of the effects represented by the financial elements we have described, several paths can be followed. In a fist study we are going to use a well-known technique that consists of separating the financial effects into two groups according to whether is more or less directly perceived. With this we arrive at a chain with certain initial causes, certain intermediary effects that at the same time are causes of certain final effects. This part of the study, on some occasions creates certain problems arising from the difficulty of clearly establishing this separation. In spite of this we are going to admit the grouping of the following elements as acceptable:

Group of initial incidence

A = volume of external financing
B = cost of outside capitals
C = market share
D = sales prices
G = production costs

Group of indirect incidence

E = financial solvency
F = financial risk
H = auto-financing
I = ration of indebtedness
J = profits
K = share quotation

A quick glance at this classification will be sufficient to bring to light its vulnerability to criticism. Indeed, it would seem somewhat risky to consider profits and auto-financing, or profits and share quotation, for example, on the same level. But drawing up a chain of n links would lead to a treatment of this problem parallel to the one described in the chapter on the use of multivalent logic. Our objective here is different. Let us not forget that we are faced with an attempt to determine the accumulated effects of first and second generation and in a later study recuperate forgotten effects.

Therefore, although conscious of the aforementioned inconveniences, we are going to continue with this process, not before making a brief mention to the fact that at the time of assigning a number to the incidence of an element over another we will take the segment $[0; 1]$, using the hendecagonal system. This means the acceptance of the fact that the influence effected by a cause on an effect is not limited to all or nothing, but that there may exist 11 "degrees" that are equivalent to the different intensities considered as possible.

Having made these observations, we will now draw up a graph, that is a diagram of arrows, which will bring to light, in a visual manner, the structure of the relations of causality (see Fig. 23.1).

An expert has valued the incidences in accordance with the already mentioned system. The result of these valuations has been placed above each arc (arrow) of Fig. 23.1. It should be understood that the absence of arcs joining two vertices means that the said expert that no relation of causality exists whatsoever.

Once the estimates made by the expert are known, we proceed to do the calculations in order to arrive at the accumulation of direct or indirect effects.

$$\frac{a \to E}{0{,}8 \wedge 0{,}8 = 0{,}8} \quad \frac{b \to E}{0{,}7 \wedge 0{,}8 = 0{,}7} \quad \frac{c \to E}{0{,}4 \wedge 0{,}8 = 0{,}4}$$

$$\frac{g \to E}{(0{,}7 \wedge 0{,}7) \vee (0{,}9 \wedge 0{,}8) = 0{,}8} \quad \frac{h \to E}{1 \wedge 0{,}8 = 0{,}8}$$

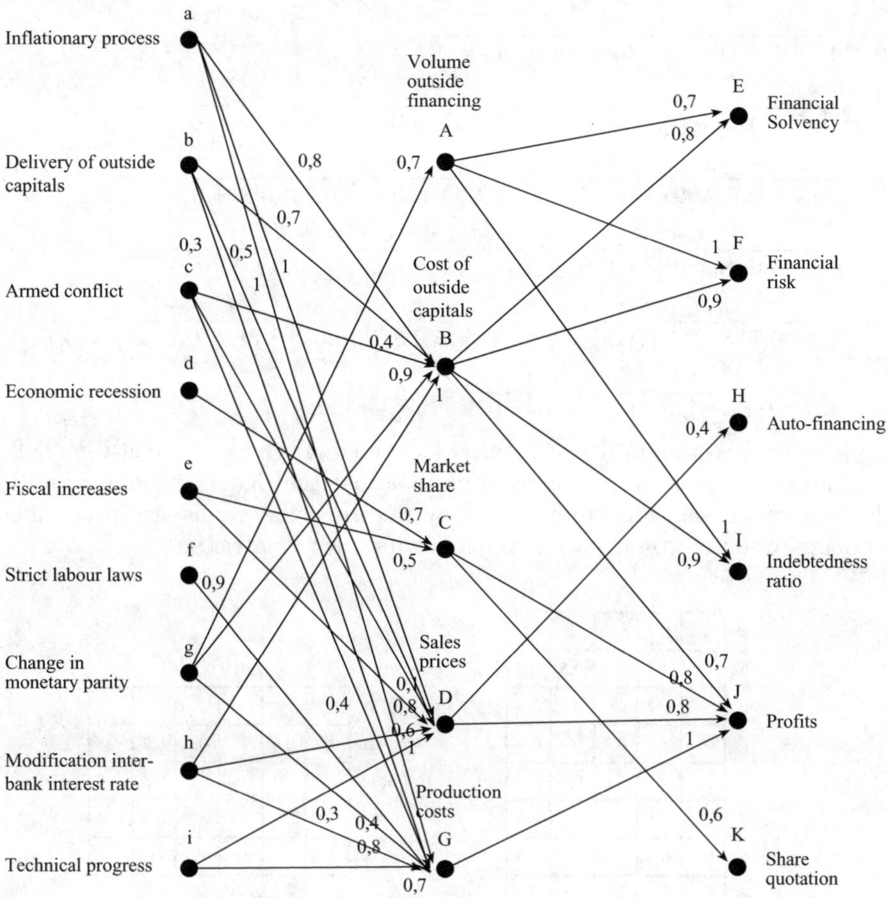

Fig. 23.1.

$$\frac{a \to F}{0{,}8 \wedge 0{,}9 = 0{,}8} \qquad \frac{b \to F}{0{,}7 \wedge 0{,}9 = 0{,}7} \qquad \frac{c \to F}{0{,}4 \wedge 0{,}9 = 0{,}4}$$

$$\frac{g \to F}{(0{,}7 \wedge 1) \vee (0{,}9 \wedge 0{,}9) = 0{,}9} \qquad \frac{h \to F}{1 \wedge 0{,}9 = 0{,}9}$$

$$\frac{a \to H}{1 \wedge 0{,}4 = 0{,}4} \qquad \frac{b \to H}{0{,}5 \wedge 0{,}4 = 0{,}4} \qquad \frac{c \to H}{0{,}1 \wedge 0{,}4 = 0{,}1}$$

$$\frac{e \to H}{0{,}8 \wedge 0{,}4 = 0{,}4} \qquad \frac{h \to H}{0{,}6 \wedge 0{,}4 = 0{,}4} \qquad \frac{i \to H}{1 \wedge 0{,}4 = 0{,}4}$$

$$\frac{a \to I}{0{,}8 \wedge 0{,}9 = 0{,}8} \qquad \frac{b \to I}{0{,}7 \wedge 0{,}9 = 0{,}7} \qquad \frac{c \to I}{0{,}4 \wedge 0{,}9 = 0{,}4}$$

$$\frac{g \to I}{(0{,}7 \wedge 1) \vee (0{,}9 \wedge 0{,}9) = 0{,}9} \qquad \frac{h \to I}{1 \wedge 0{,}9 = 0{,}9}$$

$$\frac{a \to J}{(0{,}8 \wedge 0{,}7) \vee (1 \wedge 0{,}8) \vee (1 \wedge 1) = 1}$$

$$\frac{b \to J}{(0{,}7 \wedge 0{,}7) \vee (0{,}5 \wedge 0{,}8) \vee (0{,}3 \wedge 1) = 0{,}7} \qquad \frac{f \to J}{0{,}9 \wedge 1 = 0{,}9}$$

$$\frac{c \to J}{(0{,}4 \wedge 0{,}7) \vee (0{,}1 \wedge 0{,}8) \vee (0{,}4 \wedge 1) = 0{,}4} \qquad \frac{d \to J}{0{,}7 \wedge 0{,}8 = 0{,}8}$$

$$\frac{e \to J}{(0{,}5 \wedge 0{,}8) \vee (0{,}8 \wedge 0{,}8) = 0{,}8} \qquad \frac{g \to J}{(0{,}9 \wedge 0{,}7) \vee (0{,}3 \wedge 1) = 0{,}7}$$

$$\frac{h \to J}{(1 \wedge 0{,}7) \vee (0{,}6 \wedge 0{,}8) \vee (0{,}8 \wedge 1) = 0{,}8} \qquad \frac{i \to J}{(0{,}6 \wedge 0{,}7) \vee (0{,}7 \wedge 1) = 0{,}7}$$

$$\frac{d \to K}{0{,}7 \wedge 0{,}6 = 0{,}6} \qquad \frac{e \to K}{0{,}5 \wedge 0{,}6 = 0{,}5}$$

If we express the results by means of a valued graph we arrive at Fig. 23.2.

The same result will be arrived at if we use the matrix system. Indeed, the relations of causality represented by the previous graphs are susceptible to being presented by means of the following fuzzy matrices:

$[\underset{\sim}{M}] =$

	A	B	C	D	G
a		0,8		1	1
b		0,7		0,5	0,3
c		0,4		0,1	0,4
d			0,7		
e			0,5	0,8	
f					0,9
g	0,7	0,9			0,3
h		1		0,6	0,8
i				0,6	0

$[\underset{\sim}{N}] =$

	E	F	H	I	J	K
A	0,7	1		1		
B	0,8	0,9		0,9	0,7	
C					0,8	0,6
D			0,4		0,8	
G					1	

If we do the max-min convolution $[\underset{\sim}{M}] \circ [\underset{\sim}{N}]$ we arrive at:

	A	B	C	D	G
a		0,8		1	1
b		0,7		0,5	0,3
c		0,4		0,1	0,4
d			0,7		
e			0,5	0,8	
f					0,9
g	0,7	0,9			0,3
h		1		0,6	0,8
i				0,6	0

$\circ$

	E	F	H	I	J	K
A	0,7	1		1		
B	0,8	0,9		0,9	0,7	
C					0,8	0,6
D			0,4		0,8	
G					1	

$=$

=	E	F	H	I	J	K
a	0,8	0,8	0,4	0,8	1	
b	0,7	0,7	0,4	0,7	0,7	
c	0,4	0,4	0,1	0,4	0,4	
d					0,7	0,6
e			0,4		0,8	0,5
f					0,9	
g	0,8	0,9		0,9	0,7	
h	0,8	0,9	0,4	0,9	0,8	
i			0,4		0,7	

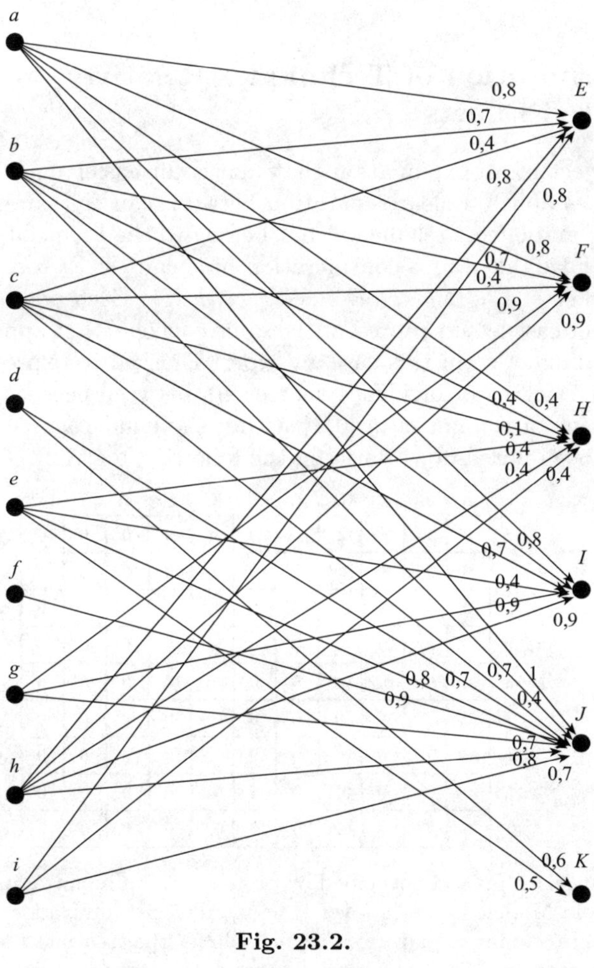

Fig. 23.2.

As can be seen the matrix system is easier and the use of the max-min convolution avoids possible errors that can arise when the calculations are done directly on the graph.

Once the problem has been technically resolved by means of the proposed method, an important question should be asked: can it be accepted that the arcs of the graph include all the possible relations of causality? More specifically, will there not be also relations of cause-effect in the original causes $\{a, b, c, d, e, f, g, h, i\}$ among themselves, as well as with the intermediary effects $\{A, B, C, D, G\}$ and final effects $\{E, F, H, I, J, K\}$ among themselves?

It appears obvious that the answer, in any event is positive and therefore although accepting the interest of the scheme we have developed we feel it is necessary to develop anew treatment for this problem that permits a better formal ordering to the reality of the economic-financial inter-connections. For this we will resort to the theory of forgotten effects[2].

23.3 Incorporation of Techniques Relative to Forgotten Effects

With the object of an explanation that makes for easier understanding, we are going to establish a clear separation between causes (external sphere of the business) and effects (elements that belong to the financial field) and, in this way we have taken into consideration nine causes: $\{a, b, c, d, e, f, g, h, i\}$ and eleven effects: $\{A, B, C, D, E, F, G, H, I, J, K\}$. Each one of these lower case and upper case letters represent the same concepts that appear under the previous heading. Also in this case an expert is asked to express an opinion relative to the degree of incidence that exists between each cause and each effect by means of a number in $[0; 1]$ within the hendecagonal system. The replies have been noted down forming the following matrix:

$[\underset{\sim}{M}] =$

	A	B	C	D	E	F	G	H	I	J	K
a	0	0,8	0	1	0	0	1	0	0	0	0
b	0	0,7	0	0,5	0	0	0,3	0	0	0	0
c	0	0,4	0	0,1	0	0	0,4	0	0	0	0
d	0	0	0,7	0	0	0	0	0	0	0	0
e	0	0	0,5	0,8	0	0	0	0	0	0	0
f	0	0	0	0	0	0	0,9	0	0	0	0
g	0,7	0,9	0	0	0	0	0,3	0	0	0	0
h	0	0	0	0,6	0	0	0,8	0	0	0	0
i	0	0	0	0,6	0	0	0,7	0	0	0	0

[2] This theory drawn up and developed by professors Kaufmann A and Gil Aluja J, in their book: Modelos para la investigación de efectos olvidados, has given rise to a group of techniques that are very suitable for the treatment of inter-actions both in the economic field and in the others spheres of science.

As can be seen, the expert has given his opinion that there are no relations between causes $a, b, \dots,$ and final effects $E, F, \dots, K$, and if these do exist they will appear as second generation effects. In some cases this is easy to accept (inflationary process $\rightarrow$ financial risk). Although a valuation could be introduced for these relations (on this point we could say even more, that is we feel that this should be so), we have not done so with the object of arriving at a "cleaner" numerical comparison, in order to be able to compare these results with the previous scheme.

The same expert is now asked to supply the degree of consequence existing between each cause $\{a, b, \dots, i\}$ and itself and with the remainder. The result has been expressed by means of the fuzzy square matrix $[\underset{\sim}{\mathbf{A}}]$ that we reproduce below:

$[\underset{\sim}{A}] =$

	a	b	c	d	e	f	g	h	i
a	1	0,7	0	0,6	0,1	0	1	0,9	0
b	0,6	1	0	0,7	0	0	1	0,4	0,2
c	0,8	1	1	0,7	0,6	0	0	0,1	0,8
d	0,8	0,9	0	1	0,4	0,4	0	0,3	0,6
e	0,7	0,9	0	0,5	1	0	0	0	0
f	0,6	0,7	0	0,8	0	1	0	0	0,3
g	0,7	1	0	0	0	0	1	0,4	0
h	0,8	1	0	0	0	0	0,2	1	0
i	0,9	0	0	0,6	0	0	0	0	1

The expert is also asked to complete the data relative to the degree of consequence of each effect $\{A, B, \dots, K\}$ with itself and with the remainder, with the object of arriving at all the relations existing between effects. The result is expressed by means of the following fuzzy square matrix $[\underset{\sim}{\mathbf{B}}]$:

$[\underset{\sim}{B}] =$

	A	B	C	D	E	F	G	H	I	J	K
A	1	1	0	0,8	0,7	1	0,8	0	1	0	0
B	0,9	1	0	0,7	0,8	0,9	0,7	0	0,9	0,7	0
C	0	0	1	0,2	0	0	0,2	0	0	0,8	0,6
D	0	0	0,9	1	0	0	0	0,4	0	0,8	0
E	1	0,6	0	0	1	0,4	0,7	0	0,5	0	0,7
F	0,9	0,6	0	0	0,6	1	0,1	0,4	0,7	0,3	0,7
G	0,2	0	0	1	0	0	1	0	0	1	0
H	0,6	0,2	0	0,4	1	1	0,7	1	1	0,7	1
I	0	0	0	0	0,8	1	0	0	1	0	0,6
J	0,7	0	0	0	1	0,9	0	1	0,7	1	1
K	0	0	0	0	0	0	0	0	0	0	1

It will be seen that matrices $[\underset{\sim}{\mathbf{A}}]$ and $[\underset{\sim}{\mathbf{B}}]$ are square by construction and also reflexive, but not symmetric.

We are now going to commence calculations for arriving at the accumulated first and second-generation effects. For this, in the first place we will do the max-min convolution between matrices $[\underset{\sim}{\mathbf{A}}]$ and $[\underset{\sim}{\mathbf{M}}]$, that is $[\underset{\sim}{\mathbf{A}}] \circ [\underset{\sim}{\mathbf{M}}]$. The result will be:

$[\underset{\sim}{\mathbf{A}}] \circ [\underset{\sim}{\mathbf{M}}] =$

	A	B	C	D	E	F	G	H	I	J	K
a	0,7	0,9	0,6	1	0	0	1	0	0	0	0
b	0,7	0,9	0,7	0,6	0	0	0,6	0	0	0	0
c	0	0,8	0,7	0,8	0	0	0,8	0	0	0	0
d	0	0,8	0,7	0,8	0	0	0,8	0	0	0	0
e	0	0,7	0,5	0,8	0	0	0,7	0	0	0	0
f	0	0,7	0,7	0,6	0	0	0,9	0	0	0	0
g	0,7	0,9	0	0,7	0	0	0,7	0	0	0	0
h	0,2	1	0	0,8	0	0	0,8	0	0	0	0
i	0	0,8	0,6	0,9	0	0	0,9	0	0	0	0

For arriving at each one of the elements of the convoluted matrix $[\underset{\sim}{\mathbf{A}}] \circ [\underset{\sim}{\mathbf{M}}]$ the max-min has been found for the row and column that determine their sub-indices. Therefore in order to arrive at $\mathcal{M}_{ij}$ consideration has been given to row i and column j. For greater detail of the calculation we will describe the way for arriving at some of the elements. For example, in order to arrive at $\mathcal{M}_{aA}$:

Row a of $[\underset{\sim}{\mathbf{A}}]$:

a	b	c	d	e	f	g	h	i
1	0,7	0	0,6	0,1	0	1	0,9	0

Column A of $[\underset{\sim}{\mathbf{M}}]$:

a	b	c	d	e	f	g	h	i
0	0	0	0	0	0	0,7	0	0

Minimum:

a	b	c	d	e	f	g	h	i
0	0	0	0	0	0	0,7	0	0
						↑		

The maximum of these minimums is 0,7 therefore $\mathcal{M}_{aA} = 0{,}7$.

In order to arrive at $\mathcal{M}_{aB}$:
Row a of $[\underset{\sim}{\mathbf{A}}]$:

a	b	c	d	e	f	g	h	i
1	0,7	0	0,6	0,1	0	1	0,9	0

Column B of $[\underset{\sim}{\mathbf{M}}]$

a	b	c	d	e	f	g	h	i
0,8	0,7	0,4	0	0	0	0,9	1	0

Minimum:

a	b	c	d	e	f	g	h	i
0,8	0,7	0	0	0	0	0,9	0,9	0
						↑	↑	

The maximum of these minimums is 0,9. Therefore $\mathcal{M}_{aB} = 0{,}9$.
In order to arrive at $\mathcal{M}_{dG}$:
Row d of $[\underset{\sim}{\mathbf{A}}]$

a	b	c	d	e	f	g	h	i
0,8	0,9	0	1	0,4	0,4	0	0,3	0,6

Column G of $[\underset{\sim}{\mathbf{M}}]$:

a	b	c	d	e	f	g	h	i
1	0,3	0,4	0	0	0,9	0,3	0,8	0,7

Minimum

a	b	c	d	e	f	g	h	i
0,8	0,3	0	0	0	0,4	0	0,3	0,6
↑								

The maximum of these minimums is 0,8. Therefore $\mathcal{M}_{dG} = 0{,}8$. And so on for all elements of matrix $[\underset{\sim}{\mathbf{A}}] \circ [\underset{\sim}{\mathbf{M}}]$.

For clearer understanding we are going to represent this max-min process by means of graphs. Thus for $\mathcal{M}_{aA}$ the Fig. 23.3 is arrived at. As can be readily seen the consequence of the inflationary process on the volume of external financing takes place in this first phase of the calculation through a change in the monetary parity.

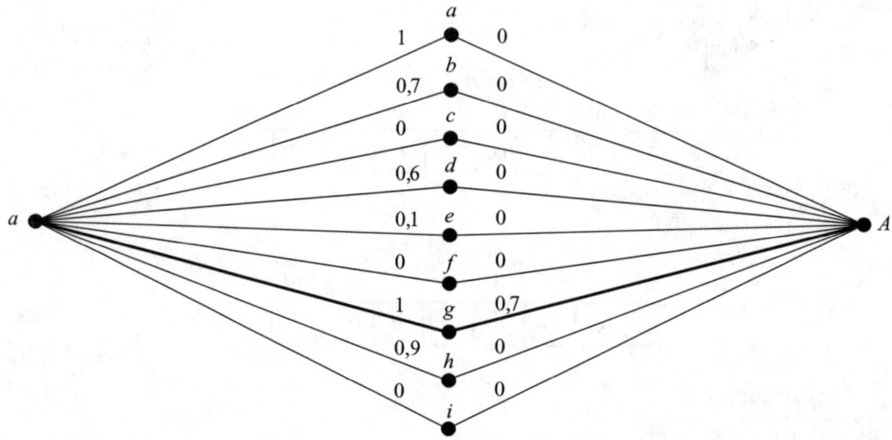

Fig. 23.3.

For arriving at $\mathcal{M}_{aB}$ we find Fig. 23.4. It will now be seen that the consequence of the inflationary process on the cost of outside capital takes place not only through the change in the cash parity but also through the modification in the inter-bank interest rate.

For $\mathcal{M}_{dG}$ the Fig. 23.5 is arrived at. On this occasion we will see that the consequence of economic recession on production costs takes place through the inflationary process.

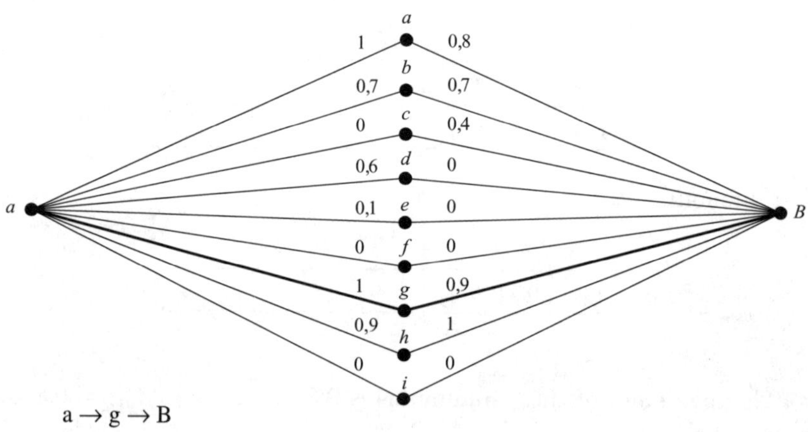

Path: $a \rightarrow g \rightarrow B$

$a \rightarrow h \rightarrow B$

Valuation: $1 \wedge 0{,}9 = 0{,}9$

$0{,}9 \wedge 1 = 0{,}9$

Fig. 23.4.

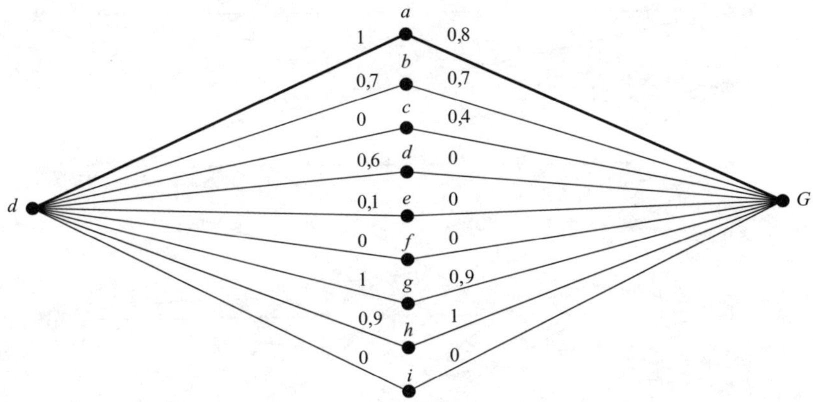

Path: d → a → G
Valuation: 0,8 ∧ 1 = 0,8

Fig. 23.5.

At this point it may be useful to compare the result we have arrived at, in this part of the process, that on using the techniques of the forgotten effects with those found in the schemes described in the second section of this chapter. The matrix and corresponding graph would now be (see Fig. 23.6):

	A	B	C	D	G
a	0,7	0,9	0,6	1	1
b	0,7	0,9	0,7	0,6	0,6
c		0,8	0,7	0,8	0,8
d		0,8	0,7	0,8	0,8
e		0,7	0,5	0,8	0,7
f		0,7	0,7	0,6	0,9
g	0,7	0,9		0,7	0,7
h	0,2	1		0,8	0,8
i		0,8	0,6	0,9	0,9

The comparison between Figs. 23.2 and 23.6 or their corresponding matrices shows that the accumulation if effects on A, B, C, D, G is very intense, in this latter process, for the greater part of the existing relations and new relations of causality occur, which did not exist when we used the previous technique, as a consequence of the indirect effects through other initial causes $a, b, c, \ldots, i$.

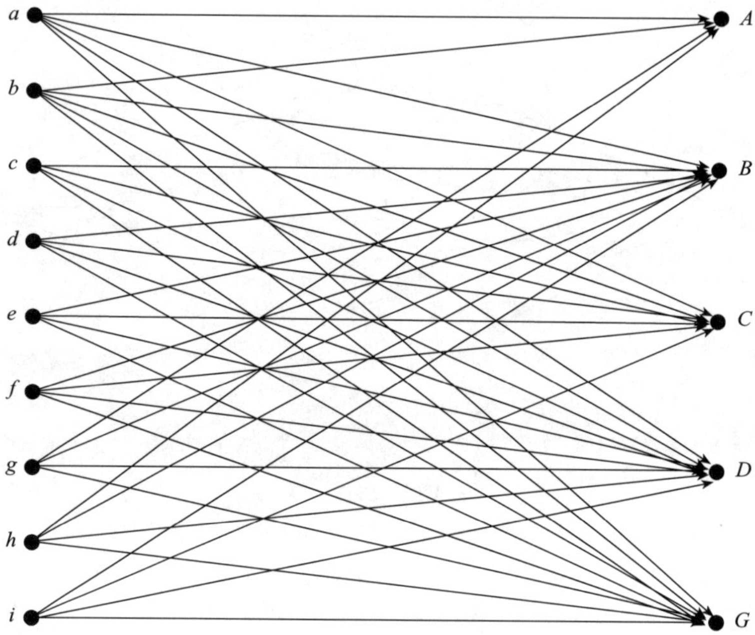

Fig. 23.6.

23.4 Presentation of the Results

With these results the first stage of this scheme has been covered. We will now move on to the second stage by doing the max-min convolution of matrix $[\underset{\sim}{\mathbf{A}}] \circ [\underset{\sim}{\mathbf{M}}]$ with matrix $[\underset{\sim}{\mathbf{B}}]$, following the same procedure as was used for arriving at $[\underset{\sim}{\mathbf{A}}] \circ [\underset{\sim}{\mathbf{M}}]$. We will therefore arrive at:

$[\underset{\sim}{A}] \circ [\underset{\sim}{M}] \circ [\underset{\sim}{B}] =$

	A	B	C	D	E	F	G	H	I	J	K
a	0,9	0,9	0,9	1	0,8	0,9	1	0,4	0,9	1	0,6
b	0,9	0,9	0,7	0,7	0,8	0,9	0,7	0,4	0,9	0,7	0,6
c	0,8	0,8	0,8	0,8	0,8	0,8	0,8	0,4	0,8	0,8	0,6
d	0,8	0,8	0,8	0,8	0,8	0,8	0,8	0,4	0,8	0,8	0,6
e	0,7	0,7	0,8	0,8	0,7	0,7	0,7	0,4	0,7	0,8	0,5
f	0,7	0,7	0,7	0,9	0,7	0,07	0,9	0,4	0,7	0,9	0,6
g	0,9	0,9	0,7	0,7	0,8	0,9	0,7	0,4	0,9	0,7	0
h	0,9	1	0,8	0,8	0,8	0,9	0,8	0,4	0,9	0,8	0
i	0,8	0,8	0,9	0,9	0,8	0,8	0,9	0,4	0,8	0,9	0,6

The results that appear in this matrix show the complete consequences of first and second generation, since it does not only include the relations of causality between the elements that constitute the causes between themselves and its effects on the effects, but also the relations of causality between the effects, included in matrix $[\underset{\sim}{\mathbf{B}}]$, the convolution of which with matrix $[\underset{\sim}{\mathbf{A}}] \circ [\underset{\sim}{\mathbf{M}}]$ provides the aforementioned accumulated consequences.

In order to isolate the indirect or second-generation consequences, resort can be made to several operators. For their simplicity and usefulness in this particular problem we propose the use of the ordinary difference between matrices $[\underset{\sim}{\mathbf{M}}^*] = [\underset{\sim}{\mathbf{A}}] \circ [\underset{\sim}{\mathbf{M}}] \circ [\underset{\sim}{\mathbf{B}}]$ and the original matrix $[\underset{\sim}{\mathbf{M}}]$. In our case we would arrive at:

$[\underset{\sim}{\mathbf{M}}^*]\,(-)\,[\underset{\sim}{\mathbf{M}}] =$

	A	B	C	D	E	F	G	H	I	J	K
a	0,9	0,9	0,9	0,1	0,8	0,9	1	0,4	0,9	1	0,6
b	0,9	0,9	0,7	0,7	0,8	0,9	0,7	0,4	0,9	0,7	0,6
c	0,8	0,8	0,8	0,8	0,8	0,8	0,8	0,4	0,8	0,8	0,6
d	0,8	0,8	0,8	0,8	0,8	0,8	0,8	0,4	0,8	0,8	0,6
e	0,7	0,7	0,8	0,8	0,7	0,7	0,7	0,4	0,7	0,8	0,5
f	0,7	0,7	0,7	0,9	0,7	0,7	0,9	0,4	0,7	0,9	0,6
g	0,9	0,9	0,7	0,7	0,8	0,9	0,7	0,4	0,9	0,7	0
h	0,9	0,1	0,8	0,8	0,8	0,9	0,8	0,4	0,9	0,8	0
i	0,8	0,8	0,9	0,9	0,8	0,8	0,9	0,4	0,8	0,9	0,6

(–)

	A	B	C	D	E	F	G	H	I	J	K
a	0	0,8	0	1	0	0	1	0	0	0	0
b	0	0,7	0	0,5	0	0	0,3	0	0	0	0
c	0	0,4	0	0,1	0	0	0,4	0	0	0	0
d	0	0	0,7	0	0	0	0	0	0	0	0
e	0	0	0,5	0,8	0	0	0	0	0	0	0
f	0	0	0	0	0	0	0,9	0	0	0	0
g	0,7	0,9	0	0	0	0	0,3	0	0	0	0
h	0	0	0	0,6	0	0	0,8	0	0	0	0
i	0	0	0	0,6	0	0	0,7	0	0	0	0

=

	A	B	C	D	E	F	G	H	I	J	K
a	0,9	0,1	0,9	0	0,8	0,9	0	0,4	0,9	(1)	0,6
b	0,9	0,2	0,7	0,2	0,8	0,9	0,4	0,4	0,9	0,7	0,6
c	0,8	0,4	0,8	0,7	0,8	0,8	0,4	0,4	0,8	0,8	0,6
d	0,8	0,8	0,1	0,8	0,8	0,8	0,8	0,4	0,8	0,8	0,6
= e	0,7	0,7	0,3	0	0,7	0,7	0,7	0,4	0,7	0,8	0,5
f	0,7	0,7	0,7	0,9	0,7	0,7	0	0,4	0,7	0,9	0,6
g	0,2	0	0,7	0,7	0,8	0,9	0,4	0,4	0,9	0,7	0
h	0,9	0	0,8	0,2	0,8	0,9	0	0,4	0,9	0,8	0
i	0,8	0,8	0,9	0,3	0,8	0,8	0,2	0,4	0,8	0,9	0,6

It will be seen that in this case a large number of second-generation effects appear and in some relations these effects are particularly intense. This is the case with relation $a \to j$, that is, the consequence that inflation exercises on profits, that initially was considered nil and which now appears as total due to the indirect effects.

Now the interesting question can be asked as to through which of the intermediary elements does an accumulated effect take place that is as important as the one already mentioned. In order to find a solution we resort to representing this by means of the following graph, in which it can be seen that the element acting as intermediary is G: production costs. The path is:

$$a \xrightarrow{1} a \xrightarrow{1} G \xrightarrow{1} j$$

where $1 \circ 1 \circ 1 = 1$ which gives rise to a maximum incidence.

The construction of this graph is very simple. We propose the following scheme:

(a) Given that we know the total accumulated consequence of 1 it will be necessary that each arc of the path has a value equal to the unit. In the relations of the elements of matrix $[\underset{\sim}{\mathbf{A}}]$ corresponding to row a there are only two relations with a valuation of 1, which are those that tie a with a and a with g.

(b) In the relations of the elements of matrix $[\underset{\sim}{\mathbf{M}}]$ corresponding to rows a and g 1 is only present in a. Therefore knot g is left alone, and the arcs continued through knot a.

(c) Of all the arcs that leave a, the only ones with a value of 1 are those arriving at D and G. The possible path will pass through D and/or G.

(d) Finally knots $A, B, \ldots, K$ are linked to knot J. It will be seen that in the relations of the elements of matrix $[\underset{\sim}{\mathbf{B}}]$ corresponding to column J there are two maximum valuations and equal to 1, that is the arc starting out from G and the one from J. Since the path can only pass through D and/or G, in this case G must be chosen.

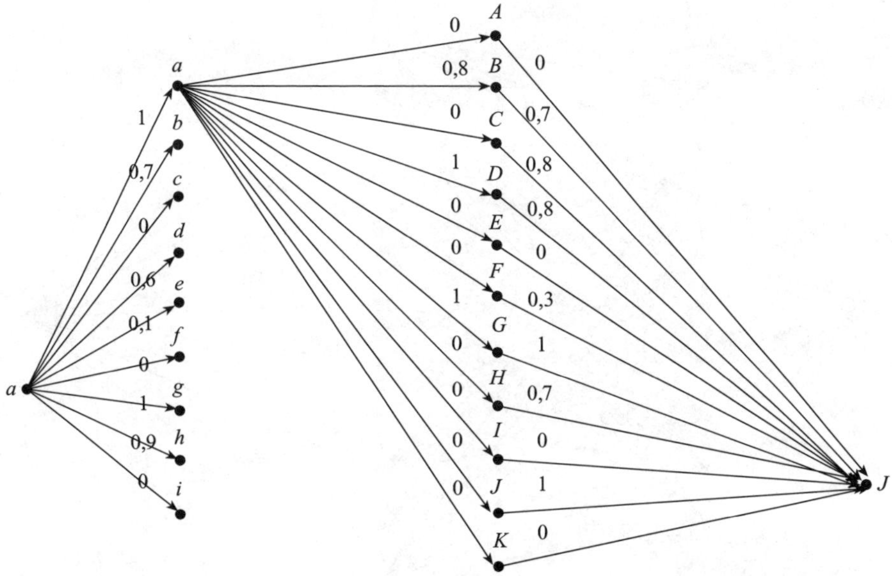

Fig. 23.7.

What we have in this way is a graphic representation that allows us to visualise a large part of the process that the mind must follow in order to connect the path followed by this linking of relations of cause-effect. And we say that the process is only partial since with this simple algorithm we have only include the path (or in the event paths) that have definitely been chosen and have not represented all the rest. Only in order to show the intricate network of connections that would be necessary to link a single cause with a single effect, below we are going to represent the complete graph corresponding to the relation $a \rightarrow j$ in Fig. 23.8.

It would seem unnecessary to insist on the fact that the graph shown in Fig. 23.8 "only" includes the connections that the mind of the expert would have to do in order to link one cause with one effect. If what was required in order to establish all the necessary links for one matrix (which on the other hand is very small) of 9 causes and 11 effects, such as the one we have been studying, this graph would have to be repeated 99 times. It would seem unnecessary to underline the fact that this is practically impossible, in the state in which the human mind is today, that we would be capable (without the help of a single algorithm) of intuitively carrying out this process. However, it will have been seen that, with the proposed technique, this was easy and also it is quite sure that there will be no error or forgetfulness.

To conclude, it would seem reasonable to establish a new comparison between the final results arrived at by using the techniques for recuperating forgotten effects and those used in the second section of this chapter. Also in

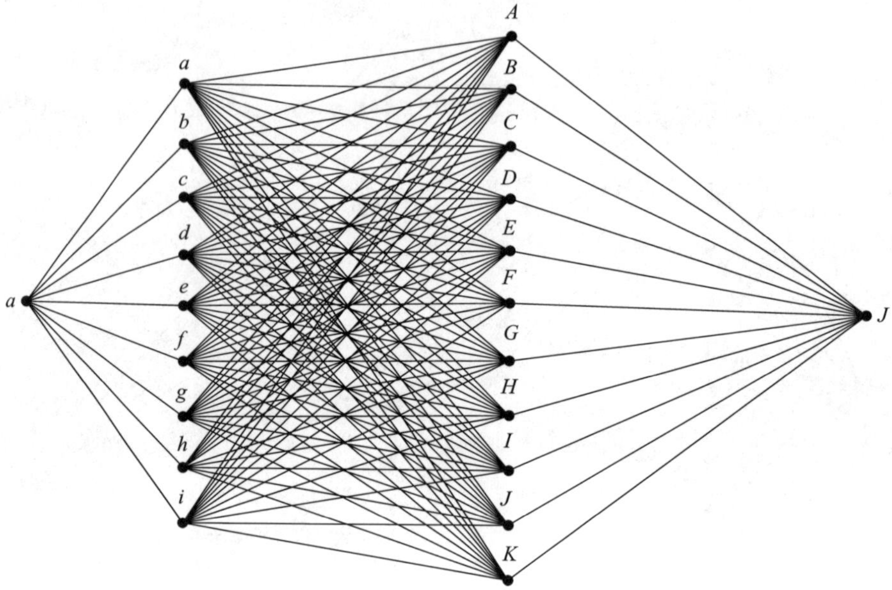

Fig. 23.8.

this case the same phenomena can be seen as those found on comparing the result of doing the initial convolution $[\underset{\sim}{\mathbf{A}}] \circ [\underset{\sim}{\mathbf{M}}]$ with that arrived at by means of the traditional technique, although on this occasion the final consequences are, in general even more intense, as the relations of causality of the effects on themselves has been included.

It is quite evident that the scheme we have proposed only constitutes an initial contribution to the many solutions that can be found for the problem we have considered. In fact, the use of Φ-fuzzy matrices, fuzzy-random and expertons, allow for generalisations that, without a doubt, would give rise to interesting works. We will leave this line of research to the future.

Bibliography

ADAMO, J.M., Une implémentation de la théorie des sous-ensambles flous. Applications a l'analyse de décisions, state doctoral thesis, Université de Lyon, March 1978.

APPLEBY, R.C., Así se dirige una empresa moderna, Salamanca, Anaya, 1971.

APOLLONI, B. And DONNARUMMA, A., "Multi-aspect decision procedures based on minimum choice fuzzy operations" Revue Busefal. L.S.I. (Univ. Paul Sabatier, Toulouse), no. 18, 1984.

ASSE, A. Aide au diagnostique industriel par des methods basées sur la théorie de sous ensambles flous, third cycle thesis, Université de Valenciennes (France), April 1985.

BACHILLER, A., LAFUENTE, A. and SALAS, V., Gestión económico-financiera del circulante, Madrid, Pirámide, 1982.

BELLMAN and DREYFUS, La Programation dinamique et ses aplications, Paris, Dunod, 1965.

BORISOV, A.N. and KRUMBERG, O.A., "A theory of possibility for decision-making", Fuzzy Sets & Syst., vol. 9, no. 1, January 1983.

BOURDIN, J., Control de Gestión, Madrid, ICE, 1978.

CAÑIBANO, L., Contabilidad, Análisis contable de la realidad económica, Madrid, Pirámide, 1988.

CARLSSON, C., "Fuzzy sets and management methodology", in Process in Cyber & Syst. Res., vol. 8, New York, Hemisphere Publ., 1980.

COHEN, E., Analyse financiére. Económica, 2nd. ed. Collection Gestión. Série: Politique générale, finance et marketing. Paris 1990.

CONCHE, B. and COURANT, P., "Applications des concepts flous aux multicritéres". In Bernard Roy Seminar, University of Paris-Dauphine, 1972.

CONSO, P., La gestión financiera de la empresa, Barcelona, Hispano Europea, 1970.

CZOGALA, E., "Application aspect of probabilistic sets on modern decision-making", in 15th International Science Seminar on Modern Methods in Decision-making, Sophia (Bulgaria), June, 1983.

CHANAS, S. and KOLODZIEJZYK, W., "Fuzzy sets in few classical operations research problems", in N.M. Gupta and E. Sánchez (pub.) Approximate Reasoning in Decision Analysis, Norh Holland, 1982.

CHOATES, S.W. and KANDEL, A., "Fuzzy resources management analysis" in 1st Chinese Nat. Symposium on Multiple Valued Logic, Changquing (People's Republic of China), November 1984.

DALE, J.D., El control de la empresa a través de los datos constables, Barcelona, Sagitario, 1961.

DALOUBEIX, J., Los presupuestos flexibles, Madrid, ICE, 1977.

DEFOSSE, G., La gestión financiera de las empresas, Barcelona, Airel, 1979, Demos Collection.

DELGADO, J.P., "A determination of subjective process by fuzzy numbers", in First IFSA Congress. Session 2–4, Palma de Mallorca (Spain), July, 1985.

DE LUCA A. and TERMINI, S., "Fuzzy sets and decision theory", Inform & Control, vol. 23, 1973.

DEPALLENS, G., Gestion financière de l'entreprise, Paris, Sirey, 1970.

DIÉZ DE CASTRO, E.P., "Presupuesto Base Cero: un Nuevo instrumento de planificación" Economía y Empresa magazine, no. 3–4.

DUBOIS, D., Modèles mathématiques de l'imprécis et de l'incertain en vue de l'application aux técniques de l'aide à la decisión, state doctoral thesis, Université de Grenoble, November 1983.

DUBOIS, D. and PRADE, H., "The management of uncertainty in expert-systems: the possibility approach", in Proc. 10th Triennial IFORS Conf. Washington, D.C., August 1984.

DURAND, D., "Cost of Debt and Equity Funds for Business: Trend and Problems of Measurement", in Conference and Research on Business Finance, New York, N.B.E.R., 1952.

ESOGBUE, A.O., "A fuzzy set for evaluation of best management practice" in Proc, International Conference Policy Analysis and Inform. Syst., Taipei, Taiwan (China), August 1981.

EUROPEAN UNION OF ACCOUNTING EXPERTS, Evaluación de empresas y partes de empresa, Bilbao, Deusto, 1962.

FERNÁNDEZ PIRLA, J.M., Teoría económica de la contabilidad, 5th Ed., Madrid, 1967.

FERON, R., Ensemble aléatoire, ensemble flou, ensemble aléatoire flou, processus aléatoires et leurs applications a l'étude du profil fiscal, Verona, Doc, Univ. Degli Studi di Pavona, Fac. de Economia e Commercio, 1979.

FERREIRO SANTANA, L.A., Introducción a la técnica del presupuesto por programas, México, Dirección General del Presupuesto por Programas, 1976, Series: study 2.

FERRY, J., "Informes del Consejo Económico y Social francés" Revista de Hacienda Pública Española, no. 11, 1971.

FUKURAMA, M; NAKAMURA, K. and ODA, M., "Fuzzy model in human decision-making" in Annual Records of J.A.A.C.E., Japan, 1972.

FUSTIER, B., Une introduction à la théorie de la demande flou, Doc. De travail no. 57, Inst. Math. Économ., Université de Dijon, 1982.

FUSTIER, M., La logique de l'arbre. Algoe, Ecully-Rhône, 1980.

GAUCHEY, Report on the 6th Annual Congress of la Chambre Des Experts Judiciaires Evaluateurs, Aix-en-Provence, November 5, 1960.

GERBER, Q.N., Management Accounting, Planning and Control, Boston, PWS-Kent Ed., 1988.

GIL ALUJA, J., Apuntes de inversiones, Barcelona, Faculty of Economic and Business Sciences, 1989, Chap. V.

GIL ALUJA, J., Inversiones, certeza, riesgo e incertidumbre, I. Faculty of Economic Sciences, Barcelona, 1991.

-, La gestión interactiva de los recursos humanos en la incertidumbre, Madrid, Ceura, 1996 (English versión by Kluwer Academic Publishers, 1998).

-, Invertir en la incertidumbre, Pirámide, Madrid, 1997 (English versión by Kluwer Academic Publishers, 1998).
-, Elementos para una teoría de la decisión en la incertidumbre, Milladoiro, Vigo, 1999 (English versión by Kluwer Academic Publishers, 1999).
GIL ALUJA, J., Modelos no numéricos de asignación en la gestión de personal. Paper presented at the II SIGEF Congress. Santiago de Compostela, November, 1995. Vol. I. Porceedings of the Congress.
GIL ALUJA, J., "La colocación óptima de los recursos financieros. The ISFUNE model (Investment Selection for Fuzzy Neuronal Networks)". Proceedings of the 1st SIGEF Congress, Reus, Spain, 16–18 November, 1997. vol. 1.
GIL ALUJA, J., GIL LAFUENTE, A.M., and TEODORESCU, N.H., "Periodicity and Chaos in Economic Fuzzy Forecasting, International fuzzy relations and Galois lattices" in X Congreso Europeo de Investigación Operativa, Aachen, 1991.
GIL ALUJA, J., TEODORESCU, N.H. and GIL LAFUENTE, A.M., "A chaotic fuzzy model for management problem", at International 92 Genève Conference "Signals and Systems", 1992.
GIL LAFUENTE, A.M., El análisis financiero en la incertidumbre, Barcelona, Ariel, 1990.
GIL LAFUENTE, A.M., " Estrategias secuenciales para la captación de medios financieros". Proceedings of the SIGEF Congress. vol. I. Buenos Aires, Argentina, November 1996.
GIL LAFUENTE, A.M., GIL ALUJA, J., TEODORESCU, N.H. and TACU, A.P., "Chaotic Fuzzy Models in Economy ", at 2nd International Conference on Fuzzy Logiz Systems and Neural Networks, Lizuka, (Japan), 1992.
GIL LAFUENTE, J., Marketing para el nuevo milenio, Madrid, Pirámide, 1997.
GORDON, P., Cadenas finitas de Markov y sus aplicaciones, Barcelona, Hispano Europea, 1967.
HANON DE LOUVET, Ch., Nouveau traité d'analyse et discussion de bilans, Brussels, Éditions Comptables, Commerciales et Financières, 1955.
IPPOLITO, C. and SAITIA, I., "Mesure d'entropie flou dans l'activation des règles dans un système de production", at 6th International Congres on Cyber & Syst, Paris, September 1984.
JEAN, W.H., Teoría analítica de la financiación, Barcelona, Ariel, 1976.
JONIO, F., PLAINDOUX, G. and LELEU, R., Control de Gestión, Barcelona, Sagitario, 1962.
KAUFMANN, A., R-expertons (expertons dans R), Note de Travail 186, La Tronche, July 1988. – Introduction à la théorie de sous-ensembles flous, Paris, Masson, vols. I, II, III and IV.
KAUFMANN, A. and GIL AJULA, J., "Grafos neuronales para la economía y gestión de empresas". Pub. Pirámide. Madrid, 1995.
- Introducción de la teoría de los subconjuntos borrosos a la gestión de las empresas, Santiago de Compostela, Milladoiro, 1986.
- Técnicas operativas de gestión para el tratamiento de la incertidumbre, Barcelona, Hispano Europea, 1987.
- Nuevas técnicas para la dirección estratégica, Barcelona, University of Barcelona, 1991.
- "Selection of affinities by means of fuzzy relations and Galois lattices", at X European Congress of Operations Investigations, Aachen, 1991.

- Técnicas de gestión de empresas. Previsiones, decisiones y estrategias, Madrid, Pirámide, 1992.
- Models per a la recerca d'efectes oblidats, Santiago de Compostela, Milladoiro, 1988.
- Las matemáticas del azar y de la incertidumbre, Madrid, Pub. Centro de Estudios Ramón Areces, 1990.
- Técnicas especiales para la gestión de expertos, Santiago de Compostela, Milladoiro, 1993.

KAUFMANN, A., GIL ALUJA, J. and GIL LAFUENTE, A.M., "La creatividad en la gestión de las empresas", Pub. Pirámide, Madrid, 1994.

KAUFMANN, A and GUPTA, M.M., Introduction to fuzzy arithmetic, Van Nostrand, Rheinhold 1985.

KHEMARKHEN, A., El control de gestión, Bilbao, Deusto, 1976.

KOLLER, Die Ermittlung von Gesanmtwert un Geschäfswert der Unternimens, Dusseldorf, 1952 (2nd. Ed.).

KOONTS, H., Management, 7th. ed. New York, McGraw-Hill, 1980.

KOVALEVCHUK, B.Y., "A method of validation of fuzzy optimisation", at Scientific Seminar on Models of Choice of Alternatives in a Fuzzy Environment, Riga (URSS), Polytech. Inst., October 1980.

KUHN, H.W., The Hungarian method for the assignment problem. Naval Research Quaterly. vol. 2, no. 1–2 March-June, 1955.

LOEB, P., El presupuesto de la empresa, Madrid, Aguilar, 1961.

MALGRANGE, Y., "Descompositions d'un graphe en sous-graphes fortement connexes maximaux", Cie. Machines Bull., 1967.

MAO, James G.T., Análisis financiero, Buenos Aires, El Ateneo, 1974.

MARTÍN RUIZ, F., El presupuesto Base Cero. Cuestiones de Economía de empresa, Universidad de Málaga, 1983.

MARTNER, G., Planificación y presupuesto por programas, México, Siglo XXI, 1967.

MATHEWS, R., Contabilidad para economistas, Madrid, Aguilar, 1984.

McCULLOCH, W.S. and PITTS, N., "A logical calculus of the ideas immanent in nervous system". Bulletin of Mathematical Biophysics no. 5, 1943.

MOLES, A. and CAUDE, R., Créativité et methods d'innovation. Collection Management. Fayard. Mane, Paris, 1972.

MUELLER, K.J., Zero-Base Budgeting in Local Government: Attempts to Implement Administrative Reform, Washington University Press of America, 1981.

NEGOITA, C.V., "On dynamics and fuzzyness in management information systems" in Proc. 3rd International Conf. On General Syst. and Cyber, Bucharest; Ed. Berlin, Springer-Verlag, 1977. NEGOITA, C.V. and RALESCU, D.A., "On fuzzy optimisation", at UNESCO Seminar, Bucharest, 1975.

NICOL, E., Los principios de la ciencia, Mexico, Fondo de Cultura Económica, 1965.

NYANG, S. And TESNIERE, J.C., Application des relations floues aux problèmes multicritères: cas des classifications des seminaries de recherche à Paris IX.

O'DONNELL, C., Management, 7th. ed. New York, McGraw-Hill, 1980.

OKUDA, K., TAMAKA, K. and ASAI, H., "Decision-making and information in fuzzy events". Bull. Univ. Osaka Sakai, Osaka), series A, vol. 2, 1974.

ORTEGA DÍAZ-AMBRONA, J.M., "Fórmulas presupuestarias de la eficacia. El PBS.", Spanish Treasury, no. 11, 1971.

ORTEGA Y GASSET, J., La rebelión de las masas, 4th. Ed. Madrid, Espasa-Calpe, 1958. Colección Austral.

PEREDA MILLÁN, M., Presupuesto Base Cero: técnicas para la racionalización y reducción del gasto. Su aplicación práctica a la empresa, Madrid, Circe, 1980.

PHILIPPATOS, G.C., Fundamentos de administración financiera, México, McGraw-Hill, 1979.

PICHAT, E., Contribution a l'algorithmique non numérique dans les ensambles ordonnés, doctoral thesis of Sciences, University of Sciences, Grenoble, 1970.

PIFARRÉ RIERA, M., Curso de introducción a la teoría de la contabilidad, Barcelona, University of Economics and Business Sciences, 1989.

PISÓN FERNÁNDEZ, I., El control de gestión en las pequeñas y medianas empresas, University of Santiago de Compostela, 1983.

PORTER, G.A., Management Accounting, Planning and Control, Boston, PWS-Kent Publ., 1988.

POURBAIX, C., Valeur de l'entreprise, Paris, Dunod, 1969.

PYHRR, P.A., "Zero-Base Budgeting", non-published text corresponding to his paper at Planning Executive Institute, New York, May 15, 1977. Reference taken from Castañeda Ordóñez, P., "Nuevas técnicas de dirección: el Presupuesto Base Cero para la reducción de gastos", Económicas y Empresariales, no. 8.

RIEBOLD, G., El cash-flow, Madrid, ICE, 1974.

RODRIGUEZ, A., Matemática de la inversion, Barcelona, Romagraf, 1983.

RODRÍGUEZ INCIARTE, M., "Fórmulas presupuestarias de la eficacia, El PBS. Hacienda Pública Española, no. 11, 1971.

ROSE, T.G., Control de gestión, Bilbao, Deusto, 1989.

SCHULTZE, C., El PBS., política y economía del gasto público, Madrid, Ministry of Finance and Institute of Fiscal Studies, 1971.

SCHWARTZ, E., "Theory of the Capital Structure of the Firm", The Journal of Finance, vol. 14, no. 1, 1959.

SEISDEDOS MARTÍNEZ, J., "Nuevas técnicas presupuestarias. El Presupuesto Base Cero", Essic-Market, no. 34, January-April 1981.

STEIN, W.E., "Probabilistic fuzzy decision-making", at ORSA/TIMS Joint Meet., Milwaukee (United States), October 1979.

SUÁREZ SUÁREZ, A., Decisiones óptimas de inversión y financiación en la empresa, Madrid, Pirámide, 1985.

SUGENO, N., Theory of fuzzy integrals and its applications, Engineering doctoral thesis, Tokyo (Japan), Institute of Technology, 1974.

TARRAGÓ SABATÉ, F., Fundamentos de la economía de la empresa, Barcelona, 6th Ed. Hispano Europea, 1986.

TEODORESCU, N.H. and GIL ALUJA, J., "Phénomènes économiques chaotiques de croissance", at International 92 Genève Conference " Signals and systems", 1992.

TUCKER, S.A., Control de gestión (método de los ratios), Barcelona, Hispano Europea, 1968.

TURKSEN, I.B., "Fuzzy sets and systems and its applications in production research", at 8th International Conference on Production Res., University of Stuttgart (Germany), August 1985.

WEIHRICH, H., Management, 7th ed., New York, McGraw-Hill, 1980.
WILAYES, D., Contribution des modèles fllous à l'analyse des systèmes, Table ronde du CNRS, Dept. de Mathématiques, Université de Lyon, July 1, 1980.
WOLK, H.I., Management, Accounting Planning and Control, Boston, PWS-Kent Publ., 1988.
WU XUEMOU, "Operation research in control and recognition. Fuzzyness and Pansystems analysis", Doc. Huazhong Inst.Techn. Wuhan. Hubei (China), 1980.
YAGER, R.R., "Application of specificity measures in decisions", at Journées d'Analyse de Problèmes Decisionnels dans un environnement incertain et imprécis, I.U.T., Reims (France), July 1985.
ZADEH, L.A., "Utilization of fuzzy evidence in decision analysis", at ORSA/TIMS Joint Meet., Houston, October 1981.